University of Ulster
LRC, JORDANSTOWN

# Quantity Surveyor's Pocket Book

WITHDRAWN
FROM THE LIBRARY OF
UNIVERSITY OF ULSTER

This second edition of the *Quantity Surveyor's Pocket Book* is fully update line with NRM1, NRM2 and JCT(11) and remains a must-have guide for dents and qualified practitioners. Its focused coverage of the data, techn , and skills essential to the quantity surveying role make it an invaluable cor - ion for everything from initial cost advice to the final account stage.

Key topics covered include:

- the structure of the construction industry;
- cost forecasting and feasibility studies;
- measurement and quantification, with NRM2 and SMM7 examp ;
- estimating and bidding;
- whole-life costs;
- contract selection;
- final account procedure.

The book includes recommended formats for cost plans, developer's budgets, financial reports, financial statements and final accounts. This is the ideal concise reference for quantity surveyors, project and commercial managers, and students of any of the above.

**Duncan Cartlidge** is a Fellow of the Royal Institution of Chartered Surveyors with more than twenty-five years involvement in construction. His experience includes private practice in both the UK and Europe, commercial management for leading European ... ering graduate and postgradaute progra... He is an Associate Tutor at the Colle... Visiting Lecturer at Glasgow Caledo... rveying and Construction Professi... www.duncancar...

# Quantity Surveyor's Pocket Book

Second edition

Duncan Cartlidge FRICS

LONDON AND NEW YORK

100598869

692.5 CAR

First edition published 2009
by Butterworth-Heinemann, an imprint of Elsevier

This edition published 2013
by Routledge
2 Park Square, Milton Park, Abingdon, Oxon OX14 4RN

Simultaneously published in the USA and Canada
by Routledge
711 Third Avenue, New York, NY 10017

*Routledge is an imprint of the Taylor & Francis Group, an informa business*

© 2013 Duncan Cartlidge

The right of Duncan Cartlidge to be identified as author of this work has been asserted by him in accordance with sections 77 and 78 of the Copyright, Designs and Patents Act 1988.

All rights reserved. No part of this book may be reprinted or reproduced or utilised in any form or by any electronic, mechanical, or other means, now known or hereafter invented, including photocopying and recording, or in any information storage or retrieval system, without permission in writing from the publishers.

*Trademark notice*: Product or corporate names may be trademarks or registered trademarks, and are used only for identification and explanation without intent to infringe.

*British Library Cataloguing in Publication Data*
A catalogue record for this book is available from the British Library

*Library of Congress Cataloging in Publication Data*
Cartlidge, Duncan P.
Quantity surveyor's pocket book / Duncan Cartlidge. – 2nd ed.
p. cm.
Includes bibliographical references and index.
1. Quantity surveying – Handbooks, manuals, etc. I. Title.
TH435.C367 2012
692'.5—dc23
2011049515

ISBN: 978–0–415–50110–1 (pbk)
ISBN: 978–0–203–11479–7 (ebk)

by Swales & Willis Ltd, Exeter, Devon

MIX
Paper from responsible sources
FSC® C004839

Printed and bound by CPI Group (UK) Ltd, Croydon, CR0 4YY

Dedication

Nicholas Jon Cartlidge 1942–2007

# Contents

# Preface

I am pleased to say that the first edition of the *Quantity Surveyor's Pocket Book* was well received and it would appear that traditional quantity surveying skills are still in great demand.

Since the publication of the first edition of the *Quantity Surveyor's Pocket Book* in 2009, the Royal Institution of Chartered Surveyors finally published its much heralded *New Rules of Measurement*. Many within the profession waited anxiously for the new rules amidst talk of a revolutionary approach to the measurement of quantities. In the end the new rules are better described as evolutionary, reflecting the changes in the approach to procurement that have gradually been evolving over the last 20 years or so.

In preparing the second edition of this pocket book I was faced with a dilemma; whether to revise the measurement chapter completely to align with NRM2 or whether to stick with examples based on SMM7. In the end, based on the fact that for some years to come the two rules of measurement will be available to quantity surveyors, I decided to include examples of measurement and estimating/cost planning based on both documents.

Duncan Cartlidge
www.duncancartlidge.co.uk

# 1

# The quantity surveyor and the construction industry

## THE UK CONSTRUCTION INDUSTRY

The UK construction industry is a unique, complex and often fragmented industry. Nevertheless, in 2011 the total turnover of the industry was close to £100 billion or the equivalent to 8.5% of the UK gross domestic product, making it an important contributor to the wealth of the nation. The industry employs approximately 190,000 contractors and 1.8 million people, a high percentage of which are self-employed with a ratio of male to female of 7:1. The construction industry is defined in accordance with Division 45 of the Revised 2003 Standard Classification to include the following:

- general construction and demolition work: establishments engaged in building and civil engineering work not specialised to be classified elsewhere;
- construction and repair of buildings: establishments engaged in construction, improvement and repair of both residential and non-residential buildings, including specialists engaged in sections of construction and repair work such as bricklaying and the erection of steel and concrete structures, etc.;
- civil engineering: construction of roads, railways, airport runways, bridges, tunnels, pipelines, etc.;
- installation of fixtures and fittings: establishments engaged in the installation of fixtures and fittings including: gas fittings, plumbing, electrical fixtures and fittings, etc.;
- building completion work: establishments engaged in work such as painting and decorating, plastering, onsite joinery, etc.

### Market drivers

Demand for construction and civil engineering work can be divided in broad terms into the public and private sectors. Public sector work is work for any public authority such as:

- government departments
- public utilities
- universities
- the National Health Service
- local authorities.

Private sector work is for a private owner or organisation or for a private client and includes:

- work carried out by firms on their own initiative;
- work where the private sector carries out the majority of the risk/gain.

Increasingly the distance between public and private sectors is disappearing with the introduction of strategies such as the Private Finance Initiative which is included in Chapter 4.

Demand for construction is influenced by the following factors:

- The industry is vulnerable to economic influences as witnessed by the downturn in the UK housing sector in 2008. Figure 1.1 illustrates that the last 40 years have seen a number of periods of 'boom and bust' associated with the economic performance of the UK as a whole. The construction industry has regularly been used by government in the recent past as a method of regulating the general economy, for example by varying interest rates in order to adjust demand for housing.

**Figure 1.1** Construction output – percentage change 1965–2012 (Source: BERR)

- Almost half of all construction works are commissioned by the public sector and therefore cutbacks in public sector spending on projects such as: schools, hospitals, roads, etc. can have the effect of cooling down an overheating economy.
- Demand comes from a variety of sources, from mega projects, such as the 2012 London Olympic Games to a single-storey kitchen extension.
- A buoyant construction market depends on the availability of reasonable cost credit.
- Further ways in which the government can manipulate demand are with the use of tax breaks for certain categories of development, e.g. Enterprise zones were established in various parts of the country in 1982. These zones offer certain types of development: lucrative tax breaks, rapid planning approvals and exemption from business rates. Enterprise zones were re-introduced in England in the 2011 budget, when twenty-one locations were named. In addition to the above incentives: high-speed broadband was added to the enterprise zone package.
- Nearly half the output of the construction industry is repairs and maintenance, often neglected in times of economic downturns.

The supply side of construction is characterised by the following factors:

- Its unique structure: statistics produced by the BERR indicate that in 2009 there were approximately 194,000 firms working in construction, however only 62 of these had more than 600 employees, with the greatest majority being small jobbing builders. This structure makes it difficult to introduce new initiatives and working practices to increase productivity and/or efficiency. It is generally only the larger organisations that have the time and resources to try to bring about change.
- 1.5% of construction firms account for approximately 60% of the workload. The UK structure is not unique and is replicated, for example, in Australia and North America.
- During the past thirty years or so there has been less reliance on traditional construction trade skills such as bricklaying and plastering: instead there has been a move towards an assembly process, for example, the extensive use of timber kits for low and medium rise structures.
- The early 1960s witnessed a movement towards the use of industrialised buildings, buildings that were constructed from mass produced, factory made, components that were assembled on site. The object was to reduce the amount of skilled expensive labour in the construction process, thereby reducing costs and increasing profits and by 1966 this accounted for 25% of all social housing starts. However, industrialised building tends

to be inflexible and the Ronan Point disaster of 1968, when a high-rise block of system-built flats partially collapsed, without warning, turned public opinion away from this type of approach.
- The time lag between the response to supply to increased demand will nearly always result in a distortion of the market. For example, increased house building in response to increased demand, triggered by lower interest rates and full employment.
- With a standing housing stock of 25 million units, in addition to office buildings, shops, etc., the repair and maintenance is also an important part of the construction industry.

## THE BUILDING TEAM

In addition to construction firms, architects, surveyors and allied professions are involved in the concept, design, finance and management of the construction process. Sir Harold Emerson remarked in the Emerson Report in 1964 that: 'In no other important industry is the responsibility for the design so far removed from the responsibilities of production.' What's more, unlike other major industries such as car manufacturing or aerospace, construction activity is carried out:

- in the open air exposed to the elements;
- at various locations with each project, to some degree being bespoke, unlike a standard model of car or computer.

These factors have contributed to some of the problems that the industry has experienced during the past 50 years or so where the construction industry has been confined to a mere assembly process, with little input from the contractor. These characteristics have led to claims that the industry is inefficient and wasteful and that clients have historically received a bad deal and poor value for money, with projects being delivered late and over budget.

### The construction supply chain

The construction supply chain is the network of organisations involved in the different processes and activities that produce the materials, components and services that come together to design, procure and deliver a building.

Figure 1.2 illustrates part of a typical construction supply chain; although in reality many more subcontractors could be involved. The problems for process control and improvement that the traditional supply chain approach produces are related to:

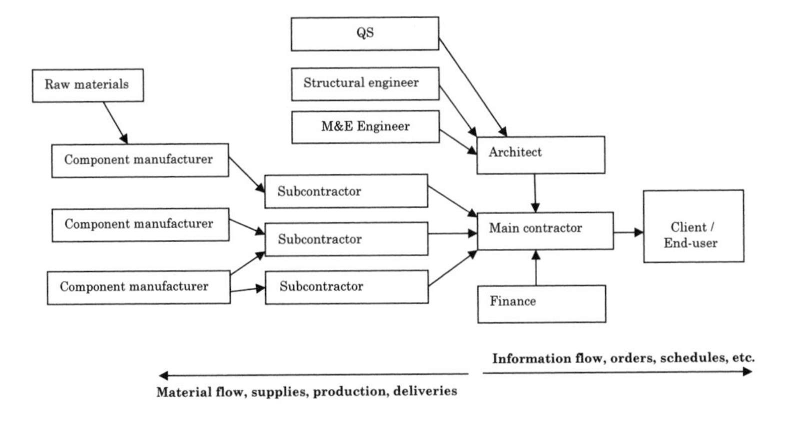


**Figure 1.2** Construction supply chain

- the various organisations which come together for a specific project disband at its completion to form new supply chains;
- communicating data, knowledge and design solutions across the organisations that make up the supply chain;
- stimulating and accumulating improvement in processes that cross the organisational borders;
- achieving goals and objectives across the supply chain; and
- stimulating and accumulating improvement inside an organisation that only exists for the duration of a project.

## Latham and Egan reports

The Latham (1994) and Egan (1998) reports were the last of a series of reports that tried to analyse the workings of the construction industry and suggest ways in which it could become more efficient and deliver better value for money. The principal messages from the reports were that the construction industry needed to concentrate and invest in:

- modernisation
- innovation, and
- mass production.

## The role of professional institutions

There are a number of professional institutions for building professionals, namely, but not in any particular order:

- The Royal Institution of Chartered Surveyors – RICS
- The Royal Institute of British Architects – RIBA
- The Institution of Civil Engineers – ICE
- The Institution of Structural Engineers – MIStructE
- Chartered Institute of Building – CIOB
- The Institution of Clerks of Works and Construction Inspectorate of Great Britain
- The Chartered Institution of Chartered Services Engineers – CIBSE.

Each of the above organisations has developed over time to regulate and further the aims of its members. Corporate membership is generally either at member or fellow grade and members must pay substantial annual fees in order to use designatory letters after their names. The main reasons for the establishment of the professional bodies are to:

- safeguard the public, for example by ensuring that all members working in private practice have adequate professional indemnity insurance;
- enforce codes of conduct;
- lobby governments; and
- train and educate.

**Quantity surveyor**

Prior to the Napoleonic Wars, Britain, in common with its continental neighbours, had a construction industry based on separate trades. This system still exists in France as 'lots séparés', and variations of it can be found throughout Europe, including in Germany. The system works like this: instead of the multi-traded main contractor that operates in the UK, each trade is tendered for and subsequently engaged separately under the co-ordination of a project manager.

The Napoleonic Wars, however, brought change and nowhere more so than in Britain – the only large European state that Napoleon failed to cross or occupy. The government of the day was obliged to construct barracks to house the huge garrisons of soldiers that were then being transported across the English Channel. As the need for the army barracks was so urgent and the time to prepare drawings, specifications, etc. was so short, the contracts were let on a 'settlement by fair valuation based on measurement after completion of the works'. This meant that constructors were given the opportunity and encouragement to innovate and to problem solve – something that was progressively withdrawn from them in the years to come. The same need for haste, coupled with the sheer magnitude of the individual projects, led to many contracts being let to a single builder or group of tradesmen 'contracting in gross', and the general contractor was born. When peace was made the Office of Works and Public Buildings, which had been increasingly concerned with the high cost of measurement and fair value procurement, in particular in the construction of Buckingham Palace and Windsor Castle, decided enough was enough. In 1828, separate trades contracting was discontinued for public works in England in favour of contracting in gross. The following years saw contracting in gross (general contracting) rise to dominate, and with this development the role of the builder as an innovator, problem solver and design team member was stifled to the point where contractors operating in the UK system were reduced to simple executors of the works and instructions (although in Scotland the separate trades system survived until the early 1970s).

Then in 1834 architects decided that they wished to divorce themselves from surveyors and establish the Royal Institute of British Architects (RIBA), exclusively for architects. The grounds for this great schism were that architects wished to distance themselves from surveyors and their perceived 'obnoxious commercial interest in construction'. The events of 1834 were also responsible for the birth of another UK phenomenon, the quantity surveyor.

For the next 150 or so years the UK construction industry continued to develop along the lines outlined above, and consequently by the third quarter of the twentieth century the industry was characterised by:

- powerful professions carrying out work on comparatively generous fee scales;
- contractors devoid of the capability to analyse and refine design solutions;
- forms of contract that made the industry one of the most litigious in Europe;
- procurement systems based upon competition and selection by lowest price and not value for money.

Some within the industry had serious concerns about procurement routes and documentation, the forms of contract in use leading to excess costs, suboptimal building quality and time delays, and the adversarial and conflict-ridden relationships between the various parties. A series of government-sponsored reports (Simon, 1944; Emmerson, 1962; Banwell, 1964) attempted to stimulate debate about construction industry practice, but with little effect.

It was not just the UK construction industry that was obsessed with navel-gazing during the last quarter of the twentieth century; quantity surveyors had also been busy penning numerous reports into the future prospects for their profession, all produced either directly by, or on behalf of, The Royal Institution of Chartered Surveyors. The most notable of which were:

- *The Future Role of the Quantity Surveyor* (1971) was the product of a questionnaire sent to all firms in private practice together with a limited number of public sector organisations. The report paints a picture of a world where the quantity surveyor was primarily a producer of bills of quantities; indeed, the report comes to the conclusion that the distinct competence of the quantity surveyor of the 1970s was measurement – a view, it should be added, still shared by many today. In addition, competitive single-stage tendering was the norm, as was the practice of receiving most work via the patronage of an architect. It was a profession where design and construct projects were rare, and quantity surveyors were discouraged from forming multidisciplinary practices and encouraged to adhere to the scale of fees charges. The report observes that clients were becoming more informed, but there was little advice about how quantity surveyors were to meet this challenge.
- *The Future Role of the Chartered Quantity Surveyor* (1983).
- *Quantity Surveying 2000: The Future Role of the Chartered Quantity Surveyor* (1991).
- *The Challenge for Change: QS Think Tank* (1998). A mere 25 years after the original report the 1998 report was drafted in a business climate driven by information technology, where quantities generation is a low-cost activity and the client base is demanding that surveyors demonstrate added value. In particular, medium-sized quantity surveying firms (i.e. between 10 and 250 employees) were singled out by this latest report to be under particular pressure owing to:
  - competing with large practices' multiple disciplines and greater specialist knowledge base;
  - attracting and retaining a high-quality work force;
  - achieving a return on the necessary investment in IT; and
  - competing with the small firms with low overheads.

Interestingly, *The Challenge for Change* report also predicts that the distinction between contracting and professional service organisations will blur – a quantum leap from the 1960s, when chartered surveyors were forced to resign from their institution if they worked for contracting organisations! The trend for mergers and acquisitions continues, although it has to be said not without its problems, with the largest quantity surveying firms developing into providers of broad business solutions.

### *The profession*

A quantity surveyor may choose to work in any number of different fields, however principally these can be divided into:

- private practice, often referred now to as project management; and
- commercial management or contracting surveying.

### *Private practice*

The conventional model for quantity surveying firms in private practice is to trade as a sole practitioner or as a partnership. A surveyor who is a partner in a partnership is jointly and severally liable for all debts and liabilities of the partnership and liable to the full extent of their personal wealth for the debts of the business. However, in 2001 the Limited Liability Partnership Act made limited liability partnerships (LLPs) available to any 'two or more persons associated for carrying on a lawful business with a view to profit'. This followed on from the RICS's decision in 1986 to remove the restrictions on limited liability. The arguments for the introduction of LLPs are as follows:

- The general partnership, which had existed since the Partnership Act of 1890, was no longer an appropriate vehicle for modern firms and for their businesses.
- Unlike a partnership, an LLP is a separate legal entity. Although some LLPs call their members 'partners', they are not partners in a partnership, their legal title is 'members of an LLP'.
- The primary purpose of an LLP is to provide additional protection for the members through limited liability; a member will not be personally liable for acts and defaults of a fellow member however, he or she may still be personally liable for his or her own negligence.

### *Commercial management*

Commercial management is generally meant to be managing the contractual and commercial aspects of projects for the supply side of the industry. Many commercial managers are members of the CIOB and the RICS.

### *The Royal Institution of Chartered Surveyors (RICS)*

The RICS was founded in 1868 and today there are approximately 130,000 members operating in over 140 countries of which approximately 35,000 are quantity surveyors.

The key roles of RICS are to:

- regulate and promote the profession;
- maintain the highest educational and professional standards;
- protect clients and consumers through a strict code of ethics;
- provide impartial advice, analysis and guidance.

### *Training and education*

Until the 1960s the principal route to becoming a quantity surveyor was to follow a course on either a full or part-time basis (some of these courses were really tests of attrition, involving attending evening classes for three hours a night, four nights a week for several years) and then sit the examinations of either the RICS or the Institute of Quantity Surveyors (IQS). These examinations had a fearsome reputation with approximately only 35% of all candidates being successful each year. However, during the 1960s the first CNAA degree and diploma courses in surveying started to be offered at universities and other institutions of higher education that granted exemption from the majority of the professional institutes' examinations. In 1982 the RICS and the Institute of Quantity Surveyors merged and the IQS ceased to exist. With the transition from examinations set by the RICS to degrees and diplomas the RICS's role changed to one of accrediting body. In 2001 the RICS radically revised its accreditation process and introduced a 'Partnership' scheme for selected universities, etc. The aims of the partnership arrangements are to:

- maintain standards;
- attract the best entrants to the profession;
- promote research;
- develop courses in response to the needs of the profession and industry;
- improve education/professional links.

One of the principal routes to becoming a member of the Royal Institution of Chartered Surveyors is as follows:

- The individual obtains a first degree awarded by an RICS partnership university. In the UK there are numerous higher education establishments that offer three-year, or four-year in the case of Scotland, degree courses.
- On completion of the first degree the graduate then typically gains employment in a private practice or contractor's organisation with a structured training framework and after a minimum of two years of work experience applies to take the Assessment of Professional Competence (APC).

- The purpose of the APC is to ensure that those applying for RICS membership are competent to practice. It is structured to provide a number of pathways to cover nineteen different areas of practice. The APC for quantity surveyors cover a number of mandatory, core and optional competencies such as:
  - commercial management of construction or design economics;
  - contract practice;
  - construction technology and environmental services;
  - procurement and tendering;
  - project financial control;
  - quantification and costing of construction studies.
- If successful, the candidate may apply for membership of the RICS (MRICS). Approximately half of all entrants to the surveying profession come via this route. Corporate membership is at two levels; members and fellows. In 2002 the RICS raised the standards for its fellowship award to reflect career achievements. Normally only MRICS members with a minimum of five years' service who are major achievers will be considered.

In addition to the first degree route, in recent years a number of other routes have become very popular. These are:

- Cognate and non-cognate degree courses available in full-time, part-time and distance learning modes. These courses are typically two to three years in duration and have been developed to attract candidates who already have a first degree in a related (cognate) or unrelated (non-cognate) subject area. These courses are intensive but have proved to be very popular with almost 50% of all entrants to surveying now coming from this route.
- Masters degrees (MSc). There are a number of second degree courses both full-time, part-time and distance learning that are recognised by the RICS for entry to the institution and enrolment to take the APC.
- In addition to the above routes to entry to the profession the RICS encourages entry from non-traditional routes and many higher education establishments retain a number of places for candidates who do not have any of the above traditional qualifications.
- Higher National Diplomas and Certificates in a cognate area with passes at a high level are also recognised as an entry qualification to an RICS first degree course at advanced level.

### AssocRICS

For many years the RICS has been promoting a 'two tier' profession, the lower technical tier being provided by technicians or Tech RICS. This clearly did not capture the imagination of the industry and take-up was always at low levels and consequently TechRICS was replaced in 2010 by AssocRICS, an experience- and competency-based route leading to Associate membership of the RICS.

One of the routes leading to AssocRICS is with the completion of the College of Estate Management's Diploma in Surveying Practice. Entry to the course is typically:

- 'A' Level at grade C or above in a relevant subject;
- HNC/HND in relevant subjects (merits and distinctions recognised in the final award);
- relevant professional qualifications or at least 10 years' relevant experience in the property and construction profession with evidence of five years' appropriate structured learning (Level 1 module exemption only).

AssocRICS can be a route to full corporate membership of the RICS.

***Continuing professional development (CPD) and Lifelong Learning (LLL)***

Since 1984 CPD has been mandatory for all corporate members and is a process by which practising surveyors can keep pace with the latest professional standards and practices whilst monitoring current levels of knowledge.

CPD can be grouped into four main categories:

1. professional work-based activities
2. personal activities outside work
3. courses, seminars and conferences
4. self-directed and informal learning.

Lifelong learning is defined as:

> a learning activity undertaken through life with the aim of improving knowledge, skills and competence within a personal and/or employer-related perspective. It is seen as a key element of CPD and an important tool in maintaining a person's employability in a rapidly changing business environment.

In addition to the quantity surveyor other professionals are involved in the design and the delivery of a construction project. Collectively they are called the design team.

## Project manager

The role of the project manager is one that has emerged during the past 30 years or so. Latterly, the term 'project manager' has been used to refer to a quantity surveyor working for the client side in private practice and the term quantity surveyor is used to refer to a quantity surveyor working for a contracting organisation. This is by no means universal, especially outside the UK where the term 'quantity surveyor' is more universally used to describe the quantity surveyor irrespective of where or for whom they work. In the UK therefore it is possible for a private practice to supply both quantity surveying and project management services for the same project and client. Project managers may be drawn from all building professionals with the appropriate training and expertise.

The project manager therefore represents the client's interests from the initial inception to the completion and commissioning of a project. For the client, the main advantage of using a project manager is the establishment of a single point of contact in the case of queries. The client simply contacts the project manager instead of having to decide which of the design team has the answer. Training and qualifications for project managers are generally at postgraduate level, typically MSc.

## Architect

Traditionally in the UK the architect has been regarded as the leader of the design team and the first person to be appointed by the client at the start of a new project. So much so, that traditional single-stage tendering is sometimes referred to still as: 'architect-led tendering'.

Until recently it was usual for the majority of a quantity surveyor's work to come via the architect although this has changed to some extent with quantity surveyors and other members of the design team winning work in their own right. Perhaps the most difficult part of the architect's role is to interpret a client's user requirements and transform them into a building. Architects can also act as contract administrators, although increasingly this role is being taken over by others. Unlike the rest of Europe most architects work within private practice, with few working for contractors or developers. The UK is home to some of the largest firms of commercial architects in the world.

The work of architects influences every aspect of our built environment, from the design of energy efficient buildings to the integration of new buildings in sensitive contexts. Architects work closely with other members of the construction industry including engineers, builders, surveyors, local authority planners and building control officers. The Royal Institute of British

Architects and The Royal Incorporation of Architects in Scotland are the professional institutes for architects in the UK.

### Building surveyor

Building surveying is a comparatively new profession, being a branch of the RICS General Practice section until the 1970s when it became a separate division within the RICS. In-depth knowledge of building pathology is key to building surveying and building surveyors can frequently be found working on historic and conservation projects. For smaller new build contracts building surveyors can also take on the design role and contract administration.

### Structural engineer

A structural engineer is involved in the design and supervision of the construction of all kinds of structures such as houses, theatres, sports stadia, hospitals, bridges, oil rigs, space satellites and office blocks. The central strength of a building lies in the framework, often hidden, that supports the shape and design concept produced by the architects and is integral to the completed project's function. To the chartered structural engineer, the considerations of strength, shape and function are paramount in their conception of the framework of a structure. Having chosen appropriate materials such as steel, brick, concrete or timber, they have to design the structure and make all the necessary checks and calculations to ensure that the foundations will be sound, that the floors and roof will not fall down, and that the construction as a whole will remain safe and serviceable for the length of its intended lifetime. The specialist skills of a structural engineer will include calculating loads and stresses, investigating the strength of foundations and analysing the behavior of beams and columns in steel, concrete or other materials to ensure the structure has the strength required to perform its function safely, economically and with a shape and appearance that is visually satisfying.

### Civil engineer

Civil engineers are involved with the design, development and construction of a huge range of projects in the built and natural environment. Their role is central to ensuring the safe, timely and well-resourced completion of infrastructure projects in many areas, including: highways construction, waste management, coastal development and geotechnical engineering.

Consulting civil engineers liaise with clients to plan, manage, design and supervise the construction of projects. They work in a number of different settings and, with experience, can run projects as project manager. Within civil engineering, consulting engineers are the designers; contracting engineers turn their plans into reality. Consulting civil engineers provide a wide range of services to clients. During the early stages of a career, work will involve taking responsibility for minor projects, but the size of the projects may increase as experience is gained. Typical work activities include:

- undertaking technical and feasibility studies and site investigations;
- developing detailed designs;
- assessing the potential risks of specific projects, as well as undertaking risk management in specialist roles;
- supervising tendering procedures and putting together proposals;
- managing, supervising and visiting contractors on site and advising on civil engineering issues;
- managing budgets and other project resources;
- managing change, as the client may change their mind about the design, and identifying, formalising and notifying relevant parties of changes in the project;
- scheduling material and equipment purchases and delivery;
- attending public meetings and displays to discuss projects, especially in a senior role;
- adopting all relevant requirements around issues such as building permits, environmental regulations, sanitary design, good manufacturing practices and safety on all work assignments;
- ensuring that a project runs smoothly and that the structure is completed on time and within budget;
- correcting any project deficiencies that affect production, quality and safety requirements prior to final evaluation and project reviews.

Infrastructure is the thing that supports our daily life: roads and harbours, railways and airports, hospitals, sports stadiums and schools, access to drinking water and shelter from the weather. Infrastructure adds to our quality of life, and because it works, we take it for granted. Only when parts of it fail, or are taken away, do we realise its value. In most countries, a civil engineer has graduated from a post-secondary school with a degree in civil engineering, which requires a strong background in mathematics, economics and the physical sciences; this degree is typically a four-year degree, though many civil engineers continue with their studies to obtain a masters, engineer, doctoral and post-doctoral degrees. In many countries, civil engineers are subject to licensure, and often, persons not licensed may not call themselves 'civil engineers'.

**Building services engineer**

Building services engineers are responsible for ensuring the cost-effective and environmentally sound and sustainable design and maintenance of energy using elements in buildings. They have an important role in developing and maintaining buildings, and their components, to make the most effective use of natural resources and protect public safety. This includes all equipment and materials involved with heating, lighting, ventilation, air conditioning, electrical distribution, water supply, sanitation, public health, fire protection, safety systems, lifts, escalators, facade engineering and even acoustics.

Whilst the role increasingly demands a multidisciplinary approach, building services engineers tend to specialise in one of the following areas:

- electrical engineering
- mechanical engineering
- public health.

Activities will vary according to the specialist area of work and whether you are employed by a single organisation or a consultancy, but tasks typically involve:

- advising clients and architects on energy use and conservation in a range of buildings and sites, aiming to minimise the environmental impact and reduce the carbon footprint;
- managing and forecasting spend, using whole-life-cycle costing techniques, ensuring that work is kept to budget;
- developing and negotiating project contracts and agreeing these with clients, if working in consultancy, and putting out tenders;
- attending a range of project groups and technical meetings;
- working with detailed diagrams, plans and drawings;
- using specialist computer-aided design (CAD) software and other resources to design all the systems required for the project;
- designing site-specific equipment as required;
- commissioning, organising and assessing the work of contractors;
- overseeing and supervising the installation of building systems and specifying maintenance and operating procedures;
- monitoring building systems and processes;
- facilities management;
- ensuring that the design and maintenance of building systems meets legislative and health and safety requirements.

The professional institution for building services engineers is the Chartered Institution of Building Services Engineers. There are a variety of grades of membership depending on qualifications and experience.

## Clerk of works

The clerk of works is the architect's representative on site and usually a tradesman with many years practical experience.

- 1882 – formed as 'The Clerk of Works Association'
- 1903 – renamed the 'Incorporated Clerks of Works Association of Great Britain'
- 1947 – became 'The Institute of Clerks of Works Great Britain Incorporated'.

The job title 'clerk of works' is believed to derive from the thirteenth century when 'clerics' in Holy Orders were accepted as being more literate than their fellows, and were left to plan and supervise the 'works' associated with the erection of churches and other religious property. By the nineteenth century the role had expanded to cover the majority of building works, and the clerk of works was drawn from experienced tradesmen who had wide knowledge and understanding of the building process.

The clerk of works, historically as well as now, is a very isolated profession on site, most easily associated with the idiom 'poacher turned gamekeeper'. The clerk of works is the person that must ensure quality of both materials and workmanship and, to this end, must be absolutely impartial and independent in their decisions and judgements. A clerk of works cannot normally, by virtue of the quality role, be employed by the contractor – only the client, and normally by the architect on behalf of the client. Their role is not to judge, but simply to report (through exhaustive and detailed diary notes) all occurrences that are relevant to the role.

Experience in the many facets of the building trade is essential and, in general terms, most practitioners will have 'come from the tools' in the first place. When originally formed the Association was to allow those that were required to operate in isolation on site a central organisation to look after the interests of their chosen profession, be it through association with other professional bodies, educational means or simply through social intercourse amongst their own peers and contemporaries. Essential to this, as the Association developed, was the development of a central body that could lobby Parliament in relation to their profession, and the quality issues that it stands for.

Although the means of construction, the training of individuals and the way in which individuals are employed have changed dramatically over the past 120 years, the principles for which the Association was originally formed remain sacrosanct.

### Site manager/agent

The site manager, often referred to as an agent, is the person in charge of a building contract and, as such, must be aware and in control of all aspects of site operations, including the planning of site progress. It is the manager/agent who has responsibility for both the profitability of operations and adherence to the agreed construction and cost plans.

Site managers/agents are employed by building and construction companies, civil engineering firms and contractors.

Typical work activities include:

- attending regular site meetings with professionals, including quantity surveyors, building services engineers, foremen, subcontractors, and the client who has commissioned the building;
- maintaining strict quality control procedures – this necessitates regular testing of materials, visual inspections of work, and frequent tours of the site;
- conducting regular site safety checks;
- ensuring the project runs to schedule and to budget, and finding solutions to problems that may cause delays, i.e. late arrival of materials.

Recent graduates are unlikely to take on a full site manager/agent role until the necessary site engineering experience is gained. However, the period of apprenticeship or training in the role of site engineer appears to be shortening, with firms forced to promote graduates earlier.

#### *Training and education*

The Chartered Institute of Building (CIOB) offers a qualifications framework for trainee and practising site managers/agents. Progression is normally to contracts management or project management. A number of site managers/agents are self-employed.

## UK PROFESSIONALS AND THE EU

UK construction professionals have always been in demand worldwide. With the establishment of the single European market in 1992 many professionals began expanding their practices into Europe, with varying degrees of success.

EU Directive 2005/36/EC on the recognition of professional qualifications allows holders of a regulated professional qualification to pursue their profession in another EU member state.

## REGULATION AND CONTROL OF THE CONSTRUCTION PROCESS

Where and what you can build in the UK is heavily controlled and before undertaking most building projects, it is first necessary to obtain planning permission and Building Regulation approval.

### Planning permission

The main statutes governing planning law are:

- Town and Country Planning Act 1990 as amended
- Planning (Listed Building and Conservation Areas) Act 1990
- Planning and Compensation Act 1991
- Planning and Compulsory Purchase Act 2004.

The purpose of the planning system is to protect the environment as well as public amenities and facilities. The planning control process is administered by local authorities and exists to 'control the development and use of land and buildings for the best interests of the community'. The levels of planning are:

- Regions set out regional policy through Regional Planning Guidance;
- Structure Plans establish broad planning policies at County Council level;
- Local Plans set out detailed policy at District Council level.

There are three types of planning permission, all of which are subject to a fee that can range from hundreds to thousands of pounds, depending on the scale of the proposed project:

1. Outline – this is an application for a development in principle without detail of construction, etc. Generally used for large-scale developments to get permission in principal.
2. Reserved matters – a follow-up to an outline application stage.
3. Full planning permission – sometimes referred to as detailed planning permission when a fully detailed application is made. Permission when granted is valid for six years.

If planning permission is refused then there is an appeals process, although appeal can only be made on certain matters, listed below. Appeals are made to the Secretary of State for the Environment, Transport and the Regions, the National Assembly for Wales or the Scottish Executive. Allowable reasons for appeal are:

- properties in conservation areas;
- non-compliance with local development plan;
- property is subject to a covenant;
- planning permission already exists;
- infringements of rights of way.

Appeal may *not* be made on the grounds of:

- loss of view;
- private issues between neighbours;
- loss of privacy;
- etc.

It is recommended that prior to a proposed development the Structure Plans are read thoroughly and understood. Buildings erected without planning permission will have a demolition order served on them and the structure will be taken down and destroyed.

In 2012 the UK Coalition Government proposed radical changes to planning legislation.

**Building Regulations**

Even when planning permission is not required, most building work is subject to the requirements of the Building Regulations. There are exemptions such as buildings belonging to the Crown, the British Airports Authority and the Civil Aviation Authority. Building Regulations ensure that new and alteration works are carried out to an agreed standard that protects the health and safety of people in and around the building. Builders and developers are required by law to obtain building control approval, which is an independent check with which the Building Regulations comply. There are two types of building control providers: the local authority and approved private inspectors.

The documents which set out the regulations are:

- The Building Act 1984.
- The Building Regulations 2000 for England and Wales, as amended.
- The Building (Scotland) Act 2003.
- The Building (Scotland) Regulations 2004.

The Building Regulations 2000, England and Wales, are a series of Approved Documents. Each Approved Document contains the relevant subject areas of the Building Regulations. This is then followed by practical and technical guidance (that include examples) detailing the regulations. The current set of approved documents is in 13 parts and includes details of areas such as: Structural, Fire Safety, Electrical Safety, etc. In Scotland the Approved Documents are replaced with Technical Handbooks.

Contravention of the Building Regulation is punishable with a fine or even a custodial sentence plus, of course, taking down and rebuilding the works that do not comply with the regulations.

There are two approaches to complying with Building Regulations:

1. Full plan application submission, when a set of plans is submitted to the local authority that checks them and advises whether they comply or whether amendments are required. The work will also be inspected as work proceeds.
2. Building notice application, when work is inspected as the work proceeds and the applicant is informed when work does not comply with the Building Regulations. The work is also inspected as work proceeds.

Once approval is given and a building notice is approved, it is valid for three years.

## Health and safety

Construction is one of the most dangerous industries in the UK. In the last 25 years, over 2800 people have died from injuries they received as a result of construction work. Many more have been injured or made ill.

Efficient site organisation is of vital importance from two aspects:

1. safety
2. efficient working.

### *Safety: key statistics*

- In 2010/11, 50 operatives were killed in construction related activities.
- There were 6,789 over-three-day injuries to employees (2008/09).
- 3.3 million working days lost per year due to injury and ill health (2009/10).

(Source: Health and Safety Executive (HSE))

It is clear from the above figures that for a number of reasons the UK construction sector has a poor record in health and safety matters. In an attempt to improve the industry's attitude to health and safety the Construction (Design & Management) Regulations were introduced in 1994 in order to comply with EU legislation. The CDM (1994) Regulations made the duties of clients and designers explicit by identifying the need to reduce risk by better co-ordination, management and co-operation. The introduction of the CDM Regulations, without doubt, led to major changes in how the industry managed health and safety although several years after their introduction there were concerns from industry and the Health and Safety Executive (HSE) that the regulations were not delivering the improvements in health and safety that were expected. The principal reasons were said to be:

- slow acceptance, particularly amongst clients and designers;
- effective planning, management, communications and co-ordination was less than expected;
- competence of organisations and individuals slow to improve;
- a defensive verification approach adopted by many, leading to complexity and bureaucracy.

During 2002–2005 extensive consultations were carried out between the HSE and the construction industry. As a result, in April 2007, CDM 2007 brought into force the characteristics which were to:

- simplify the regulations and improve clarity;
- maximise their flexibility;
- focus on planning and management;
- strengthen requirements on co-operation and co-ordination and to encourage better integration;
- simplify competence assessment, reduce bureaucracy and raise standards.

The CDM 2007 structure is as follows:

- Part 1: Introduction
- Part 2: General management duties applying to all construction projects
- Part 3: Additional duties where projects are notifiable
- Part 4: Worksite health and safety requirements
- Part 5: General
- Supported by a CDM 2007 Approved Code of Practice.

Notifiable construction works under CDM 2007 are construction projects with a non-domestic client and involve:

- construction work lasting longer than 30 days, or
- construction work involving 500 person days.

### *A client's perspective*

The definition of a client under CDM 2007 is: 'an individual or organisation who in the course or furtherance of a business, has a construction project carried out by another or by himself. This excludes domestic clients from the definition, but not necessarily domestic premises.'

A domestic client is someone who lives, or will live, in the premises where the work is carried out. However, the CDM client duties will still apply to domestic premises if the client is a:

- local authority
- landlord
- housing association
- charity
- collective of leaseholders, or
- any other trade, business or undertaking (whether for profit or not).

To some the exclusion of domestic clients was a missed opportunity given the pattern of workload in the UK.

Duties on clients can be summarised as follows:

- check competence and resources of those they appoint;
- allow sufficient time and resources;
- provide key information to designers and contractors – it is for clients to arrange for any gaps in information to be filled, e.g. commissioning an asbestos survey;
- ensure that all those involved in the work co-operate and co-ordinate their activities;
- ensure that suitable management arrangements are in place;
- ensure that adequate welfare facilities are on site;
- ensure workplaces are designed correctly and comply with Health, Safety and Welfare Regulations 1992 and ensure that construction work does not start unless there is a health and safety plan;
- appoint competent CDM co-ordinator and provide key information. For notifiable projects, where no CDM co-ordinator or principal contractor is appointed then the client will be deemed to be the CDM co-ordinator and subject to their duties.

### *The role of the CDM co-ordinator*

Many surveyors are practising CDM co-ordinators.

The CDM co-ordinator is a role new to the CDM 2007 Regulations whose responsibilities are to:

- advise the client about selecting competent designers and contractors;
- help identify what information will be needed by designers and contractors;
- co-ordinate the arrangements for health and safety of planning and design work;
- ensure that HSE is notified of the project;
- advise on the suitability of the initial construction phase plan;
- prepare a health and safety file.

However, CDM co-ordinators do *not* have the power to:

- approve the appointment of other duty holders, although they give advice;
- approve or check designs, although they should be satisfied the hierarchy is addressed;
- approve or supervise the principal contractor's construction phase plan;
- supervise or monitor work on site.

Who can be a CDM co-ordinator?

- Any person or body provided the competency requirements given in the ACOP are met.
- The duties can be carried out by a:
  - client
  - principal contractor
  - contractor
  - designer, or
  - full-time CDM co-ordinator.

## SUSTAINABILITY AND THE QUANTITY SURVEYOR

Perhaps the most important influence on construction and the professions since the turn of the millennium is the prominence given to sustainable issues.

If this book had been written 13 years ago, then sustainability would not have been an issue: 10 years ago sustainability issues were starting to be discussed, but were considered only to be of interest to 'tree-hugging' cranks.

Welcome to the second decade of the twenty-first century where sustainability and the need to be badged a green construction organisation is seen to be vital to maintain market share.

The Stern Review (2006) came to the conclusion that 'An overwhelming body of scientific evidence now clearly indicates that climate change is a serious and urgent issue. The Earth's climate is rapidly changing, mainly as a result of increases in greenhouse gases caused by human activities.' As illustrated in Figure 1.3, buildings account for nearly 50% of UK carbon emissions and it is for this reason that such importance has been placed both by the European Union and the UK governments on the introduction of Energy Performance Certificates (EPCs) and Display Energy Certificates (DECs). It is the stated long-term goal in the UK is to reduce carbon emissions by 60% by 2050.

## Legislative background

EU Directive 2002/91/EC, the Energy Performance of Buildings Directive (EPBD), became law on 4 January 2003 and made it mandatory for EPCs and Display Energy Certificates (DEC) to be available for constructed, marketed or rented buildings including non-dwellings by 4 January 2009, at the latest. The EU Directive was implemented in the UK by means of:

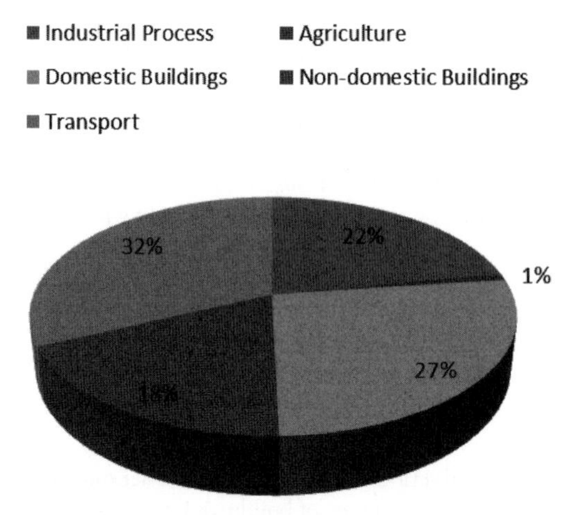


**Figure 1.3** UK carbon emissions (Source: BRE/BRECSU)

- The Housing Act 2004, Section 134.
- The Home Information Pack (No. 2) Regulations 2007.
- The Energy Performance of Buildings (Certificate and Inspections) (England and Wales) Regulations 2007/991.
- Similar enabling legislation was introduced for the devolved administrations of Scotland (The Housing (Scotland) Act 2006) and Northern Ireland.

A recast of the Directive (EPBD2) has now been agreed. The key provisions in the recast are:

- minimum energy performance requirements to be set for all new and refurbished buildings and compared against requirements calculated in accordance with cost-optimal requirements;
- energy use of technical building systems to be optimised by setting requirements relating to installation, size, etc. Covers heating, hot water, air-conditioning and large ventilation systems;
- all new buildings developed after 2020 to be nearly zero energy buildings, with an earlier target date of 2018 where the building will be owned and occupied by a public authority;
- property advertisements to include details of EPC rating;
- member states to provide details of the fiscal incentives in place (if any) which could be used to improve the energy efficiency of their buildings;
- content of EPCs to be improved by making them more specific to a particular building and including more detailed information on the cost-effectiveness of recommendations, along with the steps to be taken to implement those recommendations;
- DECs to be issued and displayed in buildings larger than $500m^2$ (current threshold is $1{,}000m^2$) that are occupied by a public authority and frequently visited by the public. This threshold will fall to $250m^2$ after five years;
- EPCs to be displayed in commercial premises larger than $500m^2$ that are frequently visited by the public and where one has previously been issued; and a statistically significant percentage of EPCs and ACRs to be checked by independent experts for quality assurance purposes;
- EPBD2 will be implemented by member states by 2012–13.

Prior to the introduction of EPCs a series of other models were developed to assess the energy performance of buildings across the UK, namely:

### *EcoHomes points*

Now superseded by the Code for Sustainable Homes, some development projects still have a requirement for this assessment.

EcoHomes assesses the green performance of houses over a number of criteria:

- reducing $CO_2$ emissions from transport and operational energy;
- reducing mains water consumption;
- reducing the impact of materials used;
- reducing pollutants harmful to the atmosphere; and
- improving the indoor environment.

### *BREEAM (Building Research Establishment Environmental Assessment Method)*

BREEAM has been developed to assess the environmental performance of both new and existing buildings over the following areas:

- management – overall management policy, commissioning and procedural issues
- energy use
- health and well-being
- pollution
- transport
- land use
- ecology
- materials
- water consumption and efficiency.

Unlike EcoHomes points, BREEAM covers a range of building types, such as: offices, industrial units, retail units, schools and even leisure centres. BREEAM measures the environmental performance of buildings by awarding credits for achieving levels of performance. The cost of having a BREEAM assessment can be considerable.

## What is sustainability?

So what is sustainability? There are many definitions, as with any new buzz term, people queue up to add their definition in order to gain their five minutes of fame! In reality it would appear to mean different things to different people in different parts of the world, depending on their circumstances. Consequently, there may never be a consensus view on its exact meaning; however, one way of looking at sustainability is 'The ways in which built assets are procured and erected, used and operated, maintained and repaired, modernised and rehabilitated and reused or demolished and recycled constitutes the complete life-cycle of sustainable construction activities.'

Why is construction significant in the sustainability big picture?

- Over 90 million tonnes of construction and demolition waste arises annually in the UK alone.
- The construction industry spends over £200 million on landfill tax each year.
- 13 million tonnes of construction and demolition waste is material that is delivered to sites but never used!
- Over 5 million tonnes of hazardous waste is produced in England and Wales, 21% of which is produced by construction and demolition.
- Construction and demolition waste forms nearly 30% of all Environment Agency recorded fly tipping incidents.
- In addition around 40% of total energy consumption and greenhouse gas emissions are directly attributable to constructing and operating buildings.

Although high on the face of it, the true cost of waste is generally around 20 times of the disposal costs due to the:

- purchase cost of materials;
- cost of storage, transport and disposal of waste;
- loss of income from selling salvaged materials.

The so-called waste hierarchy has been described as follows:

- eliminate – avoid producing waste in the first place;
- reduce – minimise the amount of waste you produce;
- re-use – use items as many times as possible;
- recover (recycling, composting, energy) – recycle what you can only after you have re-used it;
- dispose – dispose of what is left in a responsible way.

The process of getting the minimum whole-life cost and environmental impact is complex as illustrated in Figure 1.4. Each design option will have associated impacts, costs and trade-offs, e.g. what if the budget demands a choice between recycled bricks or passive ventilation?

The solution to a complex problem will be iterative.

Generally, attention to the following issues will increase the design costs, but not the costs of the building itself and will reduce whole-life costs:

- short supply chains to reduce transport costs;
- exercise waste minimisation and recycling construction;
- building orientation;
- durability and quality of building components, generally chosen to last for the appropriate refurbishment or demolition cycle;
- local sourcing of materials;
- design sensitive to local topological, climatic and community demands;
- construction type – prefabrication, wood or concrete structures.

During procurement supply chains should be aware that components should be chosen selectively to minimise:

- embodied energy; energy of production and transport;
- atmospheric emissions from boilers, etc.;
- disposal to landfill of non-biodegradable waste;
- air quality contaminants, e.g. solvents and wood preservatives continue to emit volatile chemicals long after construction, though in much smaller quantities and these have been implicated in 'sick building syndrome';
- replacement due to poor durability;
- use of finite resources, or at least promote the use of materials like wood from forests which are being replenished.

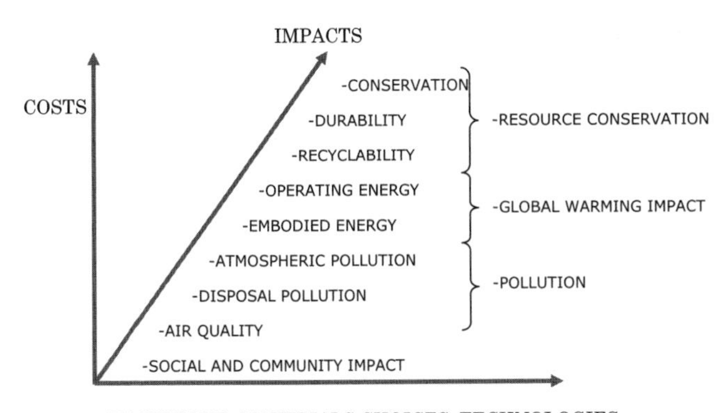


**Figure 1.4** Graph showing the minimum whole-life cost and environmental impact

## THEMES FOR ACTION DURING THE PROCUREMENT PROCESS

The following courses of action should be considered by the quantity surveyor and the procurement team when developing a procurement strategy:

- Re-use existing built assets – consider the need for new build. Is a new building really the answer to the client's needs, or is there another strategy that could deliver a more appropriate solution and add value?
- Design for minimum waste – think whole-life costs – involve the supply chain – specify performance requirements and think about recycled materials. Sustainability in design requires a broad and long-term view of the environmental, economic and social impacts of particular decisions. Design out waste both from the process and the life span. As well as the obvious definition waste can also include:
  - the unnecessary consumption of land, or
  - lower than predicted yield from asset.

Consider Lean Construction – Lean Construction, in its entirety, is for many a complex and nebulous concept; however the ethos of Lean with its focus on the following, is worthy of consideration:

- continuous improvement
- waste elimination
- strong user focus
- high-quality management of projects and supply chains
- improved communications.

### Minimise energy in construction and in use

Fully investigate the whole-life cost and life-cycle cost implication of the materials and systems that are being procured. Draw up environment profiles of components.

### Do not pollute

Understand the environmental impacts of construction and have policies to manage such impacts in a positive manner.

Construction can have a direct and obvious impact on the environment. Sources of pollution can include:

- waste materials
- emissions from vehicles

- noise
- releases into water, ground and atmosphere.

### Set targets

Use benchmarking and similar techniques discussed in Chapter 6 to monitor continuous improvement. UK construction industry KPIs are issued in June each year.

### Site waste management plans

Clearly, given the above statistics, the construction industry can address sustainability by reducing waste, for example from 6 April 2008 Site Waste Management Plans are required for all construction projects with a value of over £300,000.00. The responsibility for the production of the site waste management plan is jointly shared between the client and the contractor and should contain the following details:

- types of waste removed from the site;
- identity of the person who removed the waste and their waste carrier registration number;
- a description of the waste;
- site that the waste was taken to;
- environmental permit or exemption held by the site where the material is taken.

Without doubt, sustainable considerations will continue to be high on the list of construction industry issues for the foreseeable future.

## *The RICS New Rules of Measurement (NRM) Initiative*

In April 2012 the RICS launched the three volumes of the *New Rules of Measurement*:

- **Volume 1:** NRM1: *Order of Cost Estimating and Cost Planning for Capital Building Works*.
- **Volume 2:** NRM2: *Detailed Measurement for Building Works*.
- **Volume 3:** NRM3: *Order of Cost Estimating and Cost Planning for Building Maintenance Works* – due autumn 2012.

This suite of documents, the result of an initiative of the RICS Quantity Surveying and Construction UK Professional Group Board that began in 2003, aimed to more accurately reflect the changes that have taken place in procurement practice during the past 15 years or so, as well as, for the first time to give guidance on the preparation of cost plans and whole-life costs. The rationale for the introduction of the NRM is that they provide:

- a standard set of measurement rules that are understandable by all those involved in a construction project, including the employer, thereby aiding communication between the project/design team and the employer;
- direction on how to describe and deal with cost allowances not reflected in measurable building work; and
- a more universal approach than the UK-centric SMM7.

The impact of the launch of the NRM suite will be explained and discussed more fully in the following chapters.

## ETHICS AND THE QUANTITY SURVEYOR

Ethics is an important topic and particularly so for surveyors who operate in a sector that is generally perceived to have low ethical standards. Professions can only survive if the public retains confidence in them. Conducting professional activities in an ethical manner is at the heart of professionalism and the trust that the general public has in professions such as the chartered quantity surveyor. One of the principal reasons for construction-related institutions like the RICS is to ensure that their members operate to high ethical standards, indeed ethical standards was a top priority on the RICS Agenda for Change (1998). For quantity surveyors transparency and ethical behaviour is particularly important as they deal on a day-to-day basis with: procurement, contractual arrangements, payments, valuations and client's money.

Interestingly in a survey carried out by the Chartered Institute of Building in 2006, nearly 40% of those questioned regarded the practice of cover pricing as either 'not very corrupt' or 'not corrupt at all', regarding it as the way that the industry operates! In addition 41% of respondents admitted offering bribes on one or more occasions. One of the major issues from the CIOB survey is a clear lack of definition of corruption and corrupt practices. The industry is one that depends on personal relationships and yet a particular nebulous area in non-cash gifts that range from pens to free holidays.

Recently the RICS has published a number of guides/documents to help surveyors find their way through the ethical maze. These are:

- Professional ethics guidance note (2000).
- Professional ethics guidance note (2003) – Case studies.
- RICS Core Values (2006).
- RICS Rules of Conduct for Members (2007).
- RICS Rules of Conduct for Firms (2007).
- Fraud in Construction – Follow the money (2009).
- Fraud in Construction – RICS Guidance (2010).

In addition to the above the RICS has also published a help sheet on Maintaining Professional and Ethical Standards in which the behaviour of a chartered surveyor in their professional life is characterised as follows:

1. **Act honourably** – never put your own gain above the welfare of your clients or others to whom you have a professional responsibility.
2. **Act with integrity** – be trustworthy in all that you do, never deliberately mislead, whether by withholding or distorting information.
3. **Be open and transparent** in your dealings.
4. **Be accountable** for all your actions and do not blame others if things go wrong.
5. **Know and act within your limitations** and competencies.
6. **Be objective at all times** – never let sentiments or your own interests cloud your judgement.
7. **Always treat others with respect** – never discriminate.
8. **Set a good example** in both public and private behaviour.
9. **Have the courage to make a stand.**
10. **Comply with relevant laws and regulations.**
11. **Avoid conflicts of interest** – declare any potential conflicts of personal or professional interest.
12. **Respect confidentiality** of client's affairs.

(Source: RICS Ethics Help Sheet, 2007)

Although the list above appears to be straightforward, things are never quite that simple in practice when matters such as economic survival and competition are added into the mix. The position is even more complicated when operating in countries outside the UK where ideas of ethics may be very different to those expected by the RICS.

# 2

# Forecasting costs and value

## FORECASTING COSTS

### Cost management

The quantity surveyor has the responsibility for providing accurate and timely cost advice throughout the duration of a project to a variety of organisations including the client and architect; Tables 2.1 and 2.2 show the various stages in the pre-contract costing process.

The terms commonly associated with cost advice are detailed below.

### Element

- Part of a building that fulfils a specific function or functions irrespective of its design, specification or construction, e.g. the element *external walls* provides the external vertical envelope to a building, separating the internal and external environment irrespective of how it may be constructed.
- Each element is fully defined in order that there can be no doubt as to what is included within a particular elemental cost and that all estimates are prepared on the same basis.

The following is an extract from the BCIS standard list of elements.

#### 2.6 Windows and external doors

2.6.1 Windows
Sashes, frames, linings and trims. Ironmongery and glazing. Shop fronts. Lintels, sills, cavity damp-proof courses and work to reveals of openings.

2.6.2 External doors
Doors, fanlights and sidelights. Frames, linings and trims. Ironmongery and glazing. Lintels, thresholds, cavity damp-proof courses and work to reveals of openings.

**Table 2.1** RIBA Outline Plan of Work 2007 (Amended 2009), Copyright RIBA

| | |
|---|---|
| **Preparation** | **A Appraisal**<br>Identification of client's needs and objectives, business case and possible constraints on development.<br>Preparation of feasibility studies and assessment of options to enable the client to decide whether to proceed. |
| | **B Design Brief**<br>Development of initial statement of requirements into the Design Brief or on behalf of the client confirming key requirements and constraints.<br>Identification of procurement method, procedures, organisational structure and range of consultants and others to be engaged for the project. |
| **Design** | **C Concept**<br>Implementation of Design Brief and preparation of additional data.<br>Preparation of Concept Design including outline proposals for structural and building services systems, outline specification and preliminary cost plan.<br>Review of procurement process. |
| | **D Design Development**<br>Development of concept design including structural and building services systems, updated outline specifications and cost plan.<br>Completion of Project Brief. |
| | **E Technical Design**<br>Preparation of technical design(s) and specifications, sufficient to co-ordinate components and elements of the project. |
| **Pre-construction** | **F Production Information**<br>Preparation of detailed information for construction. |
| | **G Tender Documentation**<br>Preparation of tender documentation. |
| | **H Tender Action**<br>Identification and evaluation of potential contractors and specialists.<br>Obtaining and appraising tenders. |
| **Construction** | **J Mobilisation**<br>Letting the building contract, appointing the contractor.<br>Issuing information to the contractor.<br>Arranging handover of site. |
| | **K Construction to Practical Completion**<br>Administration of the building contract to Practical Completion.<br>Provision to the contractor of further information.<br>Review of information. |
| **Use** | **L Post Practical Completion**<br>Administration of the building contract and making final inspections.<br>Assisting building user during initial occupation period.<br>Review of project performance in use. |

**2.7 Internal walls and partitions**
Internal walls, partitions and insulation. Chimneys forming part of internal walls up to plate level. Screens, borrowed lights and glazing. Moveable space-dividing partitions. Internal balustrades excluding items included with 'Stair balustrades and handrails' (2.7.2)

**2.8 Internal doors**
Doors, fanlights and sidelights. Sliding and folding doors. Hatches. Frames, linings and trims. Ironmongery and glazing. Lintels, thresholds and work to reveals of openings.

(Reproduced with permission of the BCIS)

## Cost planning

- Cost planning came to prominence during the 1970s and was devised in an attempt to introduce more rigor and accuracy into the pre-contract costing process.
- The key publication that introduced cost planning was Building Bulletin No. 4 published by the then Department of Education and Science in 1972.
- The system, described in the following text, is based to some extent on the standard model set out in Building Bulletin No. 4 plus the standard list of elements developed by The Building Cost Information Service (BCIS) that is now the industry standard for presenting cost plans and cost advice.
- The BCIS format therefore attempts to produce all cost information in a standard format based on standard rules and parameters.
- Note that in order to maintain accuracy in the target cost, there should be no major changes in design after Stage D – Design Development, see Table 2.2.

## Cost control

- Planning cost is only half of the story and once a project commences on site there is a need to control cost targets, set in the pre-contract phase, to ensure that costs do not spiral out of control.

## Cost analysis

- Cost planning of new projects is dependent on accurate and well-documented cost information. Cost analysis is the process of analysing and recording cost data of projects once tender information has been received.

**Table 2.2** Elemental cost planning

| Design sequence | Process | Advice | Technique | NRM1 |
|---|---|---|---|---|
| A. Appraisal | Cost planning | Cost range | Interpolation | Floor area<br>Unit method<br>Elemental |
| B. Design Brief | Cost planning | Feasibility study | Interpolation | Floor area<br>Unit method<br>Elemental |
| C. Concept | Cost planning | Confirm cost limit | Single rate estimating | |
| D. Design | Cost planning Development | Cost plan | Single rate estimating | Elemental cost planning |
| E. Technical Design | Cost control | Cost checking | Approximate quantities | Elemental cost planning |
| F. Production | Cost control Information | Cost checking | Approximate quantities | Elemental cost planning |
| G. Tender | Cost control Documentation | Cost checking | Single rate estimating | Pre-tender estimate |
| H. Tender action | Cost control | Cost analysis | Elemental analysis of tender | Pre-tender estimate |

- A cost analysis is a record of how cost has been distributed over the elements of a building; it includes a brief description of the overall nature of the project and specification notes on the general level of finishes and so on.
- A cost analysis should be prepared as soon as possible after tenders have been received for a number of reasons such as, currency of the cost information and familiarity with the project.
- One of the objectives of element cost planning is to make it easier for the design team and client to see at a glance where the cost of the project has been allocated.
- The cost allocation should become more detailed as the design develops although the overall budget should remain unchanged.

As illustrated in Figure 2.1 a target cost is set at the feasibility stage and as the information becomes more detailed this sum is allocated over first the element groups and the building elements themselves.

## Cost significant elements

- The BCIS standard list of elements comprises 62 individual elements and 14 groups of elements. Some of these elements/element groups will

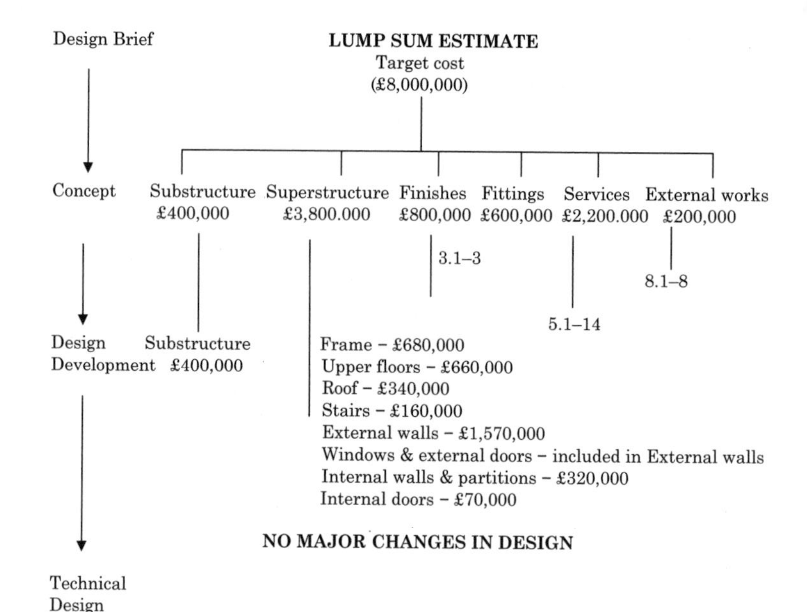


**Figure 2.1** Cost planning

be larger in value terms than others, for example an analysis of several cost analyses would show a range of values for two elements as follows:
External walls: 12–20% +
Services: 15–30% +

- It therefore makes sense to concentrate more on the cost significant elements and less on elements such as stairs and drainage, during the preparation of cost estimates. It is also good practice once the preliminary estimate is complete to check that the amount allocated to each group of elements is in an acceptable 'ball park' range for comparable building types.
- A useful guide to measurement conventions is *The Code of Measurement Practice: A Guide for Property Professionals*, RICS Property Measurement Group, see Chapter 3.

## Design risk

- An amount included in the budget at Stage B (Design Brief), after the elemental cost has been calculated, based on the perceived risk that the design may undergo changes that will impact on the budget during Stages C and D.

### Price risk

- An amount included to reflect the estimated increases in cost between the preparation of the cost plan and the tender date based on tender-based index.

## APPROXIMATE ESTIMATING TECHNIQUES

### Interpolation

Interpolation is a technique used in the early stages of the design sequence when information, drawn or otherwise, on the proposed project is in short supply. It requires a good deal of skill and experience and is the process of adding in or deducting from the cost analysis figure to arrive at a budget for a new project. Therefore in preparing a budget for a new project a cost analysis has been chosen as the basis for the estimate. However, the cost analysis will contain items that are not required for the new project and these must be deducted. For example, in the new project the client wishes to exclude the installation of air conditioning, included in the cost analysis project, from the estimate, and this will have to be deducted from the budget but on the other hand the client wishes to include CCTV throughout and the cost of providing this must be calculated and added in. It is important, as described later, to adjust costs to take account of differences in price levels. The process continues until all identified differences have been accounted for.

Other credible approaches to approximate estimating that are available to the quantity surveyor are:

- the unit and square meter methods, generally used for preliminary estimates when firm information is scarce;
- approximate quantities and elemental cost planning for later stage estimates.

Other approaches are often cited, most notably cubic metre and storey enclosure methods, but the accuracy of these approaches are somewhat dubious; they are seldom used in practice for construction projects and are not considered here.

### Unit method

The unit method is a single price rate method based upon the cost per functional unit of the building, a functional unit being, for example, a hotel bedroom. This method is often regarded as a way of making a comparison between buildings in order to satisfy the design team that the costs are reasonable in relation to other buildings of a similar nature. It is not possible to adjust the single rate price and therefore is very much a ball park approach. The unit method is suitable for clients who specialise in one type of project; for example, hotel or supermarket chains, where it can be surprisingly accurate.

Other examples where unit costs may apply are:

- schools – cost per pupil
- hospitals – cost per bed space.

**Superficial method**

The superficial method is a single price rate method based on the cost per square metre of the building. The use of this method should be restricted to the early stages of the design sequence and it is probably the most frequently used method of approximate estimating. Its major advantage is that most published cost data is expressed in this form. The method is quick and simple to use, though as in the case of the unit method, it is imperative to use data from similarly designed projects. Another advantage of the superficial method is that the unit of measurement is meaningful to both the client and the design team. Although the area for this method is relatively easy to calculate, it does require skill in assessing the price rate. The rules for calculating the area are:

- All measurements are taken from the internal face of external walls. No deduction is made for internal walls, lift shafts, stairwells, etc. – gross internal floor area.
- Where different parts of the building vary in function, then the areas are calculated separately.
- External works and non-standard items such as piling are calculated separately and then added into the estimate. Figures for specialist works may be available from subcontractors and specialist contractors.

For example:

Gross floor area for office block in Figure 2.2

10.0 × 25.0 = 250.00

Less

2/3.0 × 7.50 = 45.00

| | | |
|---|---|---|
| Area per floor | 205.00 × 5 = 1025.00 m$^2$ × £1100 m$^2$ = | £1,127,500.00 |
| Basement | 7.00 × 25.00 = 175.00m$^2$ × £1300 m$^2$ = | £227,500.00 |
| **Estimate for block** | | **£1,355,000.00** |

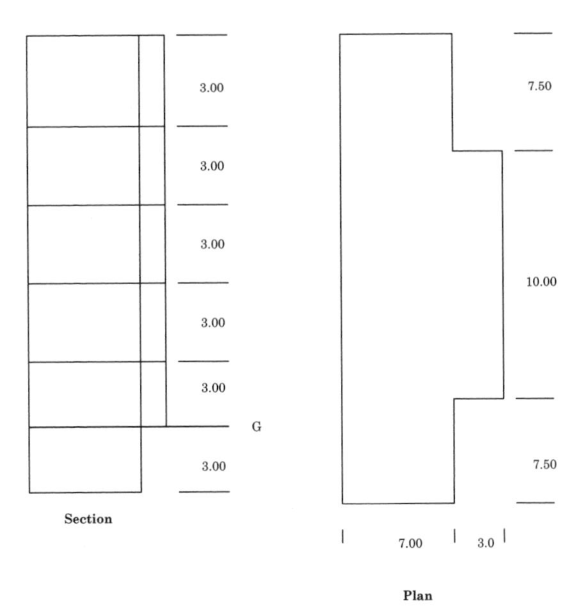


**Figure 2.2** Office block section and plan

## Approximate quantities

Regarded as the most reliable and accurate method of estimating, provided that there is sufficient information to work on. Depending on the experience of the surveyor, measurement can be carried out fairly quickly using composite rates to save time.

The rules of measurement are simple, although it must be said, not standardised and tend to vary slightly from one surveyor to another.

- One approach involves grouping together items relating to a sequence of operations and relating them to a common unit of measurement, unlike the measurement for a bill of quantities, where items are measured separately.
- Composite rates are then built up from the data available in the office for that sequence of operations.
- All measurements are taken as gross overall with the exception of the very large openings.
- Initially the composite rates require time to build up, but once calculated they may be used on a variety of estimating needs.
- Reasonably priced software packages are now available.

An example for a composite is shown below for substructure:

> Excavate trench width exceeding 0.30m maximum depth not exceeding 1.00m; earthwork support, filling to excavations, disposal of excavated material off site, in situ concrete (20N/mm$^2$ aggregate) foundation poured against earth, 275mm cavity brickwork to 150mm above ground level, bitumen-based dpc, facings externally – cost per m.

Normally, if measured in accordance with NRM these items would be measured separately as described in Chapter 3, however, when using approximate quantities a composite rate is calculated that includes a mix of units of measurement and is applied to a linear metre of trench.

### Builder's quantities

Builder's quantities are quantities measured and described from the builder's view point, rather than in accordance with a set of prescribed rules, such as SMM7 or NRM2. There can be quite a big difference between the builder's quantities and approximate quantities. For example, earthwork support may be required to be included, whereas in practice this may be omitted from the site operation with the sides of the excavation being battered back instead. A more pragmatic approach is reflected in the measurement and pricing of builder's quantities.

### Elemental cost planning

A cost plan is an estimate presented in a standard elemental format. The industry norm is the BCIS list of elements 4th Edition, referred to earlier. The estimate is based on a cost analysis of a previous project that is adjusted to suit the new project. The characteristics of elemental cost planning are as follows:

- Costs are related to elements of the building, a feature that is not possible to achieve using other methods of estimating that produce a lump sum estimate.
- The elemental cost allocation is of use to the client and the design team as it is possible to see from an early stage in the design sequence how much of the project budget has been allocated to various aspects and elements of the project.
- The estimate increases in detail as the design development progresses. Initial costs are allocated over the BCIS groups of elements:

- facilitating work
- substructure
- superstructure
- internal finishes
- fittings
- services
- work to existing buildings
- external works.

- By the design development stage it should be possible to allocate costs over the full range of BCIS elements. This transparency allows costs to be reallocated while keeping the overall target cost the same.
- Costs are based on a cost analysis of a previous similar project and adjusted for a range of factors as described below.
- Costs are presented in a variety of ways, with the cost of preliminaries shown separately and apportioned amongst the elements, for example:
- Cost of preliminaries and contingencies shown separately.

**Table 2.3** Presentation of cost analysis data

| **Element** | **Total cost of element** £ | **Cost per m$^2$ gross floor area** £ | **Element unit quantity** | **Element unit rate** £ |
|---|---|---|---|---|
| 1. Substructure | 313,000 | 62.6 | 1000m$^2$ | 313.0 |

It can be seen from Table 2.3 that the cost per m$^2$ of gross floor area is £62.60, whereas the element unit rate is £313.00 per m$^2$ of element. Given the choice, the more accurate cost information is the element unit rate as it reflects the actual cost of providing a specific element; whereas the cost per m$^2$ of floor area can be corrupted by other factors such as building plan shape, etc.

## Sources of cost information

Not all cost information has the same reputation for accuracy and reliability and care should be exercised when choosing cost data for a new estimate.

- Cost analyses as published by the BCIS. The Building Cost Information Service has since its inception in 1970 published a wealth of cost data on a wide variety of building types. The advantages of the BCIS are that the service is available, for a subscription fee, online and is published in standard cost analysis format. The BCIS is also a useful source of cost data for calculating cost forecasts, etc.

- Cost information from published price books such as Spon's or Laxton's. Price books are published annually and contain a range of prices for standard bills of quantities items. Because they are in book form the information tends to be several months old.
- Priced bills of quantities from previous projects. A useful source of information as the cost information tends to be current. As with other forms of cost information, there is a need to adjust for differences in location, etc.
- Cost analysis and cost models produced in-house. Depending on the size of organisation, perhaps the most reliable source of cost information, partly due to the fact that it is easier to ensure good quality control on the data. Also data presented in this format will be easily understood and interpreted. The disadvantage is the time and cost taken to prepare and store the information.

### *Cost planning example at the Concept and Design Development Stages (Stages C and D – RIBA Outline Plan of Work)*

The object of this stage is to allocate the budget target cost given to the client during the feasibility stage over the range of standard elements.

Cost planning is a continuing process that gradually becomes more detailed as the design process progresses as illustrated in Table 2.1. The design team has formulated a concept design and the object of the first estimate is to arrive at a cost limit that the client is happy with to which the design team can prepare their detailed design. Assuming that the client has approved the target cost the elemental cost planning can begin.

To prepare an elemental cost plan the following information should be assembled:

- a cost analysis of a previous similar building;
- sketch plans and elevations of the proposed project;
- outline specification/levels of services installation, etc. for the proposed project.

Therefore, when preparing an elemental cost plan, the first task is to select an appropriate cost analysis as discussed above. If using the BCIS, the selection can be facilitated by using their online service, where it is possible to enter the required parameters of the new project in the search facility. Several cost analyses will be selected and from these the most appropriate is chosen and used as the basis for the elemental cost plan. The selection is based on a comparison between the cost plan project and the available cost analyses over a range of parameters and should be done carefully. The criteria for selection are as follows:

- form of construction
- level of specification.

In addition cost significant elements should be examined for similarity.

Even after this process the data in the cost analysis will need to be adjusted before being used for the cost plan. The front sheet of a cost analysis contains a wealth of information relating to the analysed project that can be used in the adjustment process.

The adjustments to cost analysis data can be categorised as follows:

- price levels
- quantity
- quality.

**Price levels**

Differences in price levels between cost analysis and cost plan data are adjusted using the following:

- building cost indices
- location indices
- tender price indices.

Measures of changes in items such as location, building costs or tender prices are performed using index numbers. Index numbers are a means of expressing data relative to a base year. For example, in the case of a building cost index a selection of building materials is identified, recorded and given the index number 100. Let us say for the sake of argument that the cost of the materials included in the base index is £70.00 in January 2005. Every three months the costs are recorded for exactly the same materials and any increase or decrease in cost is reflected in the index as follows:

Building cost index January 2005 = 100
Building cost index January 2012 = 135

This therefore represents an increase of 35% in the cost of the selected materials and this information can be used if, for example, data from a 2005 cost analysis was being used as the basis for calculating costs for an estimate in January 2012.

Example:

Analysis cost for a steel-framed office project (Jan 2005) £3,500,000

$$£3,500,000 \times \frac{135}{100} = £4,725,000$$

Building cost information such as this can also be used to try to forecast how costs may alter in the future, say during the construction phase of a project.

Any variation in the cost of either of these basics will influence the cost of the works. Cost indices are an attempt to measure price variations that occur between tenders obtained at different times in differing places. The quantity surveyor is able to study the various indices, together with predicted future cost trends – be it a rise or fall and any regional variations – in order to facilitate adjustments to historical cost information. Indices reflect changes and all indices require the selection of a base period (usually this is set at 100), any increases or decreases being reflected in the indices.

### *Location indices*

Tender price levels vary according to the region of the country where the work is carried out. Generally speaking, London and the South East of England are the most expensive and the regional variations are reflected in a location index that is used to adjust prices. The BCIS annually publishes a set of location indices that cover most parts of the UK and these can be used to adjust in cases where the cost analysis building and the cost plan building are in different locations.

### *Building cost indices*

- The cost of any building is determined, primarily, by the cost of the labour and materials involved in its erection.
- Building cost indices measure changes in the cost of materials, labour and plant to the contractor. They ignore any changes in profit levels, overheads, productivity, discounts, etc. and therefore they effectively measure changes in the notional rather than actual costs.
- Building cost indices track movements in the input costs of construction work in various sectors, incorporating national wage agreements and changes in material prices as measured by government index series. They provide an underlying indication of price changes and differential movements in various work sectors, but do not reflect changes in market conditions affecting profit and overheads provisions, site wage rates, bonuses or material price discounts and premiums. In a world of global markets, building cost indices can be influenced by many factors including demands in emerging and developing markets such as China and India, for example, Davis Langdon Indices.
- Used to adjust and allow for cost increases between the date of the preparation of the estimate and the tender date.

### *Tender price indices*

- Tender price indices are based on what the client has to pay for a building as it takes into account building costs. These indices therefore reflect fluctuations in the tendering market.
- Tender price indices can be used to adjust for potential increases in cost between the date of the preparation of the cost plan and the actual date the project goes to tender.

### Other information

The front cover of the cost analysis should also contain information relating to contract type, procurement strategy, market factors, etc., that prevailed at the date the project was current. All of these factors can affect price levels and should be taken into account when preparing a cost plan:

- Contract type – there are a wide variety of contract forms available (see Chapter 6) and within these contract forms there are a variety of alternatives available. The type of contract used for a project can affect the price and should be allowed for when preparing a cost plan.
- Procurement strategy – similarly the procurement strategy can affect costs as different strategies will have different allocations of risk and this will be reflected in price levels, as discussed in Chapter 4.
- Market conditions – when market conditions are buoyant and work is plentiful contractors may choose to include high profit levels as compared to situations when work is in short supply. Once again this factor can have an influence on pricing levels and should be taken into account.

### *Differences in quantity*

This adjustment takes account of differences in the elemental quantity of the cost plan and cost analysis projects. Table 2.3 shows the information given in an elemental cost plan; elemental unit quantities and rates should be used for this adjustment.

### *Differences in quality*

The final adjustment is an attempt to allow for differences in quality, say finishes and specification levels.

#### EXAMPLE: ELEMENT 2.5 – EXTERNAL WALLS

A cost plan is being prepared for a new six-storey office project with glazed curtain walling. A cost analysis has been selected of a previous similar building and the costs are adjusted as follows:

### Data

Cost analysis

Element unit rate: £745.00/m$^2$
Element unit quantity: 6100m$^2$
Date: Jan 2005 – BCIS Index: 221
Location: West Midlands – BCIS location factor: 95

### Cost plan

Element unit quantity: 5700m$^2$
Date: November 2011 – BCIS Index 227
Location: Yorkshire & Humberside – BCIS location factor: 92

### Prices levels

Adjust for differences in tender price levels as before:

$$£745.00 \times \frac{227}{221} = £765.23$$

Adjust for location:

$$£765.23 \times \frac{92}{95} = £741.07$$

### Quantity

The updated cost can now be multiplied by the element unit quantity for the cost plan project:

£741.07 × 5700m$^2$ = £4,224,099.00

### Quality

The cost plan building is to have a higher glazing specification than the cost analysis project:

| | |
|---|---|
| | £4,224,099.00 |
| **Add** | |
| Self-cleaning glass 5700m$^2$ @ £15.00/m$^2$ | £85,500.00 |
| Target cost | £4,309.599.00 |

The process is now continued for the remainder of the elements.

As mentioned in Chapter 1, the *RICS New Rules of Measurement*, Volume 1: NRM1: *Order of Cost Estimating and Cost Planning for Capital Building Works* was published in 2009 and relaunched in 2012.

The structure of Volume 1 is:

- Part 1. This places rules of measurement in context with the RIBA Plan of Work and the Office of Government Commerce (OGC) Gateway Process* as well as explaining the definitions and abbreviations used in the rules.
- Part 2. The purpose and content of an order of cost estimate are explained as is how to prepare an order of cost estimate using three prescribed approaches; floor area, functional unit and elemental method.
- Part 3. The purpose and preparation of elemental cost plans is demonstrated.
- Part 4. The final section contains tabulated rules of measurement for formal cost plans.

### *Joined-up cost advice*

The basic premise behind the launch of NRM is that SMM7 no longer adequately addresses changes in procurement strategies and that in addition the alternative approaches to measurement being used in practice present problems for organisations like the BCIS when it comes to capture cost data, as there is no standard format.

The Building Cost Information Service Standard Form of Cost Analysis (SFCA) was first produced in 1961 when the bill of quantities was king and subsequently revised in 1969 and 2008 and has been the industry norm for the last forty years. The SFCA is presented in element format but in truth it has not really changed its format since its original launch, whereas the industry has moved on! Once the NRM becomes widely adopted there is obviously a lot of work to be done by the BCIS in order to convert its existing data to the NRM format.

Figure 2.3 illustrates a project overview of the NRM from which the various strands of the initiative can be identified.

One of the factors that has driven Volume 1 of the NRM is the lack of specific advice to quantity surveyors on the measurement of building works solely for the purpose of preparing cost estimates and cost plans. As illustrated in Figure 2.4 the process of producing a cost estimate and cost planning has been mapped against the RIBA plan of work and OGC Gateway process. It

* The OGC developed the OGC Gateway™ Process as part of the Modernisation Agenda to support the delivery of improved public services and examine programmes and projects at various key decision points throughout the life-cycle of delivery. The process is mandatory in central civil government for procurement, IT-enabled and construction programmes and projects. For the purposes of this chapter reference will be made to the RIBA Plan of Work.

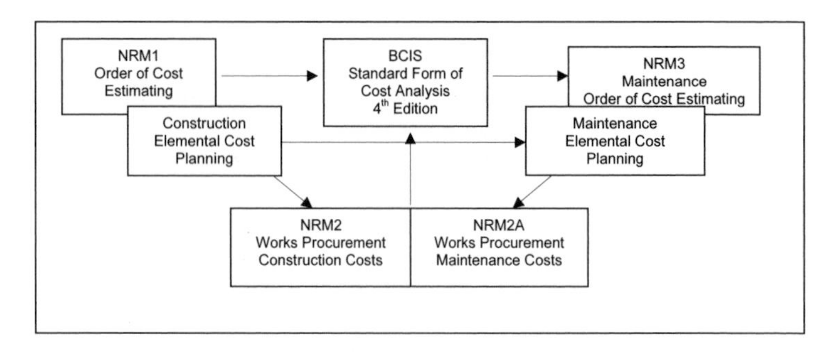


**Figure 2.3** Project overview of the NRM

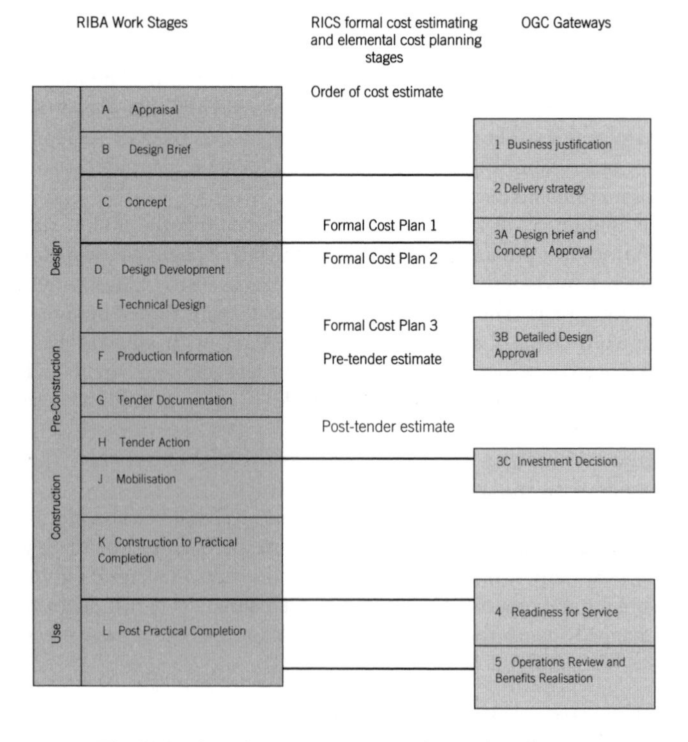


**Figure 2.4** The RICS formal cost estimation and cost planning stages in context with the RIBA Plan of Work and OGC Gateways. RIBA Plan of Work is copyright RIBA

shows that the preparation and giving of cost advice is a continuous process, that in an ideal world becomes more detailed as the information flow become more detailed. In practice it is likely that the various stages will merge and that such a clear-cut process will be difficult to achieve.

The NRM suggests that the provision of cost advice is an iterative process that follows the information flow from the design team as follows:

- order of cost estimate
- formal cost plan 1
- formal cost plan 2
- formal cost plan 3
- pre-tender estimate.

There would therefore appear to be two distinct stages in the preparation of initial and detailed cost advice:

1. Estimate – an evolving estimate of known factors. Is the project affordable? The accuracy at this stage is dependent on the quality of the information. Lack of detail should attract a qualification on the resulting figures. At this stage information is presented to the client as shown in Table 2.1.
2. Cost plan – a critical breakdown of the cost limit for the building into cost targets for each element. At this stage it should be possible to give a detailed breakdown of cost allocation as shown in Table 2.2.

In addition the NRM approach divides cost estimates and cost plans into five principal cost centres:

1. works cost estimate
2. project/design team fees estimate
3. other development/project cost estimate
4. risk allowance estimate
5. inflation estimate.

The order of cost information and cost plan stages have differing recommended formats, see Table 2.4 for Order of Cost Estimate recommended format. Compared to the BCIS (SFCA) the NRM format does provide a greater range of cost information to the client covering the following:

- building works including facilitating works
- main contractor's preliminaries
- main contractor's profit and overheads
- project/design team fees

Table 2.4 The *RICS New Rules of Measurement*: order of cost estimate format

| Ref | Item | £ |
|---|---|---|
| 0 | Facilitating Works | |
| 1 | Building Works | |
| 2 | Main Contractor Preliminaries | |
| 3 | Main Contractor Overheads and Profit | |
| | **Works Cost Estimate** | |
| 4 | Project/Design Team Fees | |
| 5 | Other Development/Project Costs | |
| | **Base Estimate** | |
| 6 | Risk Allowances: | |
| | Design Development Risk | |
| | Construction Risks | |
| | Employer Change Risks | |
| | Employer Other Risks | |
| | **Cost Limit (excluding inflation)** | |
| 7 | Inflation: | |
| | Tender Inflation | |
| | Construction Inflation | |
| | **Cost Limit (including inflation)** | |

- other development/project costs
- risk
- inflation
- capital allowances, land remediation relief and grants
- VAT assessment.

A feature of the NRM is the detailed lists of information that are required to be produced by all parties to the process; the employer, the architect, the mechanical and electrical services engineers and the structural engineer all have substantial lists of information to provide. There is an admission that the accuracy of an order of cost estimate is dependent on the quality of the information supplied to the quantity surveyor. The more information provided, the more reliable the outcome will be and in cases where little or no information is provided, the quantity surveyor will need to qualify the order of cost estimate accordingly.

The development of the estimate/cost plan starts with the order of cost estimate.

## Order of cost estimate, stages 1–3

The works cost estimate has three constituents:

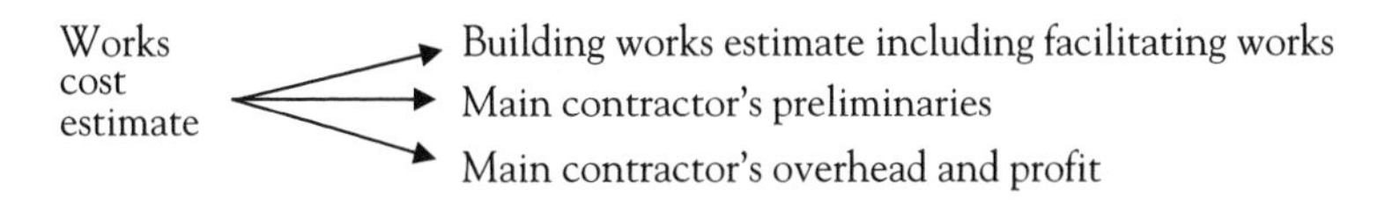


At this stage the main contractor's preliminaries and overheads and profit are included as a percentage with subcontractors' preliminaries and overheads and profit being included in the unit rates applied to building works.

Perhaps the most worthwhile feature of the NRM is the attempt to establish a uniform approach to measurement based on the *RICS Code of Measurement Practice* (6th Edition) 2007 in which there are three prescribed approaches for preparing building works estimates namely:

1. Cost per m$^2$ of floor area:
   - Gross external area (GEA)
   - Gross internal area (GIA/GIFA)
   - Net internal area (NIA).
2. Functional unit method: Building works estimate = number of functional units × cost per functional unit. A list of suggested functional units are included in the *New Rules of Measurement*; Appendix B.
3. The elemental method: Building works estimate = sum of elemental targets. Cost target (for element) = element unit quantity (EUQ) × element unit rate (EUR). The amount of detail required to be given with this approach can be seen from Figure 2.1 although the choice and the number of elements used to break down the cost of building works will be dependent on the information available. Rules for calculating EUQ are included in the NRM.

At this stage the main contractor's preliminaries and profit and overheads are recommended to be included as a percentage addition. Subcontractor's overheads and profit should be included in the unit rates applied to building works.

### Project and design fees, stage 4

In the spirit of transparency the costs associated with project and design fees are also itemised:

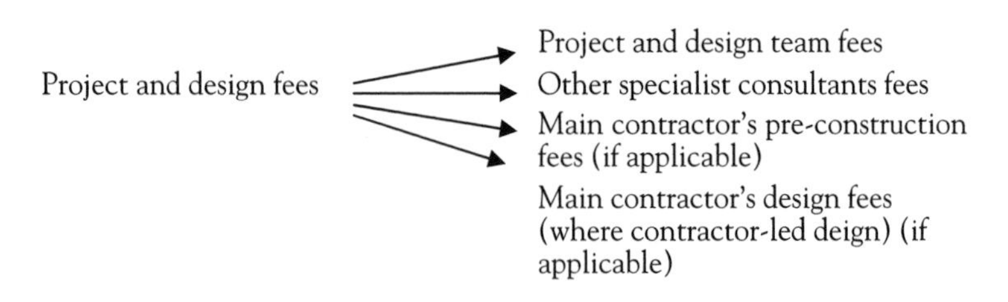


### Other development and project costs estimates, stage 5

This section is for the inclusion of costs that are not directly associated with the cost of the building works, but form part of the total cost of the building project, for example planning fees.

### Risk allowance estimate, stage 6

Risk is defined as, 'the amount added to the base cost estimate for items that cannot be precisely predicted to arrive at the cost limit'.

The inclusion of a risk allowance in an estimate is nothing new, what perhaps is new, however, is the transparency with which it is dealt with in the NRM. It is hoped therefore that the generic cover-all term 'Contingencies' will be phased out. Clients have traditionally homed into contingency allowances wanting to know what the sum is for and how it has been calculated. The rate allowance is not a standard percentage and will vary according to the perceived risk of the project. Just how happy quantity surveyors will be to be so up-front about how much has been included for unforeseen circumstances or risk is yet to be seen. It has always been regarded by many in the profession that carefully concealed pockets of money hidden within an estimate for extras/additional expenditure is a core skill.

So how should risk be assessed at the early stages in the project? It is pos-

**Table 2.5** The *RICS New Rules of Measurement*: formal cost plan format

| LEVEL 1 | LEVEL 2 | LEVEL 3 |
|---|---|---|
| **Group element** | **Element** | **Sub-element** |
| 3. Internal finishes | 1. Wall finishes | 1. Finishes to walls |
| | 2. Floor finishes | 1. Finishes to walls<br>2. Raised access floors |
| | 3. Ceiling finishes | 1. Finishes to ceilings<br>2. False ceilings<br>3. Demountable special ceilings |

sible that a formal risk assessment should take place, and this would be a good thing, using some sort of risk register. Obviously, the impact of risk should be revisited on a regular basis as the detail becomes more apparent.

Risks are required to be included under four headings:

1. design development risks – for example, design development and environmental issues;
2. construction risks – for example, site restrictions, existing services;
3. employer's change risk – for example, changes in the scope of the works or brief;
4. employer's other risk – for example, early handover, postponement/acceleration.

### Inflation estimate, stage 7

Finally, an allowance is included for inflation under two headings:

1. tender inflation – an allowance from the period from the estimate base date to the return of the tender;
2. construction inflation – to cover increases from the date of the return of tender to a mid-point in the construction process.

Inflation should be expressed as a percentage using either the retail price index or the tender price index and the BCIS building cost indices. This adjustment is of course in addition to any price adjustments made earlier in the process when adapting historic cost analysis data. In addition care should be taken not to update previous rates that were based on percentage additions, e.g. main contractor's preliminaries, main contractor's overheads and profit and project/design team fees, as these will adjust automatically when the percentages are applied.

Finally it is suggested that other advice could be included relating to:

- value added tax
- capital allowances
- land reclamation relief
- grants.

Whether this will be possible it remains to be seen, certainly giving tax advice has been a stock-in-trade for many quantity surveyors for some time, but it really needs specialist, up-to-date information. In addition, particularly with VAT, the tax position of the parties involved may differ greatly and advice should not be given lightly.

From this point on advice is given by the preparation of Formal Cost Plans 1, 2 and 3. It is anticipated that for the formal cost plan stages the elemental approach should be used and this should be possible as the quantity and quality of information available to the quantity surveyor should be constantly increasing. Table 2.5 demonstrates the degree to which detail increases during this process. At the formal cost plan stages the NRM recommends that cost advice is given on an elemental format and to this end Part 4 of the NRM contains comprehensive rules for the measurement for building works. These tabulated rules of measurement for elemental cost planning enable quantities to be measured to the nearest whole unit, providing that this available information is sufficiently detailed. When this is not possible, then measurement should be based on GIFA. From Formal Cost Plan 2 Stage cost checks are to be carried out against each pre-established cost target based on cost-significant elements. One thing that is clear is that the NRM approach, if followed, appears too labour intensive and one thought is that the cost planning stages and procurement document stages will morph so that the final cost plan becomes the basis of obtaining bids. Over the coming years it will be interesting to learn to what extent the NRM replaces the tried and trusted standard methods of measurement not only in the UK but also in overseas markets.

In the autumn of 2011 a second edition of NRM1: *Order of Cost Estimating and Cost Planning for Capital Building Works* was being prepared; the proposed changes were to more fully align the BCIS and the RICS NRM elemental breakdown structures.

## ELEMENTAL COST CONTROL

Cost planning takes place over Stages A–D: Appraisal – Design Development (see Table 2.2), after which cost control takes over. In order for the integrity of the system to be maintained it is important that no major changes in the design are now made after Stage D. At Design Development Stage each element should have a cost target; the cost control process involves checking that the cost allocation is realistic, now more detailed information is available using approximate quantities. Cost checking at the Technical Design Stage involves:

- checking the detailed design against the outline and scheme design to detect major changes;
- carrying out a cost check of each cost target using approximate quantities;
- in the case that a cost target is found to be unrealistic now detailed information is available, drawing up a 'threshold of pain' list for the client setting out the alternatives, namely:

  - work to the revised cost target, but make compensating reductions elsewhere in the building;
  - increase the target cost for the project;
- carrying out a cost check in a systematic manner with some surveyors using a pro forma to record the details.

During the construction phase the cost control process continues with the preparation of Financial Statements. These statements are produced by the quantity surveyor at either monthly or three monthly intervals and predict the final cost of the project when completed. They require the quantity surveyor to assess the financial effect of variations and other adjustments to the contract that have been issued or are expected to be issued and require a good deal of expertise to produce accurate figures and costs.

**New Office Block, Woking, Surrey**

**Financial Statement No. 3**

Date 10 March 2012

| | £ | £ |
|---|---|---|
| Contract sum | | 5,474,316 |
| Less Contingencies | | 36,000 |
| | | 5,438,316 |
| Adjust for: | | |
| Variation orders No's 1–34 | 25,000 | |
| Subcontractors | (6700) | |
| Provisional sums | 10,000 | |
| Projected variations | 36,000 | |
| Contractor's claim | Nil | 64,300 |
| Anticipated Final Account | | £5,374,016 |

Exclusions:

VAT
Professional fees

## Design and cost

The design of a building can have a major impact on costs. These can be grouped as follows:

- plan shape
- height and number of storeys

- orientation and footprint
- choice of elements/materials.

### *Plan shape*

The enclosing ratio of a building is a useful rule of thumb to assess the efficiency of a plan shape. It is found by dividing the area of the enclosing envelope by the gross floor area; the smaller the result of the alternatives being compared, the comparatively more economical the design. Figure 2.5 shows three different plan shapes that all provide 100m² of gross floor area, the height of the external envelope is 4m in all cases.

By applying the enclosing ratio to the above three alternatives the following results are obtained:

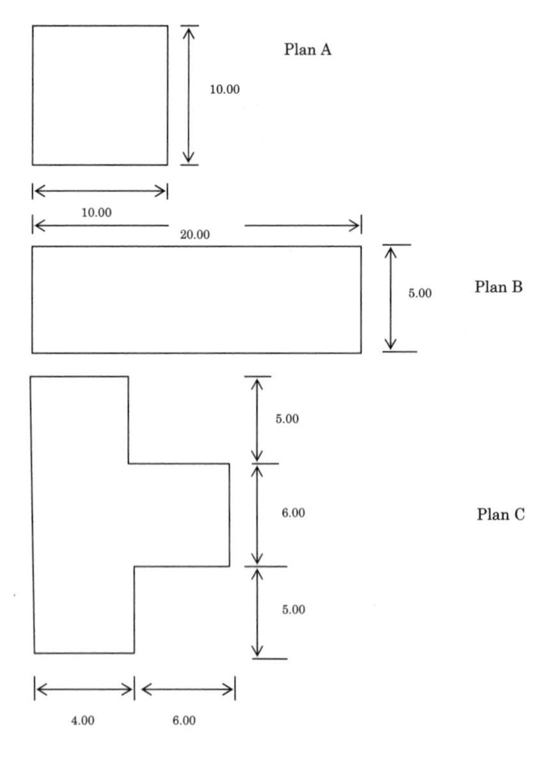


**Figure 2.5** Enclosing ratios

$$\text{Plan A} = \frac{160}{100} = 1.60$$

$$\text{Plan B} = \frac{200}{100} = 2.00$$

$$\text{Plan C} = \frac{208}{100} = 2.08$$

It is clear that Plan A gives the best, i.e. lowest, answer with 1.60 metres of enclosing envelope required per $m^2$ of gross floor area. The amount of external envelope is important as it is one of the cost significant groups of elements, containing, external walls and fenestration. It can also be seen that the more the plan shape moves away from a square, then the comparatively more expensive the design becomes. In theory the most economic design solution is a circular plan shape, a circle having the least amount of external envelope per $m^2$ of gross floor area, however, in practice other costs associated with building with a circular footprint outweigh any cost savings.

### *Height and number of storeys*

The main impact when storey height is increased is on the vertical elements, such as external and internal wall finishes. There will also be an impact on services installations with pipe and cable runs increasing.

The number of storeys in a building can affect costs in the following ways:

- Single-storey structures are comparatively expensive, as the substructure required for a two- or three-storey structure is often only marginally bigger than for a single storey.
- Buildings with more than three storeys will require a lift installation.
- Generally, multi-storey buildings require a large substructure.
- Tall buildings require areas devoted to circulation space such as lifts, escape staircases, service floors and plant rooms. These areas have to be deducted from the gross floor area to arrive at net lettable areas.
- Maintenance costs will be greater for high-rise buildings.
- Fire protection will be at a high level.

### *Orientation and footprint*

Should accommodation be provided in one large or groups of smaller buildings? Given the choice it is better to use one larger building as the cost significant elements are considerably greater when two smaller buildings are used.

Some plan shapes have become the industry norm for certain types of buildings, for example Figure 2.6 illustrates a widely used plan for hotels that maximises floor area and income with a central corridor a bedrooms on either side.

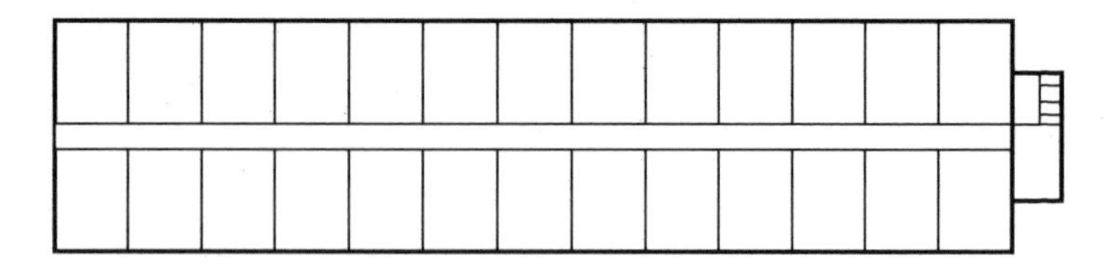

**Figure 2.6** Standard hotel floor plate layout

Note, it should be remembered that factors such as the shape of the site or topography may override the other factors.

### *Choice of elements/materials*

The choice of materials is becoming increasingly important with an increased emphasis being placed on sustainability. Care should be taken to take account of initial, running and maintenance costs as well as adaptation and disposal costs.

## FORECASTING VALUE

When attempting to forecast costs, value and income, the following tools are available.

### Sinking funds

For investments in wasting assets, e.g. a leasehold property. Given that systems are going to wear out and/or need partial replacement during the currency of a contract it is thought to be prudent to 'save for the rainy day' by investing capital in a sinking fund to meet the cost of repairs, etc. The sinking fund allowance therefore becomes a further cost to be taken into account during the evaluation process. Whether this approach is adopted will depend on a number of features including corporate policy, interest rates, etc. Sinking funds involve investing a sum of money at the end of each year to accumulate to the sum required to replace the asset. The annual sinking fund to replace £1 in 40 years at 10% is as follows:

$$\frac{i}{A - 1}$$

where: $i = \frac{R\%}{100}$

$A = (1 + i)^n$
$n$ = number of years

$$\frac{0.10}{(1+0.10)^{40}-1} = 0.0023$$

It should be noted that:

- the effect of taxation on the sinking fund should be considered. This can be done as follows:

  $$\times \frac{100}{100-P}$$ where $P$ = the client's rate of tax

- the cost of replacement will increase over time.

**Discounting appraisal techniques**

Discounting appraisal techniques are an attempt to evaluate the effect that time has over the worth of income and expenditure. Much of the financial calculation and appraisal of items that form part of a feasibility study such as income, expenditure, etc. have to be evaluated over the lifetime of the project. Discounting appraisal techniques are normally considered to be superior to conventional methods when doing this due to their implicit recognition of the time value of money. The two main discounting methods used are the net present value (NPV) and the internal rate of return (IRR).

***Net present value (NPV)***

In essence, discounted cash flow involves:

- preparation of a cash flow table showing year by year:
  - the money which is likely to flow out of the organisation as a result of creating and maintaining the investment;
  - the money which is likely to flow into the organisation from the investment;
- calculating the ultimate disposal value of the investment;
- discounting the cash flow table at a selected rate of interest, so as to bring all monies flowing into or out of the organisation, no matter when payments or receipts occurred to the same point in time, i.e. the present value.

## THE NET PRESENT VALUE APPROACH

Discounting allows the current prices of items, such as materials and components, to be adjusted to take account of the value of money of the life-cycle of the product. Discounting cash flows recognises the importance of the timing of the receipt and or payment of various cash flows by isolating differences in time between them. Discounting is required to adjust the value of costs, or indeed, benefits which occur in different time periods so that they can be assessed at a single point in time.

The choice of the discount rate is critical and can be problematic as it can alter the outcome of calculation substantially and there are two golden rules that apply:

- In the public sector follow the recommendations of *The Green Book: Appraisal and Evaluation in Central Government*, currently recommending a rate of 3.5%.
- In the private sector the rule is to select a rate that reflects the real return currently being achieved on investments.
- The discount rate can be considered almost as the rate of return required by the investor which includes costs, risks and lost opportunities. The mathematical expression used to calculate discounted present values is set out below:

$$\text{Present value (PV)} = \frac{1}{(1 + i)^n}$$

where: $(i)$ = rate of interest expected or discount rate
$(n)$ = the number of years

This present value multiplier/factor is used to evaluate the present value of sums, such as replacement costs or money that will be received or is planned to be received at say 10 or 15 year intervals in the future.

Assume that you wish to acquire *now* an investment that will produce a return of 6% in *one year's time*. What is the present day worth or value of such an investment? It can be calculated as follows:

$$\text{Present value (PV)} = \frac{100}{106} = 0.9434$$

Therefore by multiplying the expected return by the above discounting factor, the present worth can be calculated. So, if the anticipated return in one year's time is £10,000, the present value is:

$$\text{Present value (PV) @ 6\%} = £10{,}000 \times 0.9434 = \underline{£9{,}434}$$

Similarly if a benefit, with a monetary value at present values, was planned to be received by a client as part of a deal in, say, 15 years' time, the present value can be calculated as follows:

Present value of benefit × PV @ 3.5% = <u>£149,225</u>
£250,000 0.5969

Calculating the present value of the differences between streams of costs and benefits provides the net present value (NPV) of an option and this is used as the basis of comparison as follows.

Two alternative road schemes have been proposed and both are expected to deliver improvements and time savings.

- Option A requires £10 million in initial capital expenditure to realise benefits of £2.5 million per annum for the following four years.
- Option B requires £5 million in initial capital expenditure to realise benefits of £1.5 million per annum for the following four years.

**Table 2.6** Calculations of two alternative road schemes

| **Year** | 0 | 1 | 2 | 3 | 4 | **NPV** |
|---|---|---|---|---|---|---|
| Discount factor (PV £1) | 1 | 0.9962 | 0.9335 | 0.9019 | 0.8714 | |
| Option A | | | | | | |
| Costs/benefits (£) | –10.00m | 2.50m | 2.50m | 2.50m | 2.50m | |
| Present value (£) | –10.00m | 2.42m | 2.33m | 2.25m | 2.18m | 0.82m |
| Option B | | | | | | |
| Costs/benefits (£) | –5.00m | 1.50m | 1.50m | 1.50m | 1.50m | |
| Present value (£) | –5.00m | 1.45m | 1.40m | 1.35m | 1.31m | 0.51m |

The significance of the results in Table 2.6 is:

- Option A produced a negative net present value, that is to say, the costs are greater than the benefits, whereas
- Option B produced a positive net present value, that is to say the benefits are greater than the costs and it is clearly the better alternative. A marginal or zero net present value is indicative of a do nothing option.

Alternatively, for the example given in Table 2.7, the discounting factor may be calculated as follows:

$$\frac{(1 + 0.035)^4 - 1}{0.035(1 + 0.035)^4} = 3.673$$

consequently, the NPVs for Options A and B can be calculated in a single step as:

$$NPV_A = -10 + 3.673 \times 2.5 = -10 + 9.18 = -0.82\text{m}$$

$$NPV_B = -5 + 3.673 \times 1.5 = -5 + 5.51 = 0.51\text{m}$$

### Annual equivalent approach

This approach is closely aligned to the theory of opportunity costs (explained earlier), i.e. the amount of interest lost by choosing Option A or B as opposed to investing the sum at a given rate per cent, is used as a basis for comparison between alternatives. This approach can also include the provision of a sinking fund in the calculation in order that the costs of replacement are taken into account.

In using the annual equivalent approach the following equation applies:

Present value of £1 per annum (sometimes referred to by actuaries as the annuity that £1 will purchase).

This multiplier/factor is used to evaluate the present value of sums, such as running and maintenance costs that are paid on a regular annual basis.

$$\text{Present value of £1 per annum} = \frac{(1 + i)^n - 1}{i(1 + i)^n}$$

where: $(i)$ = rate of interest expected or discount rate
$(n)$ = the number of years

Previously calculated figures for both multipliers are readily available for use from publications such as *Parry's Valuation Tables*, etc.

### *Internal rate of return (IRR)*

The internal rate of return (IRR) is also commonly used in PFI contracts (see Chapter 4) to measure the rate of return expected to be earned by private sector capital in a project. The characteristics of IRR are:

- IRR is most suitable to situations where a project is predicted to produce a negative cash flow during the early years followed by positive cash flows during the later or final years.
- Many investors are as much concerned with the actual rate of interest which they are earning on their capital as they are with the total profit on any particular investment.
- IRR expresses the benefits on investing as a single rate of interest rather than an end profit.
- IRR is that rate of interest at which all the future cash flows must be discounted in order for the project's net present value to equal zero.
- IRR is defined mathematically as the discount rate which, when applied to discount a series of cash outflows followed by cash inflows, returns a net present value of zero.
- IRR can be thought of as the equivalent constant interest rate at which a given series of cash outflows must be invested in order for the investor to earn a given series of cash inflows as income; for example, a measure of the underlying return the private sector, in the case of a PPP project, expects to receive by investing in the project.
- For the purposes of calculating IRRs all funders, including lenders are considered as investors.
- IRRs can be calculated for different cash flow streams of a project, depending on:
    - which category of investor the IRR is being calculated for;
    - whether inflation is included in the underlying cash flows;
    - whether tax is included in the underlying cash flows.

A technique known as iteration is used to calculate the IRR of the profit. This involves selecting a trial discount rate and then calculating the NPV for that rate. The process is repeated for another rate until two rates are found which both have an NPV very close to and either side of zero, i.e. very low positive NPV and a very low negative NPV. By using linear interpolation the actual IRR can then be calculated to within a certain degree of accuracy. Standard software is available to carry this out and is included as part of the financial model used in investors' calculations (Figure 2.7). Unlike NPV, IRRs cannot be used to value an investment; it is purely a means of analysis and depends for its success on an initial purchase price having already been established.

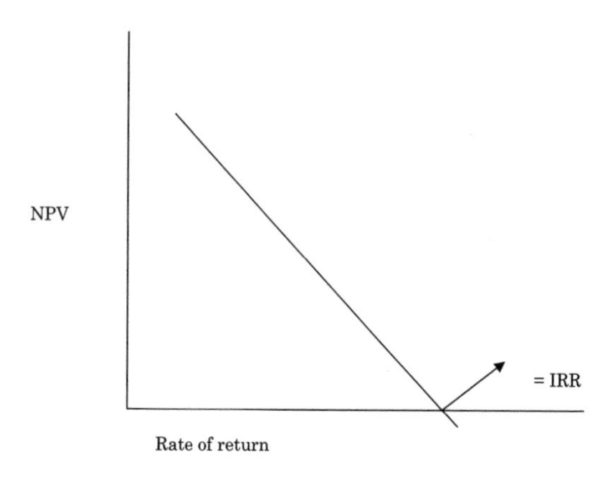


**Figure 2.7** Calculating the internal rate of return

### *NPV or IRR?*

Net present value
Discount rate – 10%

| Year | Discount rate | Option A | | Option B | |
|---|---|---|---|---|---|
| | | Cash flow | NPV | Cash flow | NPV |
| 0 | 1.000 | –1,000 | –1,000 | –1,000 | –1,000 |
| 1 | 1.100 | 340 | 309 | 200 | 182 |
| 2 | 1.210 | 305 | 252 | 235 | 194 |
| 3 | 1.330 | 270 | 203 | 270 | 203 |
| 4 | 1.464 | 235 | 161 | 305 | 208 |
| 5 | 1.611 | 200 | 124 | 340 | 211 |
| **Total** | | 350 | 49 | 350 | –2 |

NPV assesses the value of a future cash flow today, whereas IRR measures the investor's return on a project as follows:

| Year | Option A | | | Option B | | |
|---|---|---|---|---|---|---|
| | Cash flow | Discount rate | NPV | Cash flow | Discount factor | NPV |
| 0 | –1,000 | 1.0000 | –1,000 | –1,000 | 1.0000 | –1,000 |
| 1 | 340 | 1.1208 | 303 | 200 | 1.0994 | 182 |
| 2 | 305 | 1.2561 | 243 | 235 | 1.2087 | 194 |
| 3 | 270 | 1.4078 | 192 | 270 | 1.3288 | 203 |
| 4 | 235 | 1.5778 | 149 | 305 | 1.4609 | 209 |
| 5 | 200 | 1.7684 | 113 | 340 | 1.6061 | 212 |
| **Total** | 350 | | 0 | 350 | | 0 |

The discount rate that gives an NPV of 0 is the IRR = 12.08% (Option A)
9.94% (Option B)

The higher rate of return from Option A means that the cash is received earlier with this option.

Both methods have their relative advantages and disadvantages and generally speaking IRRs are not a reliable alternative to NPV-based calculations for the measurement of the value of an investment, in addition IRR has been criticised for certain implicit assumptions. The following list of assumptions is made in both forms of analysis:

- Future cash flows can be estimated with reasonable accuracy, so there is no need to consider risk or uncertainty.
- The opportunity cost of capital is known or can be estimated with reasonable accuracy.
- All project are simple investments involving an initial cash outflow followed by a series of inflows.
- Investment projects are independent of one another.

In practice, particularly during periods of inflation, risk and uncertainty often need to be taken into account. This is difficult when using NPVs, as present values are in the nature of absolute measures. However, the IRR is expressed as a rate of return so that it can easily be adjusted by a margin to allow for risk. Consequently, for its ease of understanding, its economy of presentation and its flexibility in allowing for risk IRR is preferable.

# THE PROPERTY MARKET AND DEVELOPMENT

## Taxation and property development

**Health warning:** It should be noted that taxation law changes at regular intervals.

HM Revenue and Customs seeks to classify individuals and companies that build and then dispose of property as follows.

### *Investors*

Investors are individuals or companies who build a property and retain ownership for approximately five years or so before disposal.

### *Traders*

Traders are individuals or companies who build a property and dispose of it for profit on completion.

The revenue has several rules of thumb to determine which of the above classifications a tax payer belongs to, however, suffice to say that the tax position of investors is far more favourable than traders who are treated, for taxation purposes, as any trader who produces goods or products.

## Feasibility reports

Whether in the private or public sector, when a new construction project is proposed, the first question that needs to be addressed by clients and quantity surveyors alike at an early stage in the development process is: will the proposed project prove feasible? Although public sector clients may have different interpretations of feasibility, for the purposes of this section feasibility generally means: will the capital invested in the project make the required yield or profit? This question in turn depends on many other factors and variables. Construction is a high-risk process; there are so many external influences that can derail the best laid development plans. For example:

- the general economic situation, including interest rates and employment levels;
- government intervention, for example planning issues and release of green belt land for development;
- demographics and changing needs of the market;
- new entries to the market place/increased competition;
- increase in the cost of materials;
- shortages of labour and materials.

In addition to the above, the property market is also uniquely inelastic; this means that following an increase in demand for a certain type of development it can take up to two years for supply to come on stream. The effect of this is a shortage of supply and a rise in price as companies and individuals try to secure what limited stocks are available. The quantity surveyor, when advising a client on the feasibility of a proposed new project must take all these factors into account as well as giving some indication of the construction costs – good local knowledge is an essential part of this process.

Given that there are a range of unknowns and variables that have to be taken into account when determining the feasibility of a project the most realistic way to report information to a client is to include some sort of indication that the figures that are being reported or data used in the preparation of the report are likely to be totally accurate. Various techniques for reporting probabilities will be discussed later in this section.

One of the industry standards for determining the feasibility of a project is the residual method of valuation, or as it is sometimes referred to, the developer's budget. This method of valuation is one of a set of five standard methods of valuation:

1. **Investment method.** Used extensively by general practice surveyors when determining whether the price being asked for a property, such as an office block is realistic compared to the amount of income that is generated (rent paid by the tenants).

   The value of an investment property is in no way connected to the cost of the construction or other costs, rather its income generation and is calculated as follows:

$$\text{Capital value} = \text{net income} \times \text{year's purchase}$$

$$\text{Year's purchase} = \frac{R}{100} \text{ where R} = \text{\% yield required.}$$

   - Year's purchase is a simply a multiplier but can be used also to calculate the number of years required to pay back the costs of a project.
   - The term net income is sometimes used and refers to the income after landlord's expenses have been deducted.
   - New developments can expect to have voids or empty space during the first twelve months or so, depending on the state of the rental market.

2. **Profits method.** Used to derive rental values from earnings from, for example, a hotel. It involves establishing the gross earnings for the property

and deducting from this all expenses, including profit, that are likely to be incurred by the tenant. The residual figure is the amount available for rent. It should be noted that caution is required as earnings, for a number of reasons, can be distorted.

3. **Comparative method.** The most widely used form of valuation; it uses direct comparison with prices paid for similar properties to the one being valued. The following should be noted:
   - properties compared must be similar, e.g. four-bed detached houses
   - properties must be in the same area
   - the legal status should be the same, i.e. freehold, leasehold
   - the property transactions must be recent
   - the market must be stable.

   It is important to bear in mind that two properties are rarely exactly the same and people's opinions about property are very subjective.

4. **Contractor's method.** Used for insurance purposes and for unusual or unique buildings that rarely come onto the market. It involves:
   - calculating the cost of rebuilding the property, as if new, including fees, etc.
   - applying a reduction for depreciation and wear and tear
   - adding the site value.

5. **Residual method.** Sometimes referred to as the developer's budget, this method attempts to calculate the value of a completed development. From this value all the costs associated with the development are deducted to provide a residual figure that can, among other things, equal the value of the site to the developer. A residual calculation is included later in this chapter.

Of these five methods of valuation the quantity surveyor is most often asked to use the investment and residual methods.

Figure 2.8 illustrates the building cycle. From a developer's perspective, the trick is to plan to develop a new project to come on stream to meet the building boom and not waiting until a boom before starting to develop as the average lead-in time for a new development is two years and the market may be heading for a slump by the time it is completed. Another important factor is that property is a series of markets: commercial, housing, leisure, etc., all of which have their own characteristics that peak and trough at different times.

- The starting point is a strong business cycle upturn coinciding with a relative shortage in the available supply of property, following a period of low development activity during the previous business cycle.

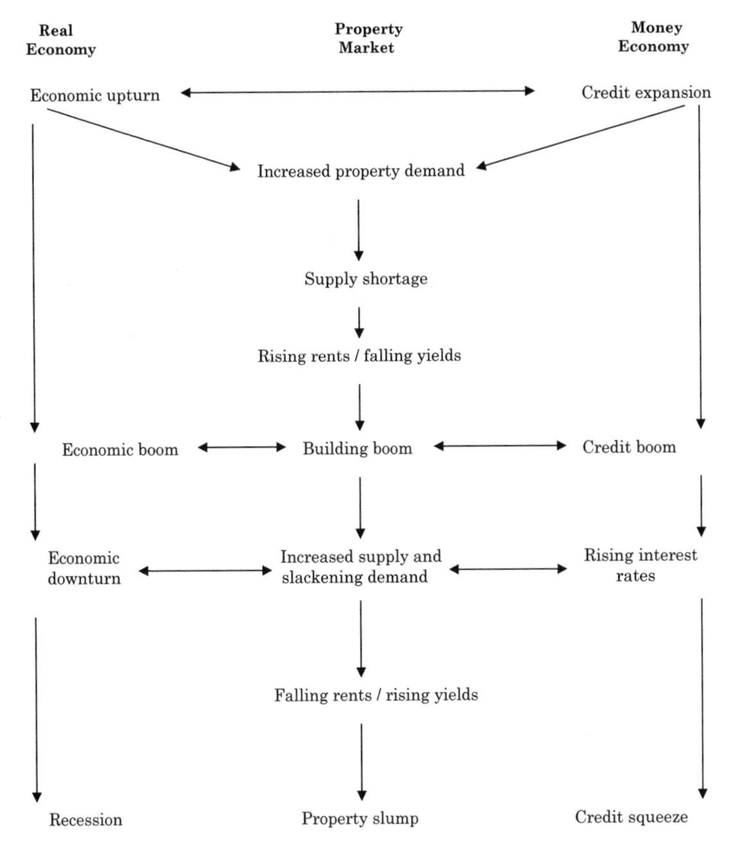


**Figure 2.8** The Barras model of the property cycle

- Strengthening demand and restricted supply cause rents and capital values to rise sharply, improving the potential profitability of development and triggering a first wave of building starts.
- If credit expansion accompanies the business cycle upturn then it can lead to a full-blown economic boom, while at the same time the banks help to fund a second wave of more speculative development activity.
- Now a major building boom is underway, but because of the inherent lags in the development process little new space has yet come through to augment supply, so that rents and values continue to rise.

- By the time that the bulk of new buildings reach completion, the business cycle has moved into its downswing, accompanied by a tightening of the money supply and an increase in interest rates to combat inflationary overheating caused by the boom.
- As the economic boom subsides the demand for property weakens, just as the new supply from the the building boom reaches its peak: the result is falling rents and values and a growing stock of vacant floor space.
- With the economy moving into recession the fall in rents and values accelerates, while the credit squeeze hits the property companies holding unlet buildings with insufficient income to cover their increased interest payments.
- The outcome is a property slump, characterised by depressed values, high levels of vacancy and widespread bankruptcies in the property sectors.

An important point to remember is that the value of a development project usually has no relation to its cost. For the majority of commercial development, value relates to yield, for example the amount of income that will be generated by the completed project. This can be the sums received from the sale of houses or the rental income from a block of flats.

### *Feasibility report checklist*

A feasibility report should comprise some or all the following sections, depending on the size of the project:

| Section | Title |
|---|---|
| A. | **Front cover.** Title of the project, name of client and professional advisers and date. |
| B. | **Contents.** List of contents and page references. |
| C. | **List of exclusions.** Leave no doubt as to which items have been excluded from the report, for example professional fees, VAT, charges for planning consent and building regulations approval, inflation, etc. It is not unknown for clients to challenge surveyors claiming that they were under the impression that items such as fees were included. |
| D. | **Executive summary**. So called because, it is intended to be a summary of the most important findings of the report in a format that can be quickly read by busy people. It is important to keep this section brief and to the point. |

E. **Main report.** This can be subdivided as follows:

*Basis of the report.* Assumptions relating to quality and specification levels, if not supplied by client, yield, development and disposal time scales.

*Developer's budget* – see below.

F. **Recommendations.** Include here also any alternative strategies if the scheme does not prove feasible in its original form. Scenario analysis.

G. **Appendices** if required and appropriate. Try not to baffle with science!

## Residual method of valuation (developer's budget)

A developer's budget or residual method of valuation is generally used to calculate the amount of money that a developer can pay for a plot of land. This is determined by firstly calculating the value of the completed development, the gross development value or GDV and then deducting from this figure all the costs that need to be expended. The residual figure is the amount available for the purchase of the land although in effect the technique can be used to determine the value of any unknown. That is to say, if the value of the land was known, how much could be allocated to construction costs. The developer's budget lends itself readily to spreadsheet applications, such as Excel, and, once set up, can quickly produce figures relating the financial feasibility of a new project.

Items to be considered:

### *Value*

- **The gross development value (GDV).** Can be total sum from sales or rental income. If rental income then it will be necessary to discount estimated future incomes.

### *Costs*

- **Site preparation and infrastructure costs.** Including demolition, contaminated land costs, roads, sewers and main services.
- **Construction costs.** A realistic estimate must be prepared by the quantity surveyor. Usually a substantial percentage of costs will be in this item therefore it is imperative that, even though information and details are limited that the forecast of construction costs is accurate. Various approaches to estimating/cost planning are discussed later.

- **Professional fees.** These will depend on the complexity of the project and whether services such as project management or structural engineering are required.
- **Finance costs.** These costs will be specific to a client and will depend on the amount of perceived risk involved in the development. In the UK, when providing funding based on an overdraft, the amount of interest charged by a High Street bank will generally follow the Bank of England's base rate, plus a percentage to reflect the perceived risk. There are many types and sources of project funding. When carrying out appraisal, and the cost and source of finance is unknown, there is a rule of thumb that is generally used as follows: interest will be calculated on the total construction costs plus fees for half the construction period. A term often used in connection with funding is LIBOR (London Inter Bank Offered Rates); that is the rate at which banks lend each other money. It can be used instead of base rate.
- **Disposal costs.** Once completed it will be necessary to dispose of a project, either by selling or letting. The costs associated with this are typically legal costs, advertising, incentives, stamp duty and agent's fees.
- **Developer's profit.** Most commercial development is carried out for profit and a client will have decided what level of profit is required for the risk of investing in a new project. Generally, the greater the risk, the higher the profit.
- Finally, after the costs have been calculated and deducted from the GDV the residual figure is the sum available to purchase the land. Of course, it may be necessary to finance the land purchase and certainly there will be legal costs involved and to cover these costs another simple rule of thumb is applied as shown in the following example.

### *Example*

A developer is considering the purchase of a city centre site that has outline planning permission for the construction of a block of offices to provide 10,000m$^2$ of gross floor area. When complete it is anticipated that, based on similar developments in the area, it will be let at £350 per square metre and that the completed development will produce a yield of 7%.

The project costs are as follows:

**Construction costs:** £1,200/m$^2$

**Infrastructure costs:** £450,000

**Professional fees**: 10% of construction costs

**Disposal costs**: 12% of first year's rent

**Interest rate**: 10% (base rate plus 9%)

**Developer's profit**: 15% of gross development value

**Construction time**: 3 years

It is usual for a project of this type to be planned and built by a developer and then once completed and fully let, sold to another party for a profit or yield. The proposed project has a gross floor area of 10,000m$^2$; this area must be converted to net lettable floor area before the process can begin. Typically, the ratio of gross to net floor area is in the range of 80–85%. It is of course important that the ratio is as high as possible in order to maximise the rental income and the value of the completed project. This can include specifying components that maximise the floor areas, for example perimeter heating or air conditioning units should be fixed at a sufficient height to allow a waste paper bin to be placed beneath them (see Figure 2.9). This way the area below the units is deemed to be usable and therefore lettable.

It is also typical with a development of this type that an allowance is made during the first year for voids; that is, for space that will remain unlet during the initial rental period. A solution to the above proposal follows.

### *Gross development value (GDV)*

The gross development value can be determined in a number of ways. If the development is to be sold then the GDV is the total amount received from sales.

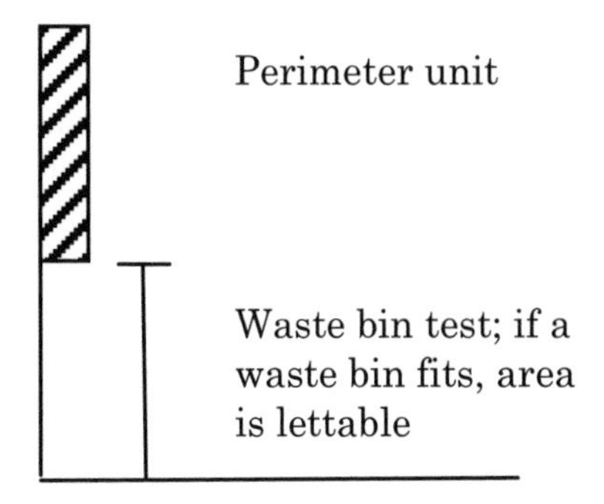


**Figure 2.9** Waste bin test

If the development is an investment property them the income from rent will form the basis of determining the value. In this example, the proposed property is for investment and therefore the capital value is calculated as follows.

To determine the value of the project when complete the following formula is used:

$$\text{Capital Value} = \text{Net Income} \times \text{Year's Purchase}$$

Year's purchase is a multiplier and is determined as follows:

$$\frac{100}{\text{Yield}} = \frac{100}{7} = 14.29$$

10,000m$^2$ gross floor area × 0.84 = 8,400m$^2$ net lettable floor area; this will then produce an income, when fully let, of:

$$8{,}400 \times £350 = £2{,}940{,}000 \text{ per annum}$$

This figure is gross income and needs to be adjusted by the deduction of landlord's expenses from net income; 5% has been taken in this case, giving £2,793,000 per annum.

For the example under consideration, it has been assumed that the developer is an investor who intends to retain ownership of the completed development for four years without rent reviews and then dispose of it. For the first year it has been assumed that there will be 40% voids, or unlet space, giving a rental income for the first year of:

$$\frac{£2{,}793{,}000}{100} \times 60 = £1{,}675{,}800$$

It is assumed that there will be no increases in rental during this period and this information is allowed for in the calculations as follows:

| | Rental income | PV £1@ 7% | |
|---|---|---|---|
| Year 1 | £1,675,800.00 | 0.9346 | £1,566,202.68 |
| Year 2 | £2,793,000.00 | 0.8735 | £2,439,685.50 |
| Year 3 | £2,793,000.00 | 0.8163 | £2,279,925.90 |
| Year 4 | £2,793.000.00 | 0.7629 | £2,130,779.70 |
| Divide by 4 | | | £8,416,593.78 |
| Capital value | | | £2,104,148.45 |

Therefore:

$$\text{Capital value} = £2,104,148.45 \times 14.29 = £30,068,280.85,$$
$$\text{say } £30,100,000.00$$

Having established the value of the completed development the cost associated with the project can be calculated as follows:

| | £ | £ |
|---|---|---|
| **Capital value** | | 30,100,000 |
| **Costs** | | |
| Construction | | |
| 10,000m² × £1,200 | 12,000,000 | |
| Infrastructure | 450,000 | |
| | 12,450,000 | |
| Professional fees | 1,245,000 | |
| | 13,695,000 | |
| Finance | | |
| As discussed above the arrangement of finance varies from client to client. For this example it will be assumed that interest will be calculated on the total construction costs for half the construction period. Fees have also been included: | 1,027,125 | |
| Disposal costs | 201,100 | |
| Developer's profit | 6,000,000 | |
| **Total costs** | **20,923,225** | **20,923,225** |
| | | **9,176,775** |
| Therefore the sum that is left is the sum available to purchase the site. However, it is usually that site purchase also needs to be financed and this is allowed for by multiplying the residual figure by the present value of £1 @ an | | 0.8163 |

appropriate percentage; in this case
7% for the duration of the
construction period: £7,491,001.43
**Say £7,500,000.00**

Therefore the amount that a developer could afford to pay for the site in this example is **£7.5 million** to achieve the level of profit and yield used in the calculation.

As discussed in Chapter 1, the costs and values of a project are usually very different. The first step in the process (Stage A – Inception) is to meet with the client and to discover exactly what the requirements are for the new project. Generally regarded as one of the most difficult and critical stages of the development process, the task can become even more difficult in the case of a client who has a number of separate departments or sections all of which perceive the project's deliverables in a different way. In this situation all kinds of hidden agendas can impinge on the process and make it difficult for the design team to determine the project objectives with accuracy.

## SOURCES OF FINANCE

Every construction project, whether large or small, requires finance. There are two broad categories where finance can be raised; these are referred to as:

1. equity, and
2. debt.

Equity refers to the practice of using the client's own resources to finance the project, whereas in the case of debt funding the project is financed by borrowing money from a variety of financial institutions. In practice most projects, particularly the larger ones, are financed from a mixture of sources. The ratio between the equity and debt funding is referred to as gearing; the higher the proportion of debt, the higher the gearing is said to be and vice versa. The gearing ratio can reflect the perceived risk of the proposed development; the higher the risk the more a developer will wish to transfer risk to the lender by borrowing as much as possible of the development costs. The accepted gearing for property development projects is around 90:10 debt to equity. Usually financing is required at two stages:

1. during the development phase, generally regarded to be the period of highest risk, as revenue will not begin to flow until completion; and

2. after completion when the developer has to repay the money lent for construction plus accumulated interest, generally regarded to be a period of lower risk.

Finance is generally referred to in terms of:

- short term – up to five years;
- medium term – five to ten years;
- long term – over fifteen years.

## Equity

The sources of equity are generally:

- from shareholders' funds or share capital;
- generating funds from internal sources such as retained profits, funds held in return for taxation, etc.

Depending on circumstances the second alternative tends to be for a short-term solution only.

### *Sources of equity*

- **Opportunity cost.** If an investor decides to use his/her own capital to fund property development then the cost of the capital needs to be considered. Opportunity cost is the amount of interest forgone by taking capital off deposit and using it to fund a new project.
- **Forward funding.** This funding method involves a financial institution; for example, an insurance company or pension fund purchases the site and provides the funds for the construction. At the completion of the project it is sold and the developer paid. From the developer's point of view the disadvantage is that the funding institution will have a major influence in the direction of the development. The advantages include the fact that the institution bears both the development and disposal risk; in return the developer's profit may be lower than the norm.
- **Joint venture.** A joint venture between two or more financial institutions may be essential for very large projects or developments and may be confined to UK institutions or cross border arrangements. The advantages include risk sharing and the potential to obtain additional expertise or local knowledge that may be essential for the successful completion of the development.

- **Lease and leaseback.** This arrangement involves the developer leasing the site to a bank or other institution for a nominal rent. The bank then arranges for the construction of the project in accordance with the developer's instructions. On completion the bank leases the building to the developer at a rate that includes construction and finance costs. This approach also has the potential for tax allowances.

### Debt finance

As has been stated previously, the big advantage to debt finance is that risk is transferred to the funder. The amount of interest charged will reflect the perceived risk of the development. The principal sources of debt finance are banks, both UK based and overseas. Banks are generally risk-averse organisations and, as such, tend to lend money on a short-term basis with the project/site usually required by the bank as security. Options include fixed rate and variable rate; variable rate is the most commonly used method:

- The developer approaches a bank and the risk is assessed based on the nature of the proposed project, market conditions, etc.
- Having assessed the risk the bank will determine the interest rate to be charged based on the London Interbank Offered Rate or LIBOR which is the rate charged between banks when they lend each other money. An additional percentage will be added to the base rate to reach the rate to be charged. The higher the perceived risk, the greater the addition.
- Variable rate finance leaves the developer exposed to increases in the base rate that can eat into profits or risk the total project viability. In the UK, the recent past has seen a period of reasonably stable base rates at historically low levels, but this has not always been the case!
- Loans are categorised into non-recourse, limited recourse and full recourse. These terms refer to the extent to which the developer guarantees the debt; in the case of a non-recourse loan the finance is secured on the development itself and if the bank has need to call in the debt, then none of the assets of the developer will be at risk. Limited recourse loans refer to the position where, as well as the development under finance, other assets of the developer are required as a guarantee. Full recourse loans refer to the situation where the funding is secured entirely on the developer's assets, and these will be at risk in the case of project failure.

Banks will, as general rule, lend up to 70% of the gross development value (GDV) depending on the status and track record of the developer.

### Mezzanine finance

Given that banks will usually lend up to 70% GDV a further cash injection is required in order to meet the required gearing ratio; this gap can be filled with so-called mezzanine finance. Mezzanine funding is rather like a second mortgage that ranks below the senior debt. This means that in the case of project failure the senior lender has first recourse to recover losses. The effect of this is to make mezzanine funding more expensive than senior debt as it reflects the developer's inability to secure adequate equity.

### Bonds

The bond market offers a source of long-dated debt and accordingly, it is now common place for many organisations to consider raising money in this market, as an alternative to the banking sector, especially if the project concerned is so big that there in insufficient liquidity in the market. This method of raising finance is ideal for individuals and organisations who:

- need to raise large sums of capital, £20 million upwards; and
- have access to regular guaranteed income for the term of the bond which can be up to 50 years.

Bonds are in effect an IOU and work as follows:

- Investors lend the bond issuer a sum of money for a fixed term.
- In return the investor receives an agreed rate of interest from the bond issuer.
- On maturity the investor receives back the original investment in full.
- Bonds may be either public or private.

### Answering the 'what if?' question

In order to carry out the developer's budget several assumptions have been made and the next step in the process is to test those assumptions against their sensitivity to change, brought about by consequences beyond the developer's control, for example, increases in interest rates, increase in construction costs due to material and/or labour shortages, etc. There are a number of techniques, ranging from the basic to the sophisticated, that test various sensitivities and scenarios and thereby attempt to answer the 'what if?' question. For example, if interest rates were to rise by 3%, the cost of financing the above scheme would increase considerably, thereby reducing the amount available for construction costs or profit and perhaps threatening the feasibility of the project. The techniques that can be used, listed in order of accuracy are:

- sensitivity analysis
- Monte Carlo simulation.

Regardless of which of the above techniques is used, the important aspect of including this item in the feasibility report is that it highlights the fact that property development has a number of inherent risks that must be correctly managed if the project is to be a success. After a budget has been finalised, market conditions rarely evolve as expected; interest rates change, materials and labour shortages occur, etc. The likelihood that these and other myriad events could negatively impact revenue is known as risk. While risk cannot be avoided, it can be mitigated and managed.

### *Sensitivity analysis*

Sensitivity analysis is a method for analysing uncertainty by changing input variables, for example, and observing the sensitivity of the result. The method can be used either on a variable-by-variable basis or by changing groups of variables at once using scenario analysis. These closely related techniques offer several advantages over other methods for examining the effects of uncertainty.

#### VARIABLE-BY-VARIABLE ANALYSIS

This approach analyses uncertainty and isolates the effect of change on one variable of the feasibility study at a time. The approach is as follows:

- List all the important factors that can affect the successful outcome of the project.
- For each factor define a range of possible values.
- Generally three ranges are proposed: Optimistic, Most Likely and Pessimistic.
- Calculate the cost–benefit ratios or net present values for each of the ranges.

A variable-by-variable analysis is based on the assumption that factors affecting a project do not operate independently of one another. Using software such as Microsoft Excel it is possible to model various scenarios to answer the 'what if?' question. For example, what will be the effect on the project feasibility if interest rates were to change from the 10% assumed in the feasibility report, during the construction period?

| | Optimistic | Most likely | Pessimistic |
|---|---|---|---|
| | 9% | 10% | 11% |
| Finance costs on £6,847,500 | £308,138 | £342,375 | £376,613 |

This process should be repeated for all key drivers.

SCENARIO ANALYSIS

A scenario analysis is based on the assumption that factors affecting cost-benefit flows do not operate independently of each other as is assumed in the variable-by-variable approach. For example, it is unlikely that increases in interest rates and lack of demand for the completed development are independent factors.

Sensitivity analysis has several advantages as it:

- shows how significant any given input variable is in determining a project's economic worth. It does this by displaying the range of possible project outcomes for a range of input values, which shows the decision makers the input values that would make the project a winner or a loser. It also helps to identify critical inputs;
- answers the 'what if?' question and is good background and preparation for defending a project. For example, if the question 'what will the outcome of the project be if operating costs increase by 20%' is posed, the answer is easily determined;
- does not require the use of probabilities;
- can be used on any measure of project worth;
- can be used where there is little information, resources or time to perform more sophisticated techniques.

The major disadvantage of sensitivity analysis is that there is no explicit probabilistic measure of risk exposure. That is, although one might be sure that one of the several outcomes might happen, the analysis contains no explicit measure of their respective likelihoods. A good way to graphically represent the outcomes of a sensitivity analysis is with the use of a spider diagram (see Figure 2.10). It is an instant snapshot of the relative importance of several uncertain variables.

### *Monte Carlo simulation*

Monte Carlo simulation was named after the town on the Côte d'Azur famous for its casinos, gambling and games of chance. Despite its exotic name, in

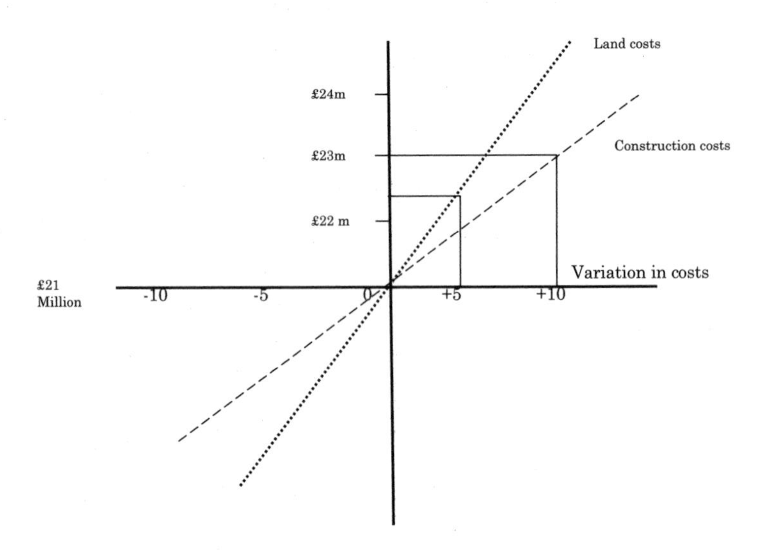


**Figure 2.10** Spider diagram

reality it is a piece of software that models or simulates the probable outcomes of a given scenario. It does this by the use of probabilistic techniques that attempt to mimic what will happen in reality. It is far superior to the sensitivity analysis described above because not only do the results illustrate the impact of an input changing, but they also take into the account the probability that any given event will happen, based upon what has happened in the past in similar projects.

Whichever of the above techniques are used, it should not be forgotten that these are simply decision-making tools and should not be used as a stand-alone set of facts. It will always be the case that the quantity surveyor will interpret and review the outcome before reporting figures to the client.

## WHOLE-LIFE COSTS

Whole-life costs include the consideration of the following factors when designing and specifying:

- **initial** or procurement costs, including design, construction or installation, purchase or leasing, fees and charges;
- **future** cost of operation, maintenance and repairs, including management costs such as cleaning, energy costs, etc.;

- **future** replacement costs, including loss of revenue due to non-availability;
- **future** alteration and adaptation costs, including loss of revenue due to non-availability;
- **future** demolition/recycling costs.

The service life of an element, product or whole building may be viewed in one or more of the following ways:

- technical life – based on physical durability and reliability properties;
- economic life – based on physical durability and reliability properties;
- obsolescence – based on factors other than time or use patterns, e.g. fashion.

Common terms used to describe the consideration of all the costs associated with a built asset throughout its life span are:

- costs-in-use
- life-cycle costs
- whole-life costs
- through-life costs
- etc.

There are a number of definitions for whole-life costing (WLC), but one currently adopted is: 'the systematic consideration of all relevant costs and revenues associated with the acquisition and ownership of an asset'.

Although whole-life costing can be carried out at any stage of the project and not just during the procurement process (see Figure 2.11), the potential of its greatest effectiveness is during procurement because:

- almost all options are open to consideration at this time;
- the ability to influence cost decreases continually as the project progresses, from 100% at project sanction to 20% or less by the time construction starts;
- the decision to own a building normally commits the user to most of the total cost of ownership and consequently there is a very slim chance to change the total cost of ownership once the building is delivered.

Typically, about 75–95% of the cost of running, maintaining and repairing a building is determined during the procurement stage.

There now follows a simple example, based on the selection of material types, illustrating the net present value and the annual equivalent approaches to whole-life cost procurement.

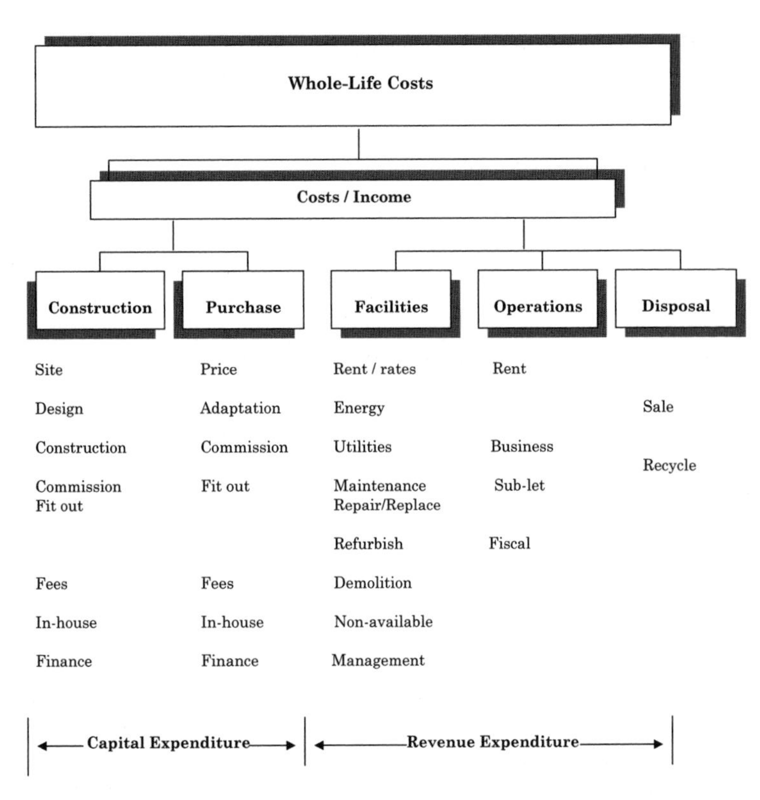


**Figure 2.11** Whole-life costs

**Table 2.7** Comparison of two different materials

| Material | Initial cost | Installation cost | Maintenance cost per day | Other maintenance costs | Life expectancy |
|---|---|---|---|---|---|
| A | £275 | £150 | £3 | £100 every 3 years for preservative treatment | 12 Years |
| B | £340 | £150 | £3 | None | 15 Years |

**Table 2.8** Results for Material A

| | | | | | | Total Discounted Costs | | |
|---|---|---|---|---|---|---|---|---|
| Year | Present value of £1 per annum (PV of £1 pa) | Present value (PV £1) | Initial cost £ | Other costs £ | Annual cost £ £3 × 365 | NPV of replacement + other + annual costs + initial costs £ | Total NPV £ | AEC £ |
| 1 | 0.943 | 0.943 | 490.00 | | 1095.00 | 1523.02 | 1523.02 | 1614.4 |
| 2 | 1.834 | 0.890 | | | 1095.00 | 974.55 | 2497.57 | 1362.25 |
| 3 | 2.673 | 0.840 | | | 1095.00 | 919.38 | 3416.95 | 1278.31 |
| 4 | 3.465 | 0.792 | | | 1095.00 | 897.34 | 4284.29 | 1236.41 |
| 5 | 4.212 | 0.747 | | | 1095.00 | 818.25 | 5102.54 | 1211.32 |
| 6 | 4.917 | 0.705 | | | 1095.00 | 771.93 | 5874.47 | 1194.65 |
| 7 | 5.582 | 0.665 | | | 1095.00 | 728.24 | 6602.71 | 1182.76 |
| 8 | 6.210 | 0.627 | | | 1095.00 | 687.02 | 7289.72 | 1173.91 |
| 9 | 6.802 | 0.592 | | | 1095.00 | 648.13 | 7937.85 | 1167.04 |
| 10 | 7.360 | 0.558 | | | 1095.00 | 611.44 | 8549.30 | 1161.58 |
| 11 | 7.887 | 0.527 | | | 1095.00 | 576.83 | 9126.13 | 1157.13 |
| 12 | 8.384 | 0.497 | | | 1095.00 | 544.18 | 9670.31 | 1153.47 |
| 13 | 8.853 | 0.469 | | | 1095.00 | 513.38 | 10183.69 | 1150.35 |
| 14 | 9.295 | 0.442 | | | 1095.00 | 484.32 | 10668.00 | 1147.72 |
| 15 | 9.712 | 0.417 | | 490.00 | 1095.00 | 661.37 | 11329.38 | 1166.50 |
| 16 | 10.106 | 0.394 | | | 1095.00 | 431.04 | 11760.42 | 1163.72 |
| 17 | 10.477 | 0.371 | | | 1095.00 | 406.64 | 12167.06 | 1161.28 |
| 18 | 10.828 | 0.350 | | | 1095.00 | 383.63 | 12550.69 | 1159.14 |
| 19 | 11.158 | 0.331 | | | 1095.00 | 361.91 | 12912.60 | 1157.24 |
| 20 | 11.470 | 0.312 | | | 1095.00 | 341.43 | 13254.02 | 1155.55 |
| 21 | 11.764 | 0.294 | | | 1095.00 | 322.10 | 13576.12 | 1154.03 |
| 22 | 12.042 | 0.278 | | | 1095.00 | 303.87 | 13879.99 | 1152.67 |
| 23 | 12.303 | 0.262 | | | 1095.00 | 286.67 | 14166.66 | 1151.45 |
| 24 | 12.550 | 0.247 | | | 1095.00 | 270.44 | 14237.10 | 1150.33 |
| 25 | 12.783 | 0.233 | | | 1095.00 | 255.13 | 14692.24 | 1149.36 |

Notes:
AEC = Annual equivalent cost.
Other cost = replacement costs every 15 years.

This problem is a classic one, which material, with widely different initial and maintenance costs, will deliver the best value for money over the life-cycle of the building (see Table 2.7). In this example, using a discount rate of 6% it is assumed that the materials are to be considered for installation in a PFI project, with an anticipated life of twenty-five years.

Table 2.8 indicates a whole-life cost calculation for Material A presented in two ways: as a net present value and also as an annual equivalent cost. The calculation is repeated for each material or component under consideration and then a comparison can be made.

A replacement expenditure profile, excluding cyclical maintenance and energy, over a range of elements during a 35-year contract period is shown in Table 2.9.

Clearly, the choice of the correct type of material or component would appear of critical importance to a client as future replacement and maintenance costs will have to be met out of future income. However, in reality, theory and practice are often very different. For example, for many public authorities, finding budgets for construction works is usually more difficult than meeting recurring running and maintenance costs that are usually included in annual budgets as a matter of course.

**Table 2.9** Replacement expenditure profile

| Element | Replacement expenditure % |
|---|---|
| Windows/doors | 22.95 |
| Kitchens | 15.79 |
| Heating | 11.82 |
| Structural | 10.63 |
| Roofs | 8.72 |
| Bathrooms | 7.79 |
| Wiring | 6.50 |
| External areas | 3.87 |
| Internal decorations | 2.48 |
| Communal decorations | 1.69 |
| Over cladding | 1.61 |
| Rainwater goods | 1.51 |
| External walls | 0.99 |
| Off-road parking | 0.82 |
| DPC | 0.73 |
| Security/CCTV | 0.60 |
| Door entry systems | 0.51 |
| Fire precautionary works | 0.50 |
| Porches/canopies | 0.44 |
| Plastering | 0.07 |

(Source: Whole Life Cost Forum)

In addition to the net present value and annual equivalent approaches described previously it has been identified that simple aggregation could sometimes be used effectively when evaluating whole-life costs.

### Simple aggregation

This method of appraisal involves adding together the costs, without discounting, of initial capital costs, operation and maintenance costs. This approach has a place in the marketing brochure and it helps to illustrate the importance of considering all the costs associated with a particular element but has little value in cost forecasting.

A similarly simplistic approach is to evaluate a component on the time required to pay back the investment in a better quality product. For example, a number of energy saving devices are available for lift installations. A choice is made on the basis of which over the life-cycle of the lift, say five or ten years, will pay back the investment the most quickly. This last approach does have some merit, particularly in situations where the life-cycle of the component is relatively short and the advances in technology and hence the introduction of a new and more efficient product is likely.

### *Criticisms of whole-life costs*

Whole-life costing is also not an exact science, as, in addition to the difficulties inherent in future cost planning, there are larger issues at stake. It is not just a case of asking 'how much will this building cost me for the next 50 years', rather it is more difficult to know whether a particular building will be required in 50 years time at all – especially as the current business horizon for many organisations is much closer to 3 years. Also, whole-life costing requires a different way of thinking about cash, assets and cash flow. The traditional capital cost focus has to be altered, and costs thought of in terms of capital and revenue costs coming from the same 'pot'. Many organisations are simply not geared up for this adjustment.

Perhaps, the most crucial reason is the difficulty in obtaining the appropriate level of information and data.

There is a lack of available data to make the calculations reliable. The Building Maintenance Information (BMI) defines an element for occupancy cost as expenditure on an item which fulfils a specific function irrespective of the use of the form of the building. The system is dependent on practitioners submitting relevant data for the benefit of others. The increased complexity of construction means that it is far more difficult to predict the whole-life cost of built assets. Moreover if the malfunction of components results in decreased yield or underperformance of the building then this is of concern to the end-user/owner. There is no comprehensive risk analysis of building components available for practitioners, only a wide range of predictions of estimated life spans and notes on preventive maintenance – this is too simplistic, there is a need for costs to be tied to risk, including the consequences of component failure. After all, the performance of a material or component can be affected by such diverse factors as:

- quality of initial workmanship when installed on site and subsequent maintenance;
- maintenance regime/wear and tear. Buildings that are allowed to fall into disrepair prior to any routine maintenance being carried out will have a different life-cycle profile to buildings that are regularly maintained from the outset;
- intelligence of the design and the suitability of the material/component for its usage. There is no guarantee that the selection of so-called high-quality materials will result in low life-cycle costs.

Other commonly voiced criticisms of whole-life cost are:

- Expenditure on running costs is 100% allowable revenue expense against liability for tax and as such is very valuable. There is also a lack of taxation incentive, in the form of tax breaks, etc. for owners to install energy efficient systems.
- In the short term, and taking into account the effects of discounting, the impact on future expenditure is much less significant in the development appraisal.
- Another difficulty is the need to be able to forecast, a long way ahead in time, many factors such as life-cycles, future operating and maintenance costs, and discount and inflation rates. WLC, by definition, deals with the future and the future is unknown. Increasingly, obsolescence is being taken into account during procurement, a factor that it is impossible to control since it is influenced by such things as fashion, technological advances and innovation. An increasing challenge is to procure built assets with the flexibility to cope with changes. Thus, the treatment of uncertainty in information and data is crucial as uncertainty is endemic to WLC. Another major difficulty is that the WLC technique is expensive in terms of the time required. This difficulty becomes even clearer when it is required to undertake a WLC exercise within an integrated real-time environment at the design stage of projects.

In addition to the above changes in the nature of development, other factors have emerged to convince the industry that whole-life costs are important.

### *Whole-life cost procurement: critical success factors*

- Effective risk assessment – what if this alternative form of construction is used?
- Timing – begin to assess WLC as early as possible in the procurement process.
- Disposal strategy – is the asset to be owner occupied, sold or let?
- Opportunity cost – downtime.
- Maintenance strategy/frequency – does one exist?
- Suitability – matching a client's corporate or individual strategy to procurement.

It is hoped that the introduction of NRM3 in the autumn of 2012 will improve the reputation of WLC advice.

### Value management/value engineering

In recent years, quantity surveyors have been using a system referred to as value management or value engineering in an attempt to make the briefing process more objective. Value management/value engineering is a structured technique that attempts to determine, at an early stage, prior to detailed design, the functions that the completed project is required to fulfil.

First developed and introduced into the manufacturing sector in America, immediately after the Second World War, the principal objective of this technique is to gather together the main stakeholders of a new project in a workshop, sometimes over a four or five day period. The process is managed by a value management practitioner, who can be a quantity surveyor. This may be the only occasion when all the stakeholders of a new project meet together to discuss the outcomes. The process involves the analysis of the required functions of a new project and then the investigation of how these functions may be achieved (see Figure 2.12). Concentration of functions allows items that do not contribute to the identified functions, referred to as waste, to be removed from the design.

The traditional approach to determining the cost of a new project is to apply costs to the product – the various parts of a new building. Value management takes a step sideways and applies costs to the required functions and then looks at ways, through design and construction, of providing functionally optimised building projects. The approach therefore can be said to be: 'don't buy product, buy function'.

Luckily for quantity surveyors, who are said to be attracted to all things standard, SAVE, the Society of American Value Engineers, has produced a standard methodology (see Figure 2.13). The SAVE methodology is based on the standard North American forty hour workshop that takes place over five days. It is broken down into three phases: the Pre-Study phase, where members of the value management team become familiar with the project parameters; the Value Study phase, which is itself broken down into six phases that take the workshop through a structured process; and the Post-Study phase.

In the Value Study phase, the characteristics of the function analysis phase are defining, classifying and establishing the worth of a function or functions which are at the heart of the value management process. The definition of function can be problematic; experience has shown that the search for a definition can result in lengthy descriptions that do not lend themselves to analysis. In addition, the definition of function has to be measurable. In order to establish some sort of hierarchy, functions need to be classified into primary (needs) or supporting function (wants). Basic functions are those that make the project work, without basic functions the final project would impact on the functionality of the completed project. Out of the list of functions emerges the highest order function that can be defined as the overall reason for the project and meets the overall needs of the client. After defining the functions of the project, the next step is to establish

**Figure 2.12** A structured technique for value management/value engineering

**PRE-STUDY**
User/Customer Attitudes
Complete Data Files

Evaluation Factors
Study Scope
Data Models
Determine team composition

**VALUE STUDY**

**Information Phase**
Complete Data Package
Finalise scope

**Function Analysis Phase**
Identify Functions
Classify Functions
Function Models
Establish Function Worth
Cost Functions
Establish Value Index
Select Functions for Study

**Creative Phase**
Create Quantity of Ideas by Function

**Evaluation Phase**
Rank & Rate Alternative Ideas
Select Ideas for Development

**Development Phase**
Benefit Analysis
Technical Data Package
Implementation Plan
Final Proposals

**Presentation Phase**
Oral Presentation
Written Report
Obtain Commitments for Implementation

**POST-STUDY**
Complete Changes
Implement Changes
Monitor Status

**Figure 2.13** Value management process (Source: Society of American Value Engineers)

a function's worth and to identify which of the functions contains a value mismatch, or in other words seems to have a high concentration to the total project cost in relation to the function. Following on from this the creative phase will concentrate on these functions. Worth is defined as 'the lowest overall cost to perform a function without regard to criteria or codes'. Having established the worth and the cost, the value index can be calculated. The formula is: value = worth/cost. The benchmark is to achieve a ratio of 1. There are a number of models, including FAST (Functional Analysis System Technique) diagrams to do this.

# 3

# Measurement and quantification

## MEASUREMENT AND QUANTIFICATION

During the recent past the bill of quantities has been criticised in some circles, as being outdated and unnecessary in the modern procurement environment. Indeed, the number of contracts based on a bill of quantities has declined sharply over the past 20 years or so. Nevertheless the bill of quantities remains unsurpassed as a model on which to obtain bids in a format that allows ease of comparison between various contractors, transparency, an aid to the quantity surveyor in valuing variations, calculating stage payments and the preparation of the final account. What's more, the ability to measure, quantify and analyse the items of labour, materials and plant necessary to construct a new project is still a much sought after skill and many would argue the core of the quantity surveying profession.

### Measurement practice

It is vitally important that measurement practice applied to buildings is both accurate and consistent. There are a number of situations that require a quantity surveyor to measure and record dimensions from both drawings as well as on site, depending on the stage of the project. In order to standardise measurement rules and conventions there are a number of standard codes and methods of measurement that are available. These are outlined below.

### The RICS *Code of Measurement Practice*, 6th Edition (2007)

According to the RICS the purpose of the code is to provide succinct, precise definitions to permit the accurate measurement of buildings and land, the calculation of the sizes (areas and volumes) and the description or specification of land and buildings on a common consistent basis. The code is intended for use in the UK only and includes three core definitions that are used in a variety of situations as follows.

### *Gross external area (GEA)*

This approach to measurement is recommended for:

- building cost estimation for calculating building costs for residential property for insurance purposes;
- town planning applications and approvals;
- rating and council tax bands.

GEA is defined as the area of a building measured externally at each floor level as follows:

| Included | Excluded |
|---|---|
| Perimeter wall thickness and external projections | External open-sided balconies |
| Areas occupied by internal walls and partitions | Canopies |
| Columns, piers, chimney breasts, stairwells, lift wells and the like | Open vehicle parking areas, roof terraces and the like |
| Atria and entrance halls with clear height above, measured at base level only | Voids over or under structural, raked or stepped floors |
| Internal balconies | Greenhouses, garden stores, fuel stores and the like in residential property |
| Structural, raked or stepped floors are to be treated as a level floor measured horizontally | |
| Mezzanine areas intended for use with permanent access | |
| Lift rooms, plant rooms, fuel stores, tank rooms, which are housed in a covered structure of a permanent nature, whether or not above the main roof level | |
| Outbuildings which share at least one wall with the main building | |
| Loading bays | |
| Areas with headroom of less than 1.5m | |
| Pavement vaults | |
| Garages | |
| Conservatories | |

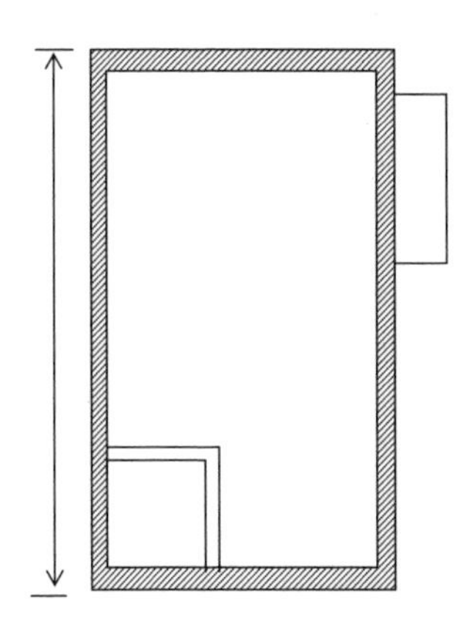

**Note:**

Areas of internal walls and partitions included.

External canopies – excluded.

Measurements taken to external face of external walls.

**Figure 3.1** Gross external area

### *Gross internal area (GIA)*

This approach to measurement is recommended for:

- building cost estimation;
- marketing and valuation of industrial buildings, warehouses, department stores;
- valuation of new homes property management – apportionment of services charges.

GIA is the area of a building measured to the internal face of the perimeter walls at each floor level.

| **Included** | **Excluded** |
|---|---|
| Areas occupied by internal walls and partitions | Perimeter wall thicknesses and external projections |
| Columns, piers, chimney breasts, stairwells, lift wells, other internal projections, vertical ducts and the like | External open-sided balconies, covered ways and fire escapes |
| Atria and entrance halls with clear height above, measured at base level only | Canopies |
| Internal open-sided balconies, walkways and the like | Voids over or under structural, raked or stepped floors |
| Structural, raked or stepped floors are to be treated as a level floor measured horizontally | Greenhouses, garden stores, fuel stores and the like in residential property |

Horizontal floors, with permanent access below structural, raked or stepped floors

Corridors of a permanent essential nature (e.g. fire corridors, smoke lobbies)

Mezzanine floor areas with permanent access

Lift rooms, plant rooms, fuel stores, tank rooms, which are housed in a covered structure of a permanent nature, whether or not above the main roof level

Service accommodation such as toilets, toilet lobbies, bathrooms, showers, changing rooms and the like

Projection rooms

Voids over stairwells and lift shafts on upper floors

Loading bays

Areas with headroom of less than 1.5m

Pavement vaults

Garages

Conservatories

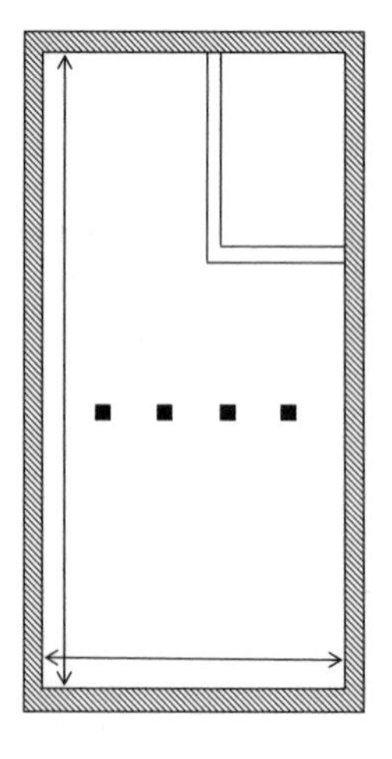

**Note:**

Areas of partitions, columns, etc. included.

Areas of canopies excluded.

Measurements taken to internal face of external walls.

**Figure 3.2** Gross internal area

**Note:** Internal face means the brick/block or plaster coat applied to the brick/block work, not the surface of internal linings installed by the occupier.

### *Net internal area (NIA)*

This approach to measurement is recommended for:

- marketing and valuation of shops, supermarkets and offices;
- rating shops;
- in property management it is used for the apportionment of services charges.

NIA is the usable area within a building measured to the internal face of the perimeter walls at each floor level.

| Included | Excluded |
|---|---|
| Atria with clear height above, measured at base level only | Those parts of entrance halls, atria landings and balconies used in common |
| Entrance halls | Toilets, toilet lobbies, bathrooms, cleaners' rooms and the like |
| Notional lift lobbies and notional fire corridors | Lift rooms, plant rooms, tank rooms (other than those of a trade process nature), fuel stores and the like |
| Kitchens | Stairwells, lift-wells and permanent lift lobbies |
| Built-in units, cupboards and the like occupying usable areas | Corridors and other circulation areas where used in common with other occupiers |
| Ramps, sloping areas and steps within usable areas | Permanent circulation areas, corridors and threshold/recesses associated with access, but not those parts that are usable areas |
| Areas occupied by ventilation/heating grilles | Areas under the control of service or other external authorities including meter cupboards and statutory service supply points |
| Areas occupied by skirting and perimeter trunking | Internal structural walls, walls enclosing excluded areas, columns, piers, chimney breasts, other projections, vertical ducts, walls separating tenancies and the like |

| Included | Excluded |
|---|---|
| Areas occupied by non-structural walls subdividing accommodation in sole occupancy | The space occupied by permanent and continuous air-conditioning, heating or cooling apparatus and ducting in so far as the space it occupies is rendered substantially unusable |
| Pavement vaults | The space occupied by permanent, intermittent air conditioning, heating or cooling apparatus protruding 0.25m or more into the usable area |
| | Areas with headroom of less than 1.5m |
| | Areas rendered substantially unusable by virtue of having a dimension between opposite faces of less than 0.25m |
| | Vehicle parking areas (the number and type of spaces noted) |

**Note:** An area is usable if it can be used for any sensible purpose in connection with the purposes for which the premises are to be used.

In addition to the *RICS Code of Measurement Practice* the two principal methods of measurement are:

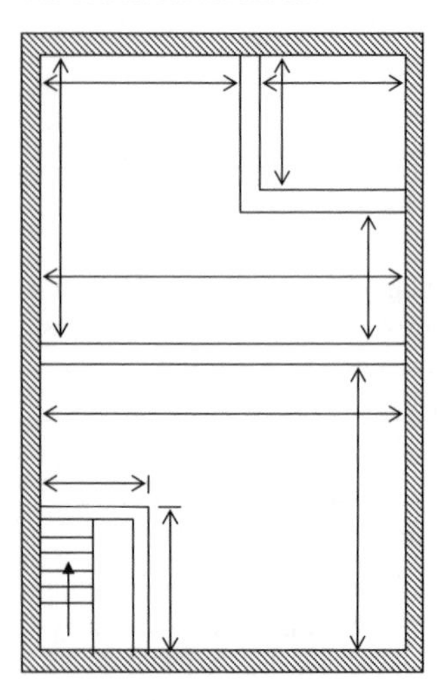

**Notes:**

Stairwells, circulation space and internal walls excluded.

Non-structural walls included.

Measurements taken to the internal face of external walls.

**Figure 3.3** Net internal area

### 1. *Standard Method of Measurement for Building Works* (SMM7)

The *Standard Method of Measurement* first appeared in 1922 and was based on 'the practice of the leading London quantity surveyors'. It was an attempt to bring uniformity to the ways by which quantity surveyors measured and priced building works. The seventh edition appeared in 1988 as a joint publication between the RICS and the Building Employers Confederation and was revised in 1998. Printing of this edition will soon cease in favour of the *RICS New Rules of Measurement*, Volume 2.

### 2. *Civil Engineering Standard Method of Measurement* (CESMM4)

Sponsored and published by the Institution of Civil Engineers the first edition of CESMM4 appeared in 1976 in order to 'standardise the layout and contents of Bills of Quantities and to provide a systematic structure'. The current (fourth) edition, was published in April 2012.

These two standard methods of measurement referred to above reflect the different approaches of the two industries, not only in the nature of the work, but also the degree of detail and the estimating conventions used by both sectors. This in turn reflects the different ways in which building and civil engineering projects are organised and carried out. In general, SMM7 and NRM2 has more emphasis on detail, whereas the CESMM4 takes a more inclusive approach to the measurement process. Building work comprises many different trades whereas civil engineering works consists of large quantities of a comparatively small range of items. For example, when measuring excavation using SMM7 it is necessary to keep excavation, earthwork support and working space as separate items, whereas when using CESMM4 all these item are included in a single item of excavation. It should be noted that NRM2 follows the CESMM4 approach and does not generally require earthwork support and working space to be measured as separate items.

Other methods of measurement are available for specific types of construction, most notably:

- Standard method of measurement for highways.
- Standard method of measurement for roads and bridges.
- Standard method of measurement for industrial engineering construction, which provides measurement principles for the estimating, tendering, contract management and cost control aspects of industrial engineering construction.
- RICS International method of measurement.

The UK *Standard Method for Measurement for Building Works* (SMM7) has been used as the basis for the preparation of methods of measurement that are used in Malaysia and Hong Kong. It is anticipated that NRM2 will similarly be adopted and adapted worldwide.

### The RICS *New Rules of Measurement: Detailed Measurement for Building Works* (NRM2)

The *Standard Method of Measurement for Building Works* (SMM), described by RICS Books as a 'landmark publication', and currently in its seventh edition, has, for as long as anyone can remember, been the standard set of rules and measurement conventions that quantity surveyors refer to when preparing bills of quantities. Traditionally the document is drawn up by both sides of the construction industry, from the client side, the RICS, and from the contracting side, the Construction Confederation. One of the ways in which NRM2 differs from SMM7 is that main contractor organisations were not consulted during its preparation, only subcontractors, reflecting the increasing importance of this side of the industry. One of the consequences in changing procurement practice is that the Building Cost Information Service is finding it increasingly difficult to obtain cost data from the industry in a form that is meaningful. In 2003 the RICS Quantity Surveying and Construction Professional Group commissioned a report 'Measurement based procurement of buildings' from the Building Cost Information Service. The report was based on a survey of practising consultants and contractors and was carried out during 2002; as with most surveys of this nature the response rate was low, with only some 20% of consultants responding and 12% of contractors. The report came to a number of conclusions; it confirmed that measurement still had an important part to play in the procurement of buildings, but perhaps not surprisingly it found that measurement was used by clients, contractors, subcontractors and suppliers in the procurement process in a variety of ways.

The report also concluded that the rise in the use of design and build procurement has encouraged the use of contractors' bills of quantities where few documents are prepared to a standard format, say for example in accordance with SMM7. Further, the allegation is that SMM7 is out of date and represents a time when bills of quantities and tender documents were required to be measured in greater detail than is warranted by current procurement practice and therefore a new approach was required. The RICS Contracts in Use Survey 2007 commented on the traditional approach to measurement:

> Bills of quantities refuse to die. It should be noted that while the use of the 'with quantities forms' have declined there are still SMM7 Bills of

Quantities being measured. It will be interesting in future years to see what is happening in the market when the RICS publishes NRM2, the procurement section of the NRM suite of documents.

The status of the rules of measurement is the same as other RICS guidance notes and as such if an allegation of professional negligence is made against a surveyor, a court is likely to take account of the contents of any relevant guidance notes published by the RICS in deciding whether or not the surveyor acted with reasonable competence.

## SMM7 and NRM2 compared

The *RICS New Rules of Measurement: Detailed Measurement for Building Works* (NRM2) was finally published in April 2012 after much speculation about the format and form of this new document. As it turned out NRM2 was far from the revolution in measurement practice that some had predicted and most quantity surveyors should quickly adapt to this new set of measurement rules as far as work procurement is concerned. For the foreseeable future the quantity surveyor will be able to choose to use SMM7 and NRM2 as alternative approaches to the measurement of quantities; whichever approach seems to be appropriate. It is expected, however, that printing and sales of SMM7 will cease shortly. The most obvious difference between the two documents is that because NRM2 has been drafted with the preparation of subcontract packages in mind there are more work sections: 41 Sections in NRM2 compared with 22 Sections in SMM7. Another important difference is that NRM2 is available as a free download in pdf format for RICS members.

**Note:** In all cases the traditional measurement conventions, for example, calculating mean girths, described later in this chapter, are identical whether using SMM7 or NRM2.

## The RICS New Rules of Measurement 2: Detailed Measurement for Building Works

NRM2 is divided into the following sections:

- Introduction.
- Part 1: General.
- Part 2: Measurement rules for *detailed measurement for building works*.
- Part 3: Tabulated rules of measurement for works procurement.
- Appendices.

**Part 1: General**

After the Foreword NRM2 opens up with a general section that outlines the big picture of the *RICS New Rules of Measurement* suite and how NRM2 fits into the whole NRM project; that is estimating and cost planning through to whole-life costs. The majority of the first section is taken up with a list of definitions.

**Part 2: Measurement rules for detailed measurement of building works**

Unlike its predecessor SMM7, NRM2 contains a lengthy section describing the purpose and benefits of bills of quantities in a rather textbook-like approach. There is a section on risk and how bills of quantities should be codified and in addition there is a typical example of a structure for use when procurement is based on work packages instead of bills of quantities. In some ways this approach is not dissimilar to CESMM4.

**Part 3: Tabulated rules of measurement for works procurement**

This part of NRM2 begins by describing the protocols and use of the tabulated rules of measurement document. This section should be carefully studied as it contains information that applies across the trade sections, for example:

**3.3.3.4 Information to be included in descriptions**

Unless otherwise stated the following information shall be given in descriptions:

- type and quality of the material;
- critical dimensions of the materials;
- method of fixing, installing or incorporating the goods or materials into the work where not at the discretion of the contractor; and
- nature or type of background.

The tabulated measurement rules are organised into two main sections:

1. Preliminaries – divided into main contractor's and work package contractor's preliminaries. The tables of information for this section are laid out over four columns:

   - **Column one** – Sub-heading 1.
   - **Column two** – Sub-heading 2.
   - **Column three** – Information requirements.
   - **Column four** – Supplementary information/notes.

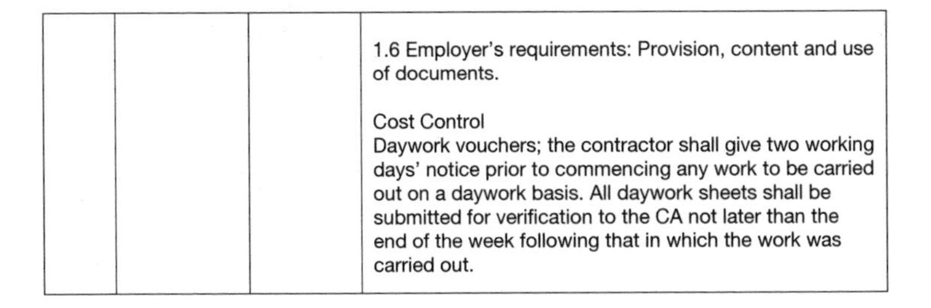

| | | | |
|---|---|---|---|
| | | | 1.6 Employer's requirements: Provision, content and use of documents.<br><br>Cost Control<br>Daywork vouchers; the contractor shall give two working days' notice prior to commencing any work to be carried out on a daywork basis. All daywork sheets shall be submitted for verification to the CA not later than the end of the week following that in which the work was carried out. |

**Figure 3.4** Preliminaries

2. Building components/items: comprising the rules of measurement for building components/items. Generally each of the work sections starts with lists of information as follows:

   - drawings that must accompany the section of measurement
   - mandatory information that is to be provided
   - minimum information that must be shown on the drawings or any other documents that accompany the section of measurement
   - works and materials that not measured, but are deemed to be included in the building components/items measured in each work section.

The tables of information for the work sections are laid out over six columns and three levels and are used, unless otherwise stated, as follows:

- **Column one** describes the item or work to be measured.
- **Column two** gives the unit in which the item or work is to be measured.
- **Column three** lists the information, including critical sizes or dimensions that must be given in the bill description.
- **Column four** lists supporting information that must be given in the bill descriptions.
- **Column five** lists further supporting information including any additional dimensional requirements.
- **Column six** (notes, comments and glossary) explains what work is deemed to be included in specific building components.

Each description will usually contain information from columns one, two and three and as many items from columns four and five as are applicable to the item being measured and when relevant, information from column six.

For example, in the case of the measurement of a one brick wall in common bricks, the description would be framed as follows (Section 14 of NRM refers):

Column one – Walls one brick thick or 225mm
Column two – Measured in $m^2$
Column three – Brickwork
Column four – Not applicable
Column five – Not applicable
Column six – Referred to as applicable.

This information will be augmented from the following mandatory information:

- type, quality and size of brick/block/stone units
- type of finish/facings to each side
- bond
- composition and mix of mortar
- type of pointing.

An NRM2 description therefore for a typical half-brick wall in facings would be:

| | | | | |
|---|---|---|---|---|
| | 23.00<br>2.00<br>56.00<br>2.00<br>78.00<br>2.00<br>65.00<br>2.00 | | Half-brick wall in facings, stretcher bond in gauged mortar Type A pointed one side as work proceeds | In practice this description is very similar to the SMM7 version. |

**Figure 3.5** Description for a typical half-brick wall in facings

A further example again from Section 14 Masonry is:

Columns 1 & 6 – Brick on end band, 225mm high on face
Column 2 – m
Column 3 – Flush
Column 4 – Horizontal
Column 5 – Entirely of stretchers.

Giving the description:

| | | | | |
|---|---|---|---|---|
| | 23.00 | | Brick on end flush horizontal band, 225mm high on face, entirely of stretchers. | |

**Figure 3.6** Masonry description

The measurement examples in this chapter will reflect measurement approaches of both SMM7 and NRM2.

### *Managing the measurement process*

The bill of quantities, when completed, is traditionally presented in trade format, that is, the same order as SMM7, for example:

- demolition and alteration
- ground work
- concrete work
- masonry
- etc.

In the case of NRM2 it will be:

- off-site manufactured materials, components and buildings
- demolitions
- alterations, repairs and conversion
- excavation and filling
- ground remediation and soil stabilisation
- etc.

There are various approaches to measurement for bills of quantities and these are as follows.

### *The group method*

When using the group approach it will be the responsibility of the person in charge of the measurement to allocate the taking off amongst the surveyors and prepare a taking-off schedule. This method is probably the most popular and is based on the following measurement groups:

- substructure
- frame
- upper floors

- external walls
- windows
- staircases
- roof
- internal doors
- doors
- internal finishes
- services
- drainage
- external works.

The taking-off schedule, as well as allocating various tasks also helps to identify missing information. On larger contracts the measurement will be divided between several surveyors and it is important that double measurement or omission of items do not occur. For example, the surveyor measuring the external walls will measure the various items gross, in other words, without deduction for any door or windows. However, the surveyor measuring the doors and windows must deduct the area occupied by doors and windows, as well as measuring the work to the sides (the reveals), the sill and the head of the door or window opening. Similarly, the surveyor measuring the roof must know where the demarcation of the roof and the external walls are and that roof measurement includes not only the roof structure, but also the coverings and the rainwater goods. The taking-off schedule will also include the start and anticipated completion date of the various groups.

When using this approach the measurement and bill process has three stages:

1. Measurement stage – the measurement and description of items in accordance with the *Standard Method of Measurement for Building Works* – 7th Edition (SMM7). SMM7 provides: 'a uniform basis for measuring building works and is to ensure that the bills of quantities fully describe and accurately represent the quantity and quality of the works to be carried out' (SMM7 – General rules 1.1). In NRM2 this has been adapted to read: 'the document provides rules of measurement for the preparation of bills of quantities and schedule of works (quantified)'. On completion of the measurement the quantities are calculated ready for the next stage.
2. Abstract stage – the process by which the measured items together with the associated quantities are transferred from the dimension paper to an abstract where all like items are grouped together and they are arranged in the order that they will appear in the bill of quantities. During these first two stages it has been traditional for quantity surveyors to use abbreviations in order to reduce time and cut down the amount of paper work, for example:

| | | | |
|---|---|---|---|
| deduct | Ddt | manhole | mh |
| extra over | EO | ground level | gl |
| ditto | do | foundations | fdns |
| softwood | sw | cast iron | ci |
| hardwood | hw | damp-proof course | dpc |
| excavate | exc | tongue & groove | t&g |
| not exceeding | ne | brickwork | bkwk |
| half brick | hb | mild steel | ms |
| as before | ab | reinforced concrete | rc |
| as described | ad | galvanised | galv |

3. Draft bill stage – using the traditional system the items are now transferred from the abstract to the draft bill, where items are written out in full. This draft bill now has to be printed in the final bill of quantities format.

It can been seen from the above notes that the traditional approach is labour intensive and therefore it is little wonder that various systems, increasingly based on IT solutions, have been developed to speed up the process. Nevertheless the ability to measure and abstract items is a skill still used by the modern quantity surveyor throughout pre- and post-contract stages of a project and one that is highly sought after by the industry on both demand and supply sides.

### *Trade-by-trade*

This method involves organising the measuring of the quantities in the same order as the final presentation. That is, each surveyor has the responsibility for measuring a complete trade and is issued with a complete set of drawings and other information relating to the project. The principal advantage of this approach is that it removes the need for the abstract stage, thereby speeding up the process and reducing the manpower needed to produce the bill. In addition, as each section is completed it can be processed. There may still be offices in the UK that use the so-called 'northern approach' dimension paper when using the trade-by-trade approach!

## Presentation of the bills of quantities

In addition to the approach to the measurement of quantities, there are also various approaches to the presentation of bills of quantities and these are:

- trade-by-trade
- elemental.

Over the years there have been other bill formats developed, such as operational, but these have never been widely adopted by the UK construction industry.

Scotland developed its own system for measurement and presentation where the approach was to let each trade as a separate contract rather similar to the lots séparés system used in France; however, during the past twenty years or so, this is used less and less.

Even a medium-sized project will generate a large amount of drawn documentation that is passed to the quantity surveyor for measurement; this can be in the form of:

- architect's drawings;
- door/window, fittings and finishing schedules, etc.;
- engineer's drawings detailing structural elements;
- specifications of materials and workmanship;
- specialists' drawings relating to piling, mechanical installations, drainage, etc.

During the measurement period it can be expected that the information on drawings and schedules will be revised and updated many times, it is therefore important that the quantity surveyor always uses the most up-to-date information when preparing the bill of quantities. Almost every day will bring an updated set of information. It is of course impossible to keep on including revisions in bills of quantities indefinitely, or otherwise the bills would never be produced, therefore the quantity surveyor has at some point to draw a line in the sand and base the measurement on what information is then available. As information arrives at the quantity surveyor's office it should be logged in the drawing register as follows:

**Table 3.1** Drawing register

| Number | Description | Date received | | Revisions | |
|---|---|---|---|---|---|
| | | A | B | C | D |
| 0094 | Substructure | 24/04/12 | | 26/04/12 | 28/04/12 |
| 0095 | Section A – A | 24/04/12 | | | |

The significance of recording the drawings is that it is possible to identify what drawn information was used to prepare the bills of quantities. In addition a list of the drawings together with the revision that is used is included in the bill of quantities under the heading:

**List of drawings from which these bills of quantities have been prepared**

The drawings included in this list will be available for inspection by the contractor at the architect's office. Applications to inspect the drawings shall be made to ABC Architects:

Architectural drawings
0094/B
0095

In this way there can be no doubt as to what information was used. Any subsequent revisions to drawings, schedules, etc. have to be dealt with at the post-contract stage with the issue of an architect's instruction.

## MEASUREMENT CONVENTIONS

Today, the majority of bills of quantities are produced using proprietary software packages, each system having its own format for inputting dimensions and formulating descriptions. However, in order to fully understand and appreciate the potential problems in the measurement process a thorough knowledge of measurement conventions is essential as measurement is not just used by the quantity surveyor during the bills of quantity stage, but also during measurement and preparation of interim valuations and the final account. The traditional approach to measurement starts with recording dimensions on traditional dimension or taking-off paper (see Table 3.2).

**Table 3.2**

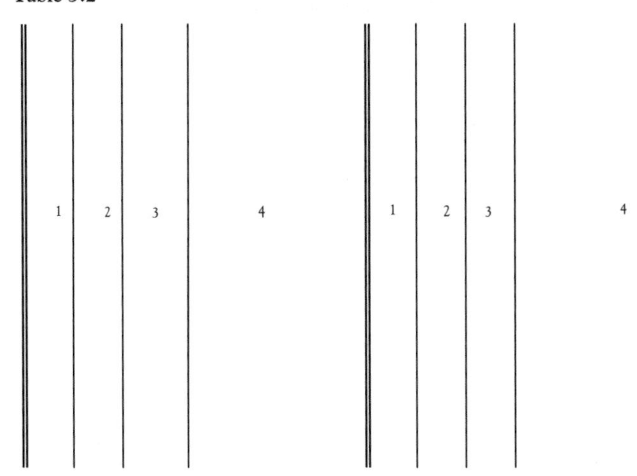

Table 3.2 illustrates traditional dimension paper. It comprises A4 paper, either printed on one or both sides, with two rows of identical columns. These columns are referred to as:

- Column 1 – the timesing column; this column is useful on the occasions where there are identical or repeat items.
- Column 2 – the dimension column; this column is used to record dimensions.
- Column 3 – the squaring column; this column is used to record the computed quantities.
- Column 4 – the description column; the widest column is used to compose the description of the measured items in accordance with SMM7 or NRM2. This description will appear in the final bills of quantities.

Finally, as illustrated later, the right-hand side of the dimension column is reserved for what is referred to as the waste calculations. They are not waste at all really, they are the 'workings out' behind the figures entered into column 3.

One of the most important disciplines for a quantity surveyor to develop is to 'signpost' or annotate the dimensions so that other people can easily see where and how the dimensions have been calculated. A well signposted set of dimensions can save a lot of time in circumstances where work has to be remeasured at the final account stage.

When taking off quantities in accordance with SMM7 or NRM2 the units of measurement will be: cubic metres, square metres, linear metres and numbered items. In addition there are some items that are required to be recorded as items, which are descriptions without a quantity, for example testing drainage.

When recording dimensions on the taking-off paper the order in which the dimensions are recorded is always length, width and depth as noted in Figure 3.7.

One of the cardinal rules when measuring quantities is to always use ink, never pencil. It may be tempting, particularly if learning the process, to use pencil, for ease of correction; however, this can cause embarrassment if the accuracy of the dimensions is called into question at some future date. If an error is made during the measurement process then there is a procedure that should be followed (see Figure 3.8).

**Anding-on**

There are quite a few occasions during the measurement process where more than one item has the same set of dimensions. In order to avoid repetition, a

| | | | | Comment |
|---|---|---|---|---|
| | 23.00<br>2.38<br>0.89 | $m^3$ | Length<br>Width<br>Depth | Note the use of the dimension column; the dimension is always underlined. SMM7/NRM2 determines whether the items are measured as cubic, square or linear metres. |
| | 23.00<br>2.38 | $m^2$ | Length<br>Width | |
| | 23.00 | m | Length | Note: NRM2 – General Rules 3.3 now gives guidance on how dimensions are to be set out. |
| | 1 | No | Numbered | |
| 2/ | 23.00<br>2.38<br>0.89<br>23.00<br>2.38<br>0.89<br>16.00<br>5.78<br>2.90 | 114.63 | If there is more than one identical item, the item can be multiplied by the number of the items, in this case two, instead of writing it again. The timesing column is used for this as shown. It is not necessary to once times items. | Dimensions are recorded to two places of decimals in the dimension column. At the end of the taking-off stage the dimensions are squared and the result recorded in the squaring column and underlined. Once again two places of decimals is used. Note: NRM2 states that 'dimensions used in calculating quantities shall be to the nearest 10mm'. |
| | | | The following description is composed in accordance with NRM2. Shorthand is often used, for example, | When there is more than one dimension that relates to a single description, the dimensions should be bracketed together. |
| | | | **Calculations** **Waste** | |
| | 23.00<br>2.38<br>0.89 | | Bulk excavation 21.000<br>not exceeding 2 metres Add<br>deep commencing 2/ 1.000 2.000<br>at reduced level Length 23.000 | The waste calculation column is not strictly delineated, but the right side of the description column is used when required. Waste calculations are calculated to three places of decimals, and then rounded off for the dimension column. |
| | 23.00<br>2.38<br>0.89 | | Bulk exc ne 2m dp commencing rl<br><br>1 | In shorthand this will be:<br><br>Column/page numbers |

**Figure 3.7**

description that refers to the same dimensions can be anded-on as illustrated in Figure 3.9.

Anding-on can also be used in the situation illustrated in Figure 3.10.

In this situation although the two items anded-on have different units ($m^2$ and $m^3$) the two items can be anded-on and the description column used to make the necessary adjustments.

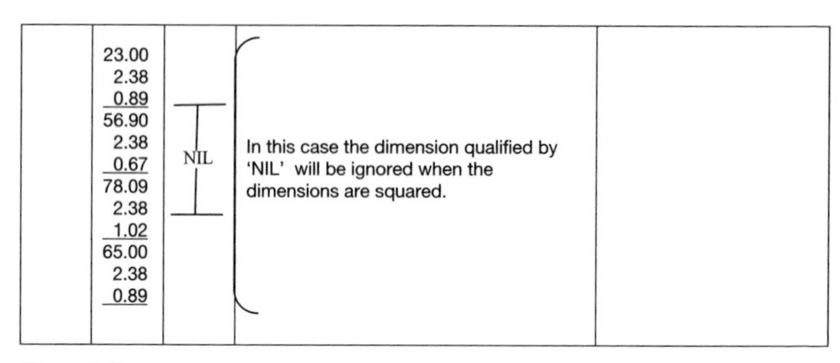


Figure 3.8

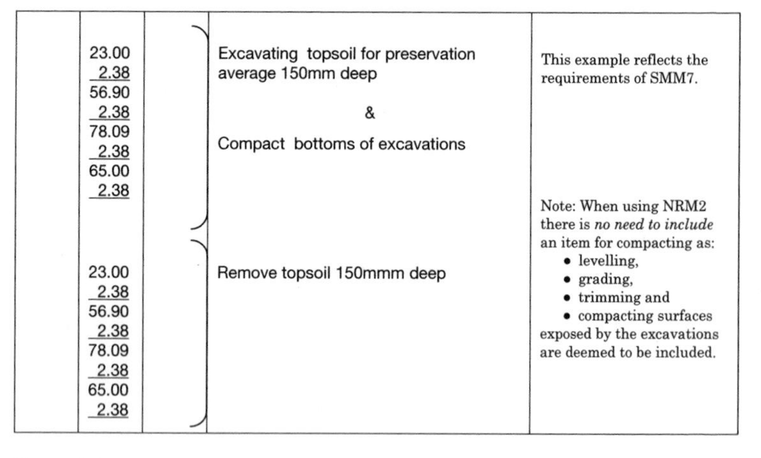


Figure 3.9

## Adds and deducts

Figured dimensions, that is dimensions shown on the drawings, and scaled dimensions, that is dimensions scaled off drawings using a scale rule, can be used when measuring quantities; however, figured dimensions always take preference (see Figure 3.11).

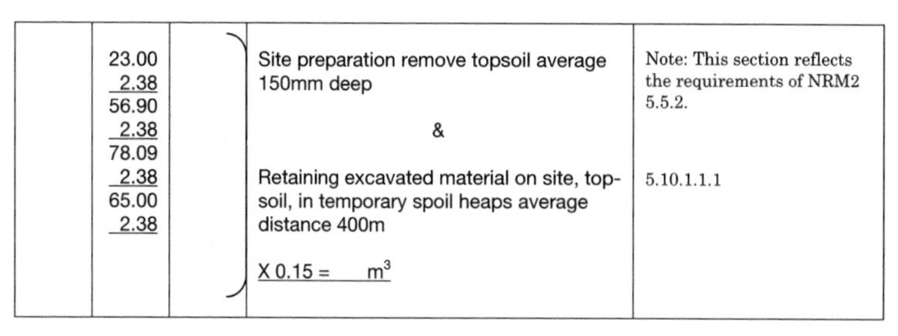

| | | | | |
|---|---|---|---|---|
| | 23.00<br>2.38<br>56.90<br>2.38<br>78.09<br>2.38<br>65.00<br>2.38 | | Site preparation remove topsoil average 150mm deep<br><br>&<br><br>Retaining excavated material on site, top-soil, in temporary spoil heaps average distance 400m<br><br>X 0.15 = m³ | Note: This section reflects the requirements of NRM2 5.5.2.<br><br>5.10.1.1.1 |

Figure 3.10

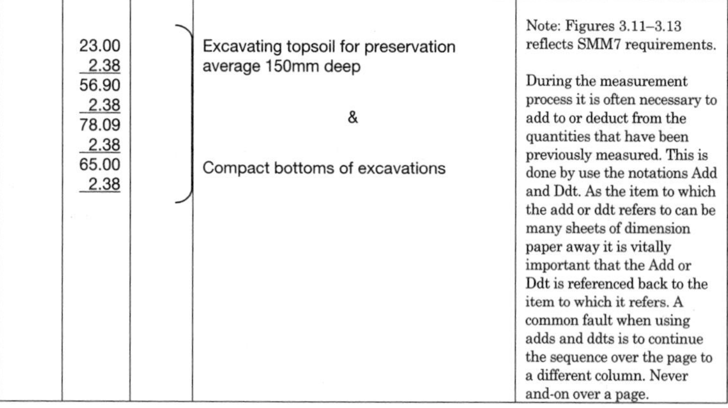

| | | | | |
|---|---|---|---|---|
| | 23.00<br>2.38<br>56.90<br>2.38<br>78.09<br>2.38<br>65.00<br>2.38 | | Excavating topsoil for preservation average 150mm deep<br><br>&<br><br>Compact bottoms of excavations | Note: Figures 3.11–3.13 reflects SMM7 requirements.<br><br>During the measurement process it is often necessary to add to or deduct from the quantities that have been previously measured. This is done by use the notations Add and Ddt. As the item to which the add or ddt refers to can be many sheets of dimension paper away it is vitally important that the Add or Ddt is referenced back to the item to which it refers. A common fault when using adds and ddts is to continue the sequence over the page to a different column. Never and-on over a page. |

Figure 3.11

## Timesing

Timesing is a technique used by takers-off in the situation where dimensions are repeated. Frequently this will occur when making adjustments for items such as windows and door openings, as illustrated in Figure 3.12.

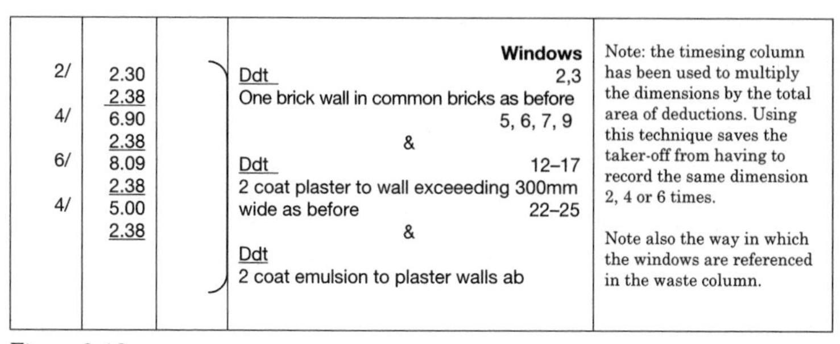

| | | | | |
|---|---|---|---|---|
| 2/<br><br>4/<br><br>6/<br><br>4/ | 2.30<br>2.38<br>6.90<br>2.38<br>8.09<br>2.38<br>5.00<br>2.38 | | **Windows**<br>Ddt 2,3<br>One brick wall in common bricks as before 5, 6, 7, 9<br>&<br>Ddt 12–17<br>2 coat plaster to wall exceeeding 300mm wide as before 22–25<br>&<br>Ddt<br>2 coat emulsion to plaster walls ab | Note: the timesing column has been used to multiply the dimensions by the total area of deductions. Using this technique saves the taker-off from having to record the same dimension 2, 4 or 6 times.<br><br>Note also the way in which the windows are referenced in the waste column. |

Figure 3.12

## Dotting on

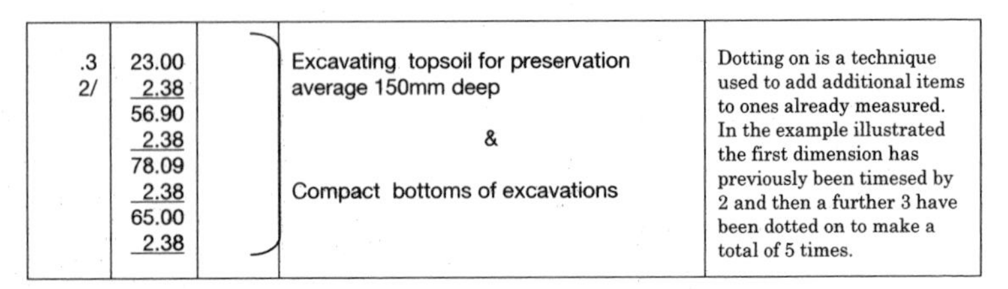

| | | | |
|---|---|---|---|
| .3<br>2/ | 23.00<br>2.38<br>56.90<br>2.38<br>78.09<br>2.38<br>65.00<br>2.38 | Excavating topsoil for preservation average 150mm deep<br><br>&<br><br>Compact bottoms of excavations | Dotting on is a technique used to add additional items to ones already measured. In the example illustrated the first dimension has previously been timesed by 2 and then a further 3 have been dotted on to make a total of 5 times. |

**Figure 3.13**

## Centre lines and mean girths

A technique used by takers-off when measuring items in the substructure and the external walls is the calculation of a centre line or mean girth. In Scotland a different approach is sometimes used, known as 'over and between'.

Figure 3.14 is a plan of a simple building where the mean girth approach would be applicable. Depending on circumstances the mean girth can be calculated from the internal or external face; in the example that follows (Figure 3.15) the internal dimensions will be used.

## *Curved work*

NRM2 (3.3.2) states that the radius of any curved work should be the mean radius measured on the centre line.

Figure 3.16 has the same overall dimensions as Figure 3.14 except that there is a 4m × 1.5m inset as shown.

## Making a start

Measurement is one of the key skills demanded of a quantity surveyor and goes hand in hand with an in-depth knowledge of construction technology.

Having mastered the basic measurement conventions it is now time to start taking off; but where to start? It is assumed for the purposes of this pocket book that a complete set of drawn information is available, whereas in practice the starting point may be dictated by the information at the surveyor's disposal. It is assumed that the group method is to be used to manage the taking-off process, as described earlier in this section, starting off with substructure.

Substructure involves the measurement of all work up to and including the damp-proof course. It includes a variety of trades such as excavation, concrete work and masonry.

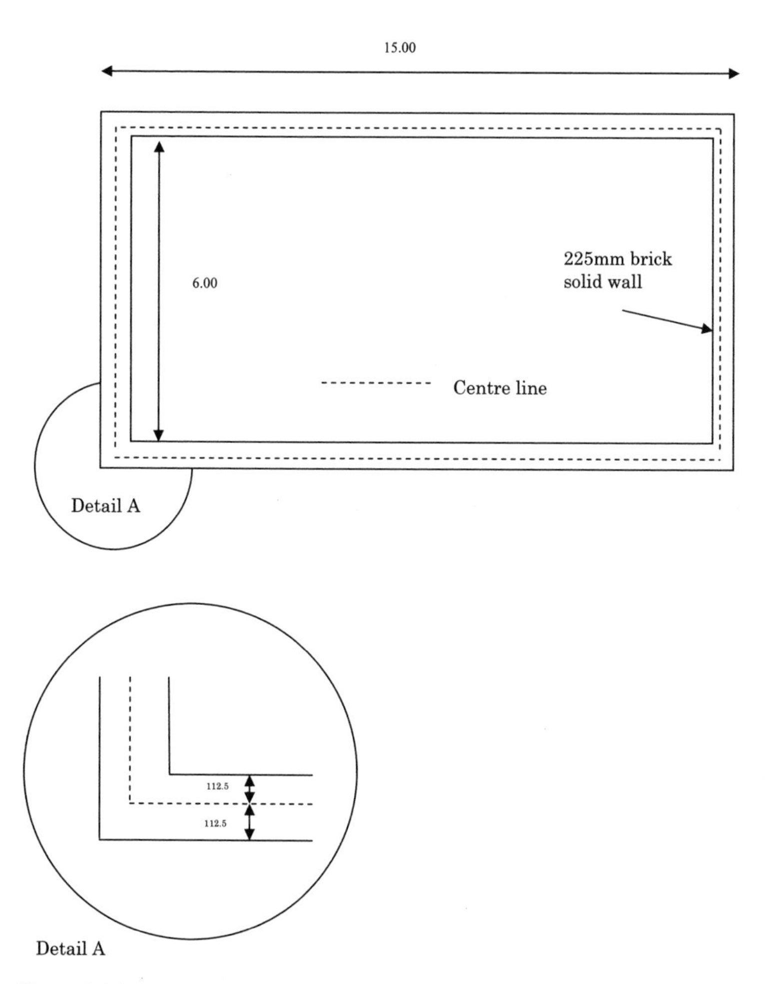


**Figure 3.14**

## Example 1: Substructure

As with most measurement exercises it is good practice to start with a taking-off list containing all the items that have to be included. Note that certain items apply to SMM7 only.

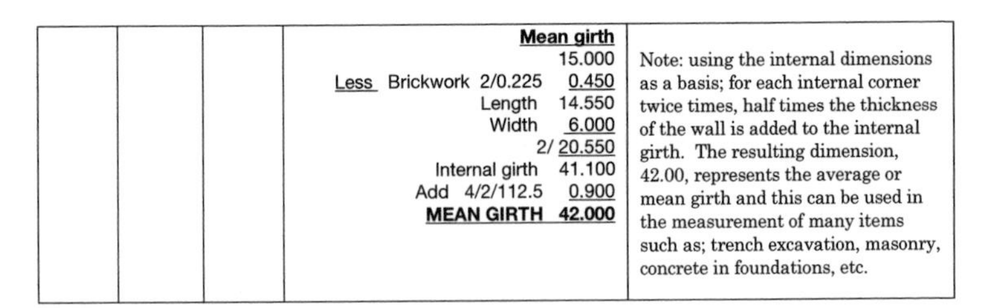

| | | | | |
|---|---|---|---|---|
| | | | **Mean girth**<br>15.000<br>Less Brickwork 2/0.225 0.450<br>Length 14.550<br>Width 6.000<br>2/ 20.550<br>Internal girth 41.100<br>Add 4/2/112.5 0.900<br>**MEAN GIRTH 42.000** | Note: using the internal dimensions as a basis; for each internal corner twice times, half times the thickness of the wall is added to the internal girth. The resulting dimension, 42.00, represents the average or mean girth and this can be used in the measurement of many items such as; trench excavation, masonry, concrete in foundations, etc. |

Figure 3.15

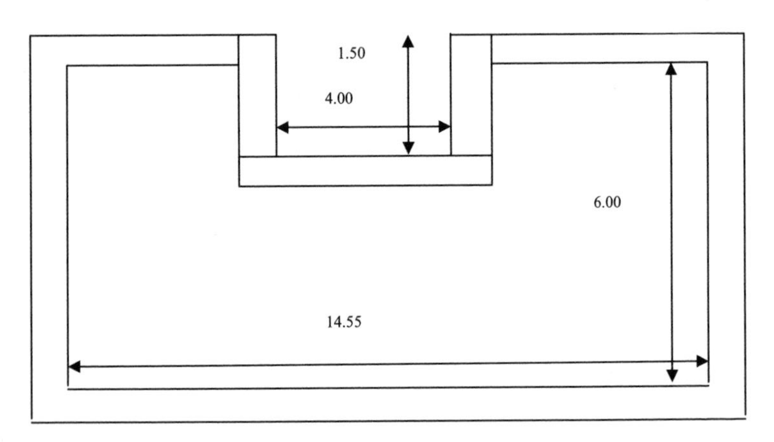


Figure 3.16

| | | | | |
|---|---|---|---|---|
| | | | **Mean girth**<br>Length 2/14.550 29.100<br>Width 2/ 6.000 12.000<br>41.100<br>Add Inset 2/1.500 3.000<br>44.100<br>Add 4/2/112.5 0.900<br>**MEAN GIRTH 45.000** | In this example again the mean girth length will be calculated. As far as calculating the mean girth is concerned the internal and external angles in the inset cancel each other out, therefore to calculate the mean girth 4/2/ the thickness of the wall needs to be either added to the internal girth or deducted from the external girth. |

Figure 3.17

**Substructure: taking-off list**

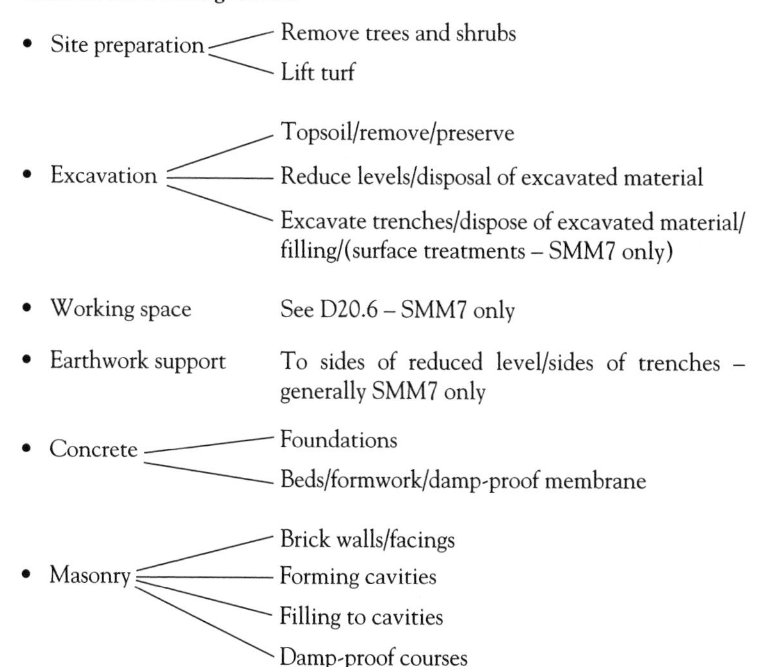


***Site levels***

Virgin sites will almost certainly be covered with a layer of vegetation that has to be removed prior to excavation commencing and stored separately or removed from site. Topsoil cannot be used for backfilling as it would, over time, cause damage to the substructure. The usual, default depth, for topsoil is 150mm although it could be more than this and a test pit may be dug to accurately determine the actual depth.

Figure 3.18 shows a 5 metre grid of a survey of levels taken on a proposed site.

The site is required to be reduced to a level of 35.62 and in order to calculate the volume of excavation required the average level of the site must be determined. This can be quite easily done by weighting the levels as follows starting at the top left-hand corner of the site (35.90) and working from left to right:

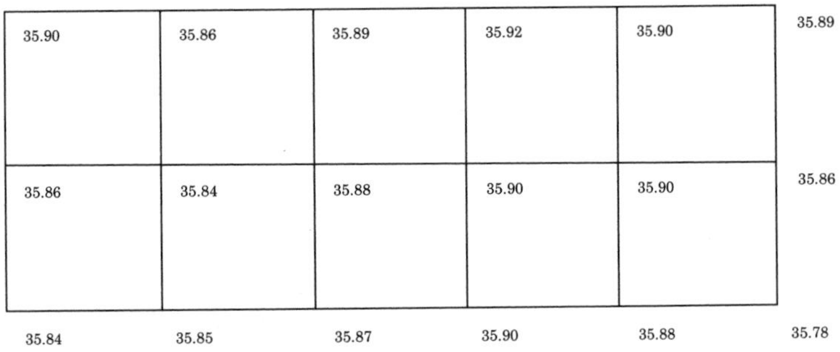

| | | | | | |
|---|---|---|---|---|---|
| 35.90 | 35.86 | 35.89 | 35.92 | 35.90 | 35.89 |
| 35.86 | 35.84 | 35.88 | 35.90 | 35.90 | 35.86 |
| 35.84 | 35.85 | 35.87 | 35.90 | 35.88 | 35.78 |

**Figure 3.18**

| | |
|---|---|
| 35.90 | 35.90 |
| 35.86 × 2 | 71.72 |
| 35.89 × 2 | 71.78 |
| 35.92 × 2 | 71.84 |
| 35.90 × 2 | 71.80 |
| 35.89 | 35.89 |
| 35.86 × 2 | 71.72 |
| 35.84 × 4 | 143.36 |
| 35.88 × 4 | 143.52 |
| 35.90 × 4 | 143.60 |
| 35.90 × 4 | 143.60 |
| 35.86 × 2 | 71.72 |
| 35.84 | 35.84 |
| 35.85 × 2 | 71.70 |
| 35.87 × 2 | 71.74 |
| 35.90 × 2 | 71.80 |
| 35.88 × 2 | 71.76 |
| 35.78 × 2 | 71.56 |
| × 41 | 1470.85 = 35.87 |

| | |
|---|---|
| Average site level | 35.87 |
| Less | |
| Reduced level | 35.62 |
| Excavation | 0.25 |

In the above case the grid levels have been multiplied by the number of grid squares that are affected by a level, the resultant total is then divided to provide an average depth of 0.25m.

Figure 3.19 shows a cross-section through the trench and reduced level excavation required for the external wall in Example 1. Note that the levels

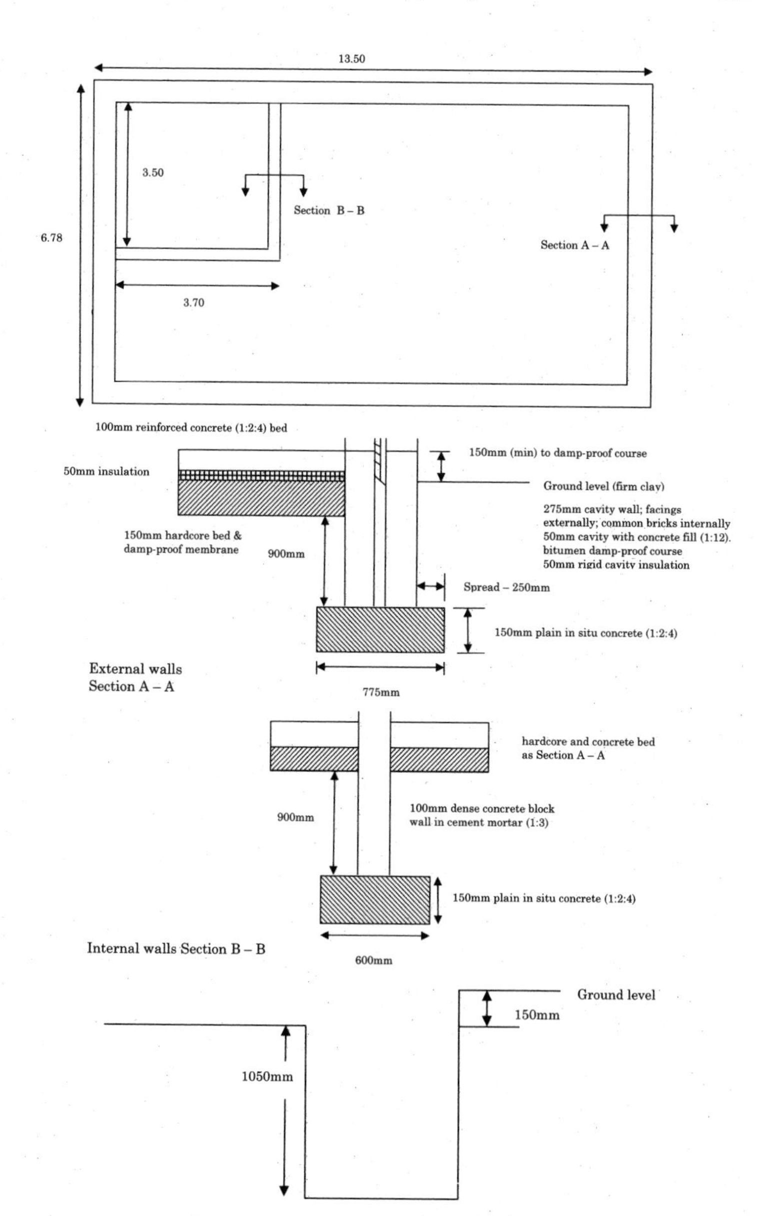
13.50
3.50
Section B – B
6.78
Section A – A
3.70
100mm reinforced concrete (1:2:4) bed
150mm (min) to damp-proof course
50mm insulation
Ground level (firm clay)
275mm cavity wall; facings externally; common bricks internally
50mm cavity with concrete fill (1:12).
bitumen damp-proof course
50mm rigid cavity insulation
150mm hardcore bed & damp-proof membrane
900mm
Spread – 250mm
150mm plain in situ concrete (1:2:4)
External walls Section A – A
775mm
hardcore and concrete bed as Section A – A
100mm dense concrete block wall in cement mortar (1:3)
900mm
150mm plain in situ concrete (1:2:4)
Internal walls Section B – B
600mm
Ground level
150mm
1050mm

**Figure 3.19**

have been reduced internally by 150mm to allow for a 150mm-thick bed of hardcore. The top of the hardcore bed when compacted will be covered or blinded with sand to prevent the damp-proof membrane, a layer of polythene sheet with a minimum thickness of 0.30mm, being perforated by the hardcore. It is important that the material used as hardcore is inert and free from chemicals, vegetable or other deleterious matter. It is a requirement of the Building Regulations that insulation is incorporated into the floor construction and in this case 50mm-thick rigid insulation board has been used. The decision as to whether to position the boards above or below the slab is one of personal choice. The bottom of the trench excavation when completed will be compacted prior to the concrete being poured; this is to prevent the soil being incorporated into the concrete and weakening the mix. This is particularly important when reinforced concrete is being used, where it is common to blind the bottom of the excavation with a weak mix concrete before the reinforcement is placed in position.

It is a requirement of SMM7 that earthwork support is measured to faces of excavation over 250mm high. However, SMM7 also adds the caveat '*whether or not required*' (D20.7), placing the risk on the contractor as to whether or not to use earthwork support. There are clear health and safety issues here for the contractor, who must decide, having priced the item in the bills of quantities, whether he is actually going to use earthwork support on site.

NRM2 takes a different approach to excavation in that the following items are deemed to be included:

- disposal of all surface water;
- earthwork support, unless expressly stated;
- working space;
- levelling, grading, trimming and compacting surfaces exposed by the excavations.

Principal differences between SMM7 and NRM2 when measuring substructure/earthworks:

- Earthwork support and working space, regarded by the industry as risk items, are no longer required to be included unless specifically required, see Example 1.
- Three new sections – 6, Ground remediation and soil stabilisation, 9, Diaphragm walls and embedded retaining walls and 10, Crib walls, gabions and reinforced earth – have been added as ground works to NRM2.

| | | | | |
|---|---|---|---|---|
| | | | **Example 1 – Substructure** | Note: This example is based on NRM2. |
| | | | Oversite excavation | |
| | | | 13.500 | |
| | | | Add spread | |
| | | | 2/0.250 0.500 | |
| | | | LENGTH 14.000 | |
| | | | Width 6.780 | |
| | | | Add spread | |
| | | | 2/0.250 0.500 | |
| | | | WIDTH 7.280 | |
| | 14.00<br>7.28<br>0.15 | | Bulk excavation n.e. 2m deep commencing formation level | Reduced level excavation is classified as 'bulk excavation'. However, NRM2 does allow certain types of bulk excavation to be classified separately. |
| | | | & | |
| | | | Disposal of excavated material off site | |
| | | | **Mean girth** | The mean girth has been calculated from the outside face of the external wall. To determine the mean girth deduct four times the thickness of the external wall. |
| | | | Length 13.500 | |
| | | | Width 6.780 | |
| | | | 2/ 20.280 | |
| | | | 40.560 | |
| | | | Ddt 4/2/ ½/0.275 1.100 | |
| | | | Mean girth 39.460 | |
| | | | Internal partition | |
| | | | 3.500 | |
| | | | 3.700 | |
| | | | 7.200 | |
| | | | Ddt | |
| | | | Spread to external walls | |
| | | | 2/0.250 0.500 | |
| | | | 6.700 | |
| | 39.46<br>0.78<br>1.05<br>6.70<br>0.60<br>1.05 | | Foundation excavation n.e. 2m deep. | At this point it is assumed that all of the trench will be backfilled. When concrete and brickwork are measured later the quantities of filling will be adjusted. |
| | | | & | |
| | | | Filling obtained from excavated materials, final thickness exceeding 500mm deep | Surface treatment of excavations is deemed to be included unless not at the discretion of the contractor. |
| | | | 1 | |

Figure 3.20

| | | | | |
|---|---|---|---|---|
| | 39.46<br>0.78<br>0.15<br>6.70<br>0.60<br>0.15 | | Plain in situ concrete (1:2:4) horizontal work in structures n.e. 300mm thick poured on or against earth<br>&<br>Ddt<br>Filling to excavation avg thickness exc 0.50m ab<br>&<br>Add<br>Disposal of excavated mats off site ab | Note that NRM2 does not require earthwork support to be measured unless not at the discretion of the contractor.<br>Horizontal includes a number of classes of concrete such as: blinding, foundations, pile caps, etc. NRM2 allows for each category to be kept separate if required.<br>Now that the concrete has been measured the filling can be adjusted together with the disposal of excavated material. |
| | | | **Brickwork**<br>Foundations 0.900<br>Hardcore 0.150<br>Insulation 0.050<br>Concrete bed 0.100<br>Height 1.200 | |
| 2/ | 39.46<br>1.20 | | Walls half-brick thick skins of hollow walls in common bricks laid strctcher bond in cement mortar (1:3) | Note: NRM2 states that walls are measured on the centre line unless otherwise stated. |
| | | | **Internal partitions** | |
| | 7.20<br>1.20 | | Walls in dense concrete blocks 100mm thick skins of hollow walls in cement mortar (1:3) | |
| | 39.46<br>1.20 | | Forming cavities in hollow walls 50mm wide with and including stainless steel twisted wire wall ties @ 5 per square metre. | Wall ties are generally spaced at 900mm horizontal and 450mm vertically for cavities between 50 and 75mm wide. The cavity filling is slightly less than the item for forming cavities to take account of the splay to the top edge. Note cavity fill should be stopped at least 225mm below the base of the dpc.<br>Generally, the volume of each type of vertical work may be aggregated or given separately, for example; filling to hollow walls. |
| | 39.46<br>0.08<br>1.15 | | Plain in situ concrete (1:12) vertical work not exceeding 300mm thick in structures | |
| | | | 2 | |

Figure 3.20 (continued)

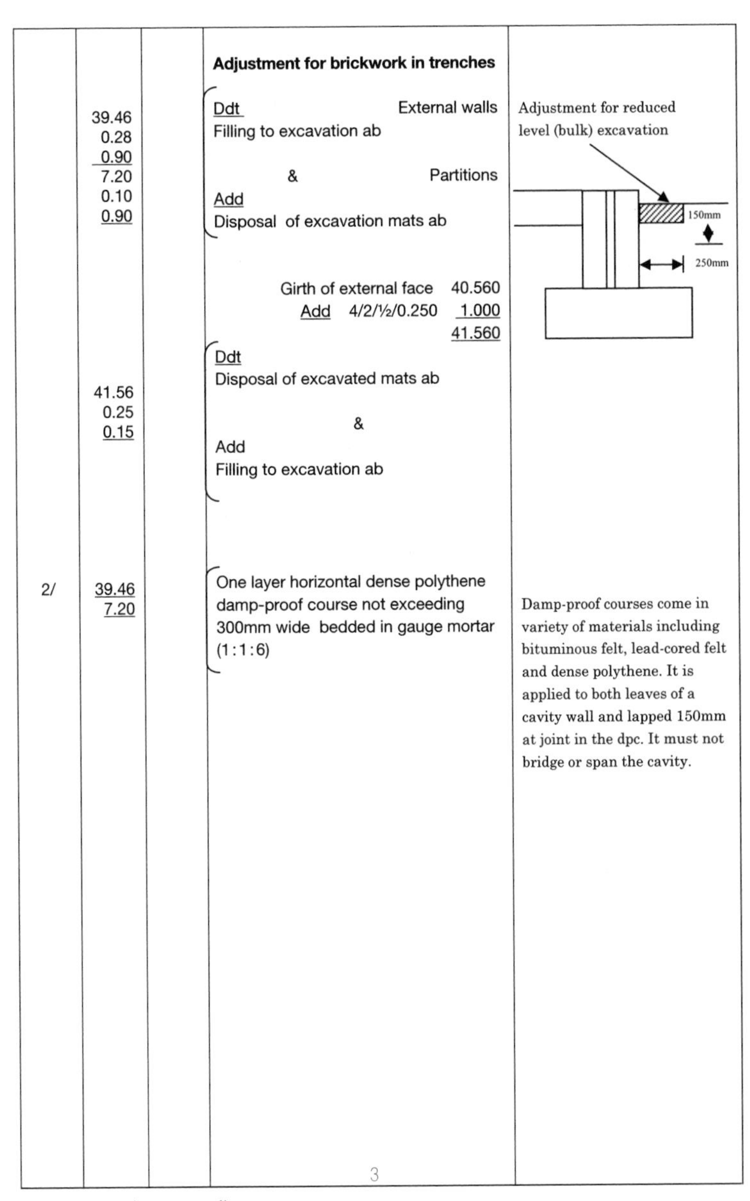

| | | | | |
|---|---|---|---|---|
| | | | **Adjustment for brickwork in trenches** | |
| | 39.46<br>0.28<br>0.90<br>7.20<br>0.10<br>0.90 | | Ddt External walls<br>Filling to excavation ab<br>& Partitions<br>Add<br>Disposal of excavation mats ab | Adjustment for reduced level (bulk) excavation<br>150mm<br>250mm |
| | | | Girth of external face 40.560<br>Add 4/2/½/0.250 1.000<br>41.560 | |
| | 41.56<br>0.25<br>0.15 | | Ddt<br>Disposal of excavated mats ab<br>&<br>Add<br>Filling to excavation ab | |
| 2/ | 39.46<br>7.20 | | One layer horizontal dense polythene damp-proof course not exceeding 300mm wide bedded in gauge mortar (1:1:6) | Damp-proof courses come in variety of materials including bituminous felt, lead-cored felt and dense polythene. It is applied to both leaves of a cavity wall and lapped 150mm at joint in the dpc. It must not bridge or span the cavity. |
| | | | 3 | |

Figure 3.20 (continued)

| | | | | |
|---|---|---|---|---|
| | | | **Ground floor** 13.500<br>Less Extl walls 2/0.275 0.550<br>12.950<br>6.780<br>Less Extl walls 2/0.275 0.550<br>6.230 | |
| | 12.95<br>6.23 | | Reinforced in situ concrete (1:2:4) horizontal work in structures n.e. 300mm thk<br>× 0.10 = m³<br><br>&<br><br>Mesh reinforcement Ref A252 weighing 3.95 kg/m² with 150mm minimum side and end laps.<br><br>&<br><br>Imported hardcore bed over 50mm but not exceeding 500mm thick, level and to falls, 150mm finished thickness<br>× 0.15 = m³<br><br>&<br><br>1200 gauge polythene horizontal damp-proof membrane exceeding 500mm wide<br><br>&<br><br>50mm thick horizonal rigid sheet insulation laid on concrete | Note: The damp-proof membrane and damp-proof course are lapped on the inner skin of the external wall. |
| | 7.20<br>0.10 | | Ddt Internal partition<br>Last 5 items | |
| | | | 4 | |

Figure 3.20 (continued)

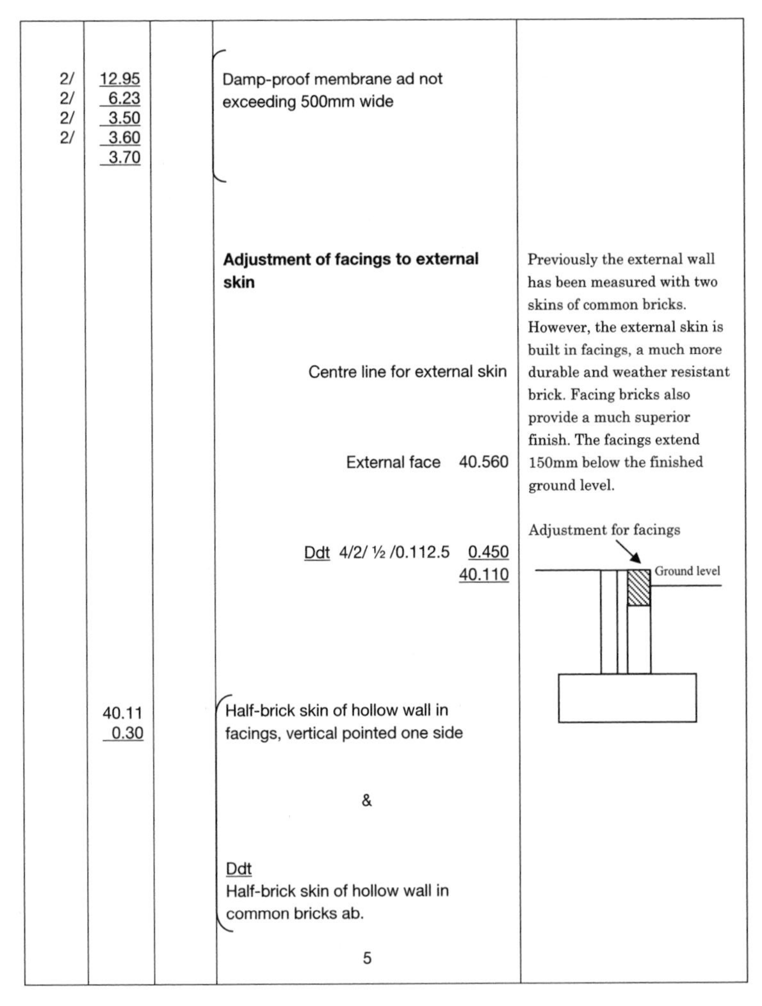

| | | | | |
|---|---|---|---|---|
| 2/<br>2/<br>2/<br>2/ | 12.95<br>6.23<br>3.50<br>3.60<br>3.70 | | Damp-proof membrane ad not exceeding 500mm wide | |
| | | | **Adjustment of facings to external skin**<br><br>Centre line for external skin<br><br>External face 40.560<br><br>Ddt 4/2/ ½ /0.112.5 0.450<br>40.110 | Previously the external wall has been measured with two skins of common bricks. However, the external skin is built in facings, a much more durable and weather resistant brick. Facing bricks also provide a much superior finish. The facings extend 150mm below the finished ground level.<br><br>Adjustment for facings |
| | 40.11<br>0.30 | | Half-brick skin of hollow wall in facings, vertical pointed one side<br><br>&<br><br>Ddt<br>Half-brick skin of hollow wall in common bricks ab.<br><br>5 | |



Figure 3.20 (continued)

## Excavation: SMM7 sundry items

There are some other commonly occurring excavation items required by SMM7 only that need to be discussed.

### *Working space*

Clause D20.6 allows working space to be measured in circumstances where workmen have to operate in situations that require them to work in trenches below ground level, for example when working with formwork, rendering, tanking or protection. It is measurable as a superficial item where there is less than 600mm between the face of the excavation and the work; all additional earthwork support, disposal, backfilling and breaking out are deemed to be included with the working space item. This is another contractor's risk item as they must decide and price what space they think is required as illustrated in Figure 3.21.

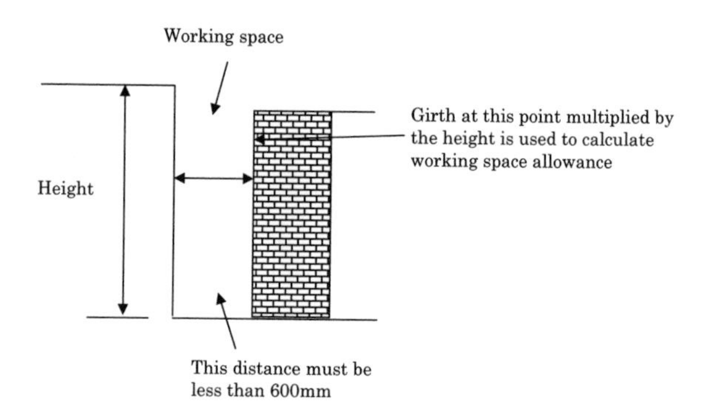


**Figure 3.21** Section through trench illustrating working space allowance

### *Earthwork support*

SMM7 requires that earthwork support shall be measured. In the case of Example 1 – Substructure this would be as shown in Figure 3.22:

| | | | | |
|---|---|---|---|---|
| 2/ | 39.46 | | Earthwork support maximum depth not exceeding 1.00m and distance between opposing faces not exceeding 2.00m | |
| | 1.05 | | | |
| | 42.56 | | | |
| 2/ | 0.15 | | | |
| | 6.70 | | Earthwork support to RL | |
| | 1.05 | | Intl 14.000 | |
| | | | Partition 7.280 | |
| | | | 2/21.280 | |
| | | | 42.560 | |

**Figure 3.22**

### *Extra over items*

Both NRM2 and SMM7 stipulate that certain types of excavations have to be measured as extra over. These include excavating below ground water level and breaking out rock, reinforced concrete, etc.

## Example 2: Walls from damp-proof course to wall plate

This section includes work from the damp-proof course up to eaves level and includes both internal and external walls. Work is measured over all doors, windows and other openings these being adjusted at a later stage. The theory is that it is better to overmeasure and then make deductions and adjustments later, rather than measure net. In this way, if items are missed then at least something has been included!

SMM7 Section F/Section 14 NRM2 deals with the measurement of Masonry and includes brick and block walling, the unit of measurement is generally the square metre.

### Brickwork

The standard size for a brick is accepted to be 215 × 102.5 × 65mm thick, however when the mortar is included this becomes 225 × 112.5 × 75mm thick. Therefore a one-brick wall will be assumed to be 225mm thick and a two-brick wall 450mm thick. Bricks are delivered to site pre-packed on wooden pallets for easy unloading and bought by the thousand, except for specially made units.

Bricks can be broadly categorised as follows.

**Common bricks:** these are suitable for general building work where the finish is not important. Common bricks are made from clay and they are cheaper than the other alternatives.

**Facing bricks:** facings come in a wide variety of finishes, colours, strengths and prices and are used typically in the external skin of cavity walls. Also made from clay, facing bricks are weather resistant and generally finished with pointed joints.

**Engineering bricks:** these bricks have low water absorption properties and a high compressive strength. Typically used in retaining walls, bridges and manholes.

**Calcium silicate bricks:** sometimes referred to as sand lime facings, these are of the more easily recognisable forms of facings made from sand, crushed flint and lime.

Walls of one brick thick and over can be built in common brickwork with facings on one side. In this case, when using SMM7, the facing brickwork is measured as extra over.

### *Expansion joints*

In order to avoid excessive cracking, 16mm-thick vertical expansion joints are required in brickwork, generally at 12 metre centres, whereas when using calcium silicate bricks the joints must be at 7.5–9.0 metre intervals and 10mm thick. The material used for expansion joints can be:

- flexible cellular polyethylene
- cellular polyethylene
- foam rubber.

The joint is finished with a sealant at least 10mm deep.

## Blockwork

Blockwork comes in larger units than brickwork with the standard size being 440 × 215 × 100mm, as well as a variety of thicknesses. As with brickwork, once the mortar is included this becomes 450 × 225 × 100mm, therefore six standard bricks are equal in size to one block; an important fact, as bricks and blocks are used for inner and outer skins of hollow walls.

### *Dense blockwork*

Dense blocks are suitable for above and below ground situations and are made from cement, sand and crushed gravel.

### *Lightweight blockwork*

Lightweight blocks include lightweight aggregates and are generally used for the internal skins of external walls or where a high degree of thermal insulation is required. They are lighter and easier to handle on site than dense concrete blocks with poor sound insulation qualities.

## Mortar

There are a variety of commonly used mortar mixes that should match the type, location and strength of the masonry. As a general rule the mortar should not be as strong as the brick or block, thereby allowing any cracking to take place in the joint and not the masonry. Therefore, a mortar mix of cement:sand (1:3) would be classed as a strong mix whereas a mix of cement:lime:sand (1:1:6) would be classified as a weaker mix. Liquid plasticisers can be used in place of lime to improve mortar workability.

## Walls: taking-off list (SMM7 and NRM2)

- External walls — External skin / Internal skin
- Cavities — Forming cavity / Insulation
- Internal partitions — Blockwork / Stud partitions

The external and internal walls contain windows and doors and the adjustment for these elements will be included in the measurement of the doors and windows (see Figures 3.23 and 3.24).

### *Stud partitions*

Timber stud partitions are generally non-load-bearing internal partitions comprising a framework of timber struts covered both sides with plasterboard

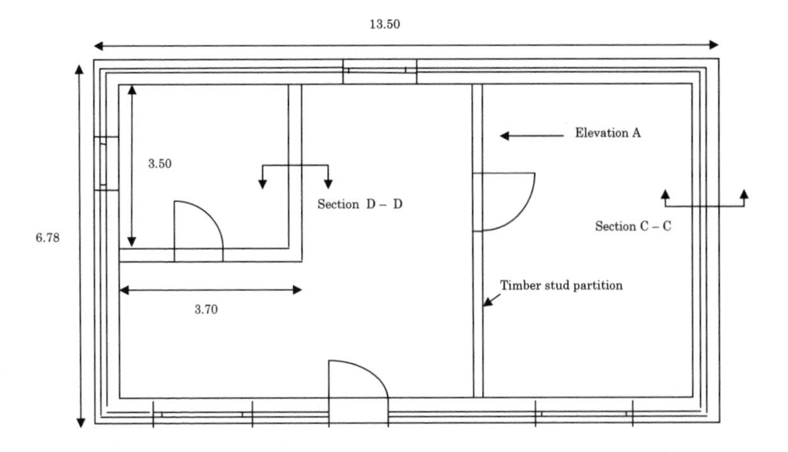


**Figure 3.23** Plan showing external and internal walls

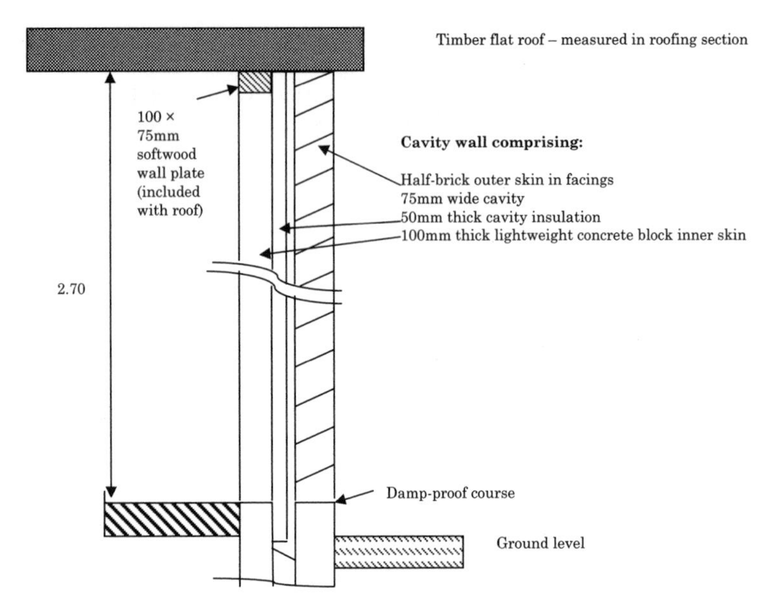


**Figure 3.24** External wall Section C – C

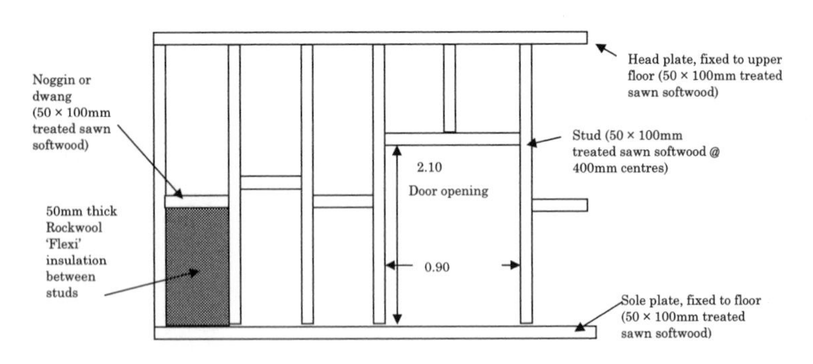


**Figure 3.25** Elevation A – A – plasterboard not shown

(see Figure 3.25). It is common practice to insulate the partitions to reduce sound transmission. The joints are formed with nails.

In addition to timber, stud partitions can be made from proprietary metal systems.

| | | | | |
|---|---|---|---|---|
| | | | **Example 2 – Walls** | Note: This example has been measured in accordance with SMM7. |
| | | | **External walls** | |
| | | | Mean girths | |
| | | | Facings 13.500 | |
| | | | 6.780 | |
| | | | 2/ 20.280 | |
| | | | 40.560 | |
| | | | Ddt 4/2/½/112.5 0.450 | |
| | | | Mean girth 40.110 | |
| | 40.11 | | Half-brick wall in facings, vertical pointed one side ab | |
| | 2.70 | | | |
| | | | Blockwork | |
| | | | 13.500 | |
| | | | Less 2/0.275 0.550 | |
| | | | 12.950 | |
| | | | 6.780 | |
| | | | Less 0.550 | |
| | | | 6.230 | |
| | | | 12.950 | |
| | | | 2/19.280 | |
| | | | 38.560 | |
| | | | Add 4/2/½ /.100 0.400 | |
| | | | 38.960 | |
| | | | Height | |
| | | | 2.700 | |
| | | | Less Wall plate 0.075 | |
| | | | 2.625 | |
| | 38.96 | | Walls in lightweight concrete blocks 100mm thick in gauge mortar (1 : 1 : 6) | |
| | 2.63 | | | |
| | | | 12.950 | |
| | | | Add 2/0.10 0.200 | |
| | | | 13.150 | |
| | | | 6.230 | |
| | | | 0.200 | |
| | | | 6.430 | |
| | | | 13.150 | |
| | | | 2/19.580 | |
| | | | 39.160 | |
| | | | Add 4/2/½ /0.08 0.320 | |
| | | | Cavity centre line 39.480 | |
| | 39.48 | | Forming cavities in hollow walls 75mm wide with and including stainless steel twisted wire wall ties @ 5 per square metre. | |
| | 2.70 | | | |
| | | | 38.560 | |
| | | | Add 4/2/½ /0.150 0.600 | |
| | | | 39.160 | |
| | | | 1 | |

Figure 3.26

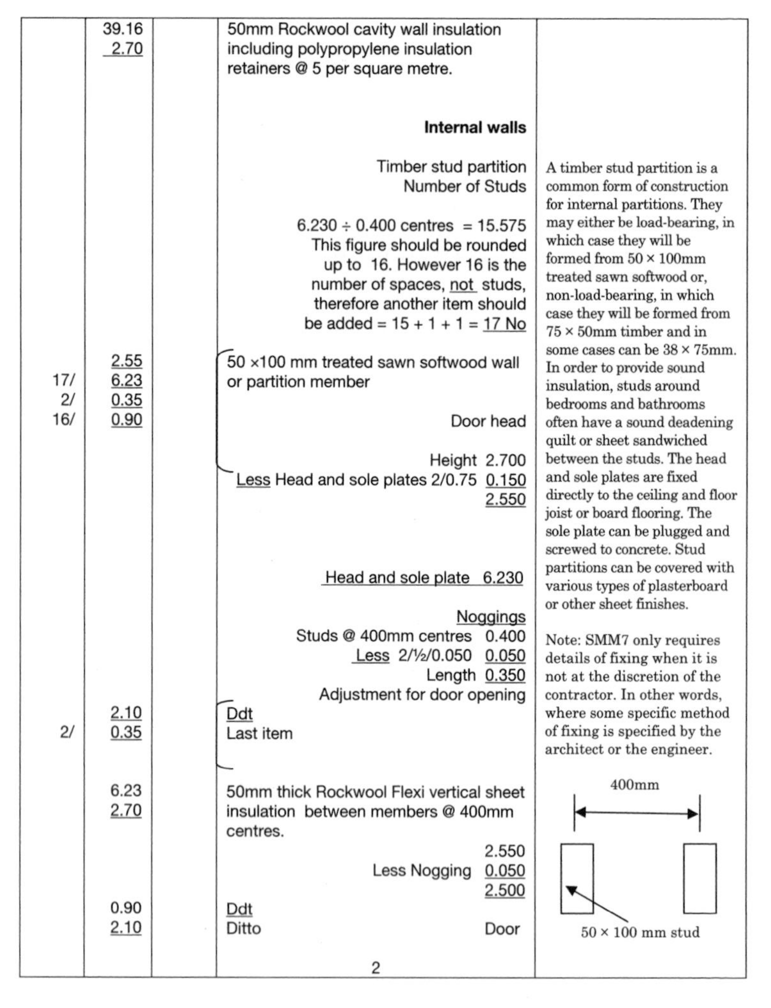

| | | | | |
|---|---|---|---|---|
| | 39.16<br>2.70 | | 50mm Rockwool cavity wall insulation including polypropylene insulation retainers @ 5 per square metre. | |
| | | | **Internal walls** | |
| | | | Timber stud partition<br>Number of Studs<br><br>6.230 ÷ 0.400 centres = 15.575<br>This figure should be rounded up to 16. However 16 is the number of spaces, not studs, therefore another item should be added = 15 + 1 + 1 = 17 No | A timber stud partition is a common form of construction for internal partitions. They may either be load-bearing, in which case they will be formed from 50 × 100mm treated sawn softwood or, non-load-bearing, in which case they will be formed from 75 × 50mm timber and in some cases can be 38 × 75mm. In order to provide sound insulation, studs around bedrooms and bathrooms often have a sound deadening quilt or sheet sandwiched between the studs. The head and sole plates are fixed directly to the ceiling and floor joist or board flooring. The sole plate can be plugged and screwed to concrete. Stud partitions can be covered with various types of plasterboard or other sheet finishes. |
| <br>17/<br>2/<br>16/ | 2.55<br>6.23<br>0.35<br>0.90 | | 50 ×100 mm treated sawn softwood wall or partition member<br><br>Door head<br><br>Height 2.700<br>Less Head and sole plates 2/0.75 0.150<br>2.550<br><br>Head and sole plate 6.230<br><br>Noggings<br>Studs @ 400mm centres 0.400<br>Less 2/½/0.050 0.050<br>Length 0.350<br>Adjustment for door opening | |
| <br>2/ | 2.10<br>0.35 | | Ddt<br>Last item | Note: SMM7 only requires details of fixing when it is not at the discretion of the contractor. In other words, where some specific method of fixing is specified by the architect or the engineer. |
| | 6.23<br>2.70 | | 50mm thick Rockwool Flexi vertical sheet insulation between members @ 400mm centres.<br><br>2.550<br>Less Nogging 0.050<br>2.500 | 400mm<br>50 × 100 mm stud |
| | 0.90<br>2.10 | | Ddt<br>Ditto Door | |
| | | | 2 | |

Figure 3.26 (continued)

## Sundry items of masonry (SMM7 and NRM2)

The following items are commonly met by the quantity surveyor when measuring masonry.

Apart from half-brick skins in hollow walls, brickwork generally comes in three forms:

1. Common brickwork, where the wall of whatever thickness is built entirely of common brickwork. The brickwork will have an 'as built' finish and will be used where the appearance of the finished work is unimportant.
2. Common brickwork with facework one side can be used in any situation where the brickwork thickness is of one brick and over and is built with an 'as built' finish on one face, whereas the other face can be built with a more expensive facing brick with pointing (Figure 3.27).

| | | | | |
|---|---|---|---|---|
| | | | One-brick wall in common bricks, English bond in gauged mortar (1 : 1 : 6) facework and pointed one side in Rustic facings with recessed joint as the work proceeds. | |

**Figure 3.27**

3. Brick facework both sides, where both sides of the brickwork are finished fair with facings or a fair face and pointing. It is therefore possible to have a two-brick thick wall, built from common bricks with facings both sides (Figure 3.28).

| | | | | |
|---|---|---|---|---|
| | | | Two-brick wall in common bricks, Flemish bond in gauged mortar (1 : 1 : 6) facework and pointed both sides in Rustic facings with recessed joint as the work proceeds | |

**Figure 3.28**

With the exception of closing cavities and bonding new work to existing, all labour such as rough and fair cutting, throats, mortices, etc. is included and does not have to be measured separately.

## *Projections*

Brick walls are often built with piers (attached projections) beyond the face of the wall as a means of adding structural stability.

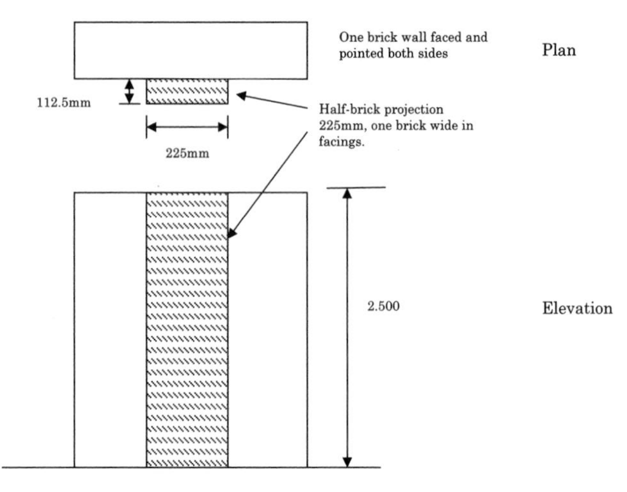


Figure 3.29

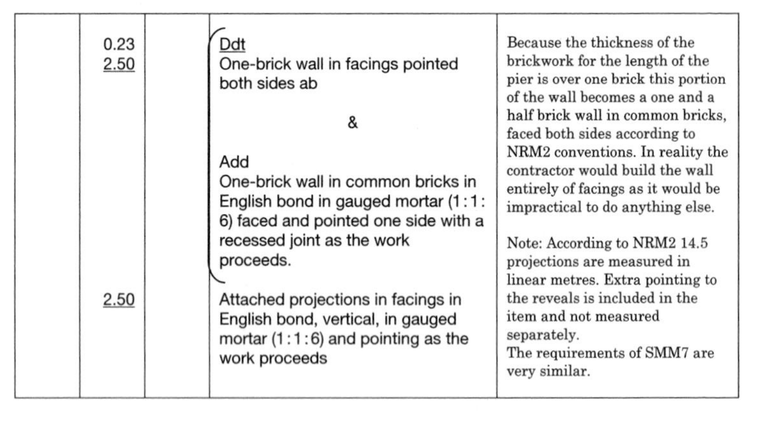

| | | | | |
|---|---|---|---|---|
| | 0.23<br>2.50 | | Ddt<br>One-brick wall in facings pointed both sides ab<br>&<br>Add<br>One-brick wall in common bricks in English bond in gauged mortar (1 : 1 : 6) faced and pointed one side with a recessed joint as the work proceeds. | Because the thickness of the brickwork for the length of the pier is over one brick this portion of the wall becomes a one and a half brick wall in common bricks, faced both sides according to NRM2 conventions. In reality the contractor would build the wall entirely of facings as it would be impractical to do anything else. |
| | 2.50 | | Attached projections in facings in English bond, vertical, in gauged mortar (1 : 1 : 6) and pointing as the work proceeds | Note: According to NRM2 14.5 projections are measured in linear metres. Extra pointing to the reveals is included in the item and not measured separately.<br>The requirements of SMM7 are very similar. |

Figure 3.30

## *Isolated piers*

According to SMM7 and NRM2 a wall becomes an isolated pier when the length on plan is less than four times its thickness. Conversely when the length on plan exceeds four times the thickness then the classification is wall (see Figures 3.29 and 3.30).

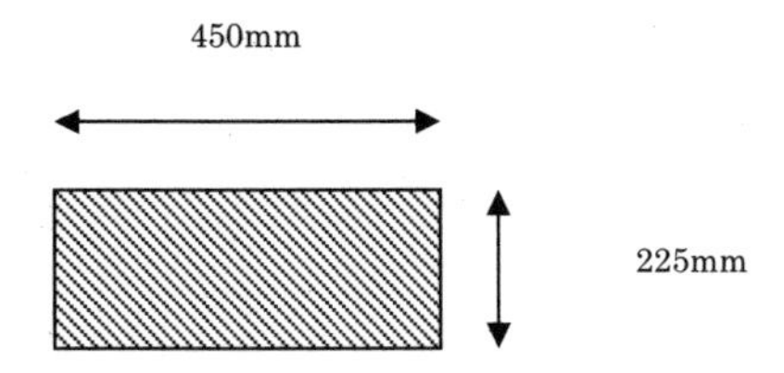


Figure 3.31

| | | | | |
|---|---|---|---|---|
| | 0.45<br>2.50 | | Isolated pier, one-brick thick vertical in common bricks laid English bond in gauged mortar ab. | Isolated piers are measured in square metres for SMM7. |
| | 2.50 | | 450 × 225mm Isolated pier, vertical in common bricks laid English bond in gauged mortar ab | Isolated piers are measured in linear metres in NRM2 (14.4). |

Figure 3.32

## Tapering walls

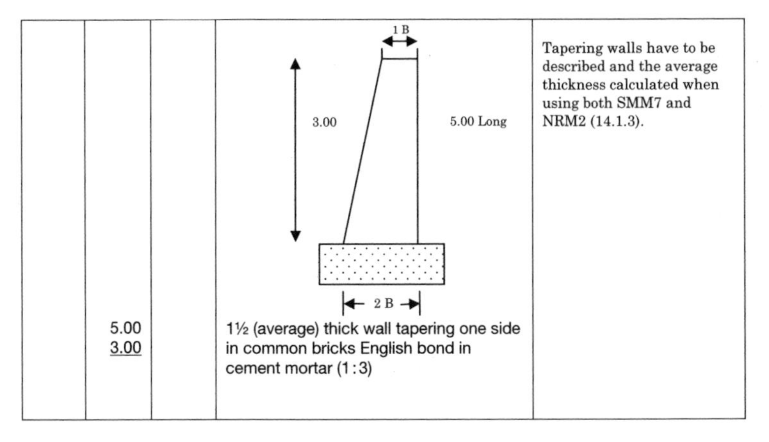


| | | | | |
|---|---|---|---|---|
| | 5.00<br>3.00 | | 1½ (average) thick wall tapering one side in common bricks English bond in cement mortar (1:3) | Tapering walls have to be described and the average thickness calculated when using both SMM7 and NRM2 (14.1.3). |

Figure 3.33

**Example 3: Floors**

Traditionally, ground and upper floors in domestic construction have been constructed from timber because it is:

- readily available;
- cheap; and
- easy to work with and fix on site.

Floor covering for most domestic construction is now provided with flooring grade tongued and grooved (t&g) chipboard as opposed to softwood boarding. Chipboard is cheaper and quicker to lay as it comes in large sheets although a disadvantage is that maintenance and access to the floor space can be more difficult compared with board flooring. Concrete and steel are just as suitable for floor construction although these materials are mainly used in medium and high-rise buildings. SMM7 / NRM2 requires that timbers in excess of 6m are kept separate and described as such, is as timber in such lengths as this is difficult to obtain and handle. To stop floor joists twisting and deforming it is usual to insert a line of strutting fixed at right angles to the joists. Solid or herring bone strutting can be used. Floor joists are generally spaced at 400mm centres and can either be built into the inner leaf of the cavity wall provided a good seal can be provided to the cavity (less common practice) or supported in a galvanised steel joist hanger built into the wall. In certain situations, galvanised or stainless steel straps are required every 2 metres to provide restraint and prevent movement of the external walls. It may be necessary to incorporate sound insulation into the floor space.

At openings in upper floors for stairs etc. the floor joists are trimmed with larger timbers to provide extra support.

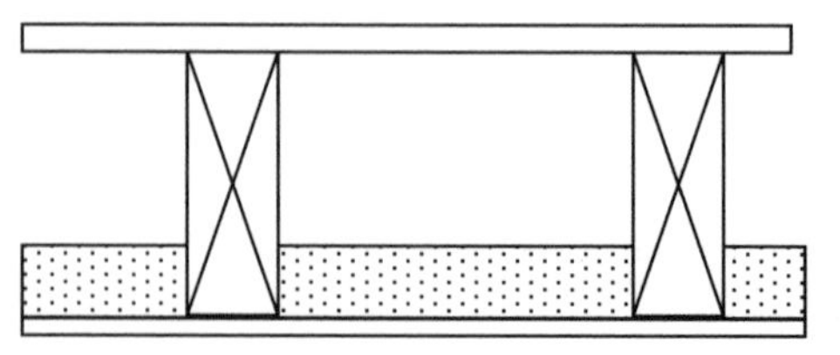


**Figure 3.34** Section through timber upper floor

Figure 3.35 Galvanised mild steel joist hanger

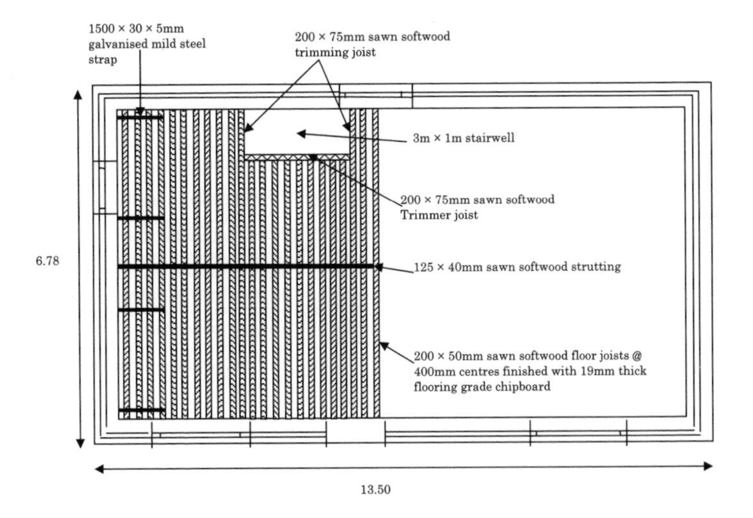


Figure 3.36 Plan showing upper floor joists

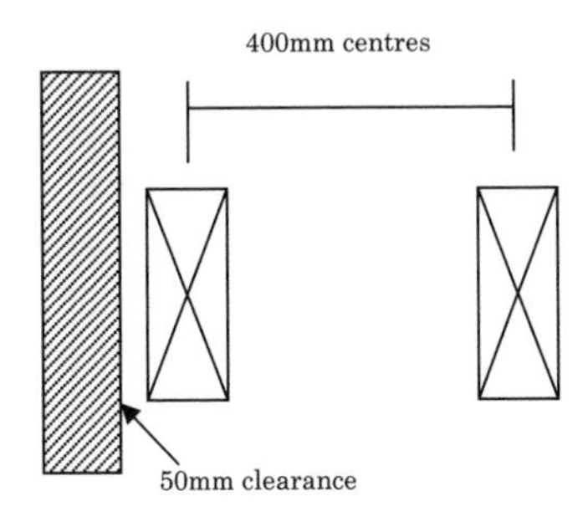


Figure 3.37

## Upper floors: taking-off list (SMM7 and NRM2)

- Joists
  - Main flooring
  - Trimming joists to stairwells, etc.
- Fixings
  - Joist hangers
  - Straps
- Finishes
  - Boarding/chipboard
  - Labour trimming – SMM7 only

| | | | | |
|---|---|---|---|---|
| | | | **Example 3 – Upper floors** | Note: This example is measured in accordance with NRM2. |
| | | | Numbers of joists<br>13.500<br>Less 2/ 0.275 0.550<br>12.950<br>Ddt first & last joist 2/0.050 0.100<br>Ddt to centre lines 2/½ /0.050 0.050 0.150<br>12.800<br>÷ 0.400<br>32 Spaces<br>+1 = 33 Joists | The numbers of floor joists are calculated by dividing the centres into the required length. |
| | | | Length<br>6.780<br>Less 2/ 0.275 0.550<br>6.230 | Note that the first joist is not placed against the internal skin of the external wall, this is to reduce the chance of the timber becoming wet and rotting and allows space to run services, therefore the first joist is located 50mm from the masonry. (See Figure 3.37) |
| 33/ | 6.23 | | 50 × 200mm treated softwood floor joist over 6m long in one length | NRM states that all timber is sawn unless stated. |
| 9/<br>2/ | 1.08<br>6.23 | | Ddt<br>Last item<br>Stairwell<br>1.000<br>Add Trimmer 0.075<br>1.075 | |
| 2/ | 3.00<br><br>6.23 | | 75 × 200mm treated sw ditto<br><br>Over 6m long in one length | Note that unlike SMM7 timbers over 6m long in one length do not have to be kept separate. |
| | | | 1 | |

Figure 3.38

| | | | | |
|---|---|---|---|---|
| 2/ | 33 | | 2mm Galvanised mild steel joist hanger for 50 × 200mm joist built into blockwork | |
| | | | Less Trimming joists 2<br>Trimmed joists 9 | |
| 2/ | 2<br>9 | | Ddt<br>Last item | |
| 2/ | 2 | | 2mm Galvanised ms joist hanger for 75 × 200mm joist built into blockwork | These hangers are for the ends of the trimmer and trimmed joists. |
| | 9 | | 2mm Galvanised mild steel joist hanger for 50 × 200mm joist built fixed to softwood | |
| | 2 | | 2mm Galvanised mild steel joist hanger for 75 × 200mm joist built fixed to softwood | |
| 2/ | 4 | | 30 × 5 × 1500mm long galvanised mild steel restraint strap once bent | Where joists run parallel to the external wall restraining straps are used. They should extend over at least three joists and are fixed at 2m centres. One end is turned down and fixed into the cavity. |
| | 12.95 | | 38mm thick treated softwood floor joist strutting | |
| | 12.95<br>6.23 | | 19mm thick flooring grade tongued and grooved chipboard horizontal flooring exceeding 600mm wide | Strutting is measured over the joists and not between them. |
| | 3.00<br>1.00 | | Ddt Stairwell<br>Last item | |
| | | | 2 | |

**Figure 3.38 (continued)**

## Example 4: Roofs (pitched and flat)

The majority of roofs found in domestic construction are pitched and covered with tiles or slate. For medium and high-rise buildings it is more usual to have a flat roof covered with a high-performance sheet finishing. Flat roofs are generally more problematic from a maintenance and life-cycle cost point of view. Traditionally, roof structures were constructed in situ from individual timbers, as detailed in Figure 3.39, however, most domestic roofs are today constructed from pre-fabricated timber roof trusses that are delivered to site by lorry and hoisted into place with a crane. Increasingly, eaves boards and fascias, traditionally made from softwood and plywood, are being manufactured in plastic, to reduce maintenance.

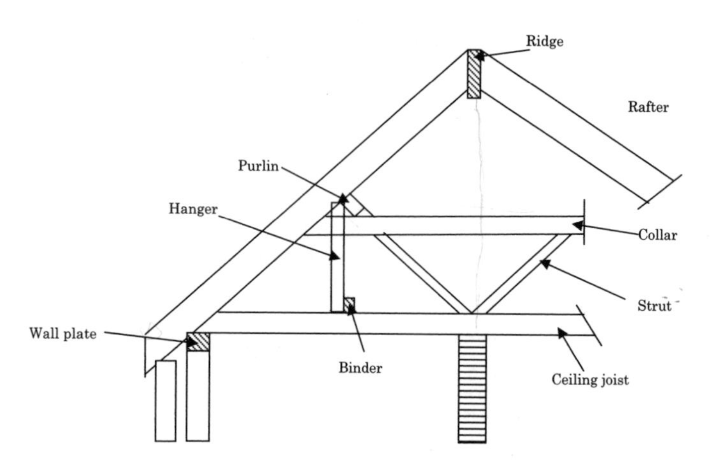


**Figure 3.39** Pitched roof members

## Pitched roofs: taking-off list (SMM7 and NRM2)

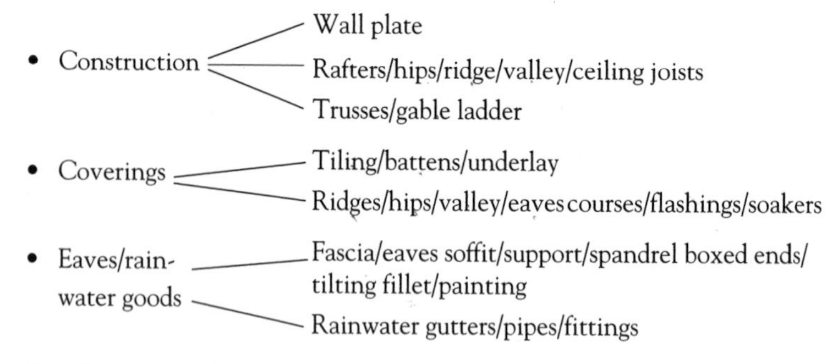


### *Double-pitch roofs*

#### Trussed rafters

Most modern roofs (1970 onwards) are constructed using trussed rafters, the most common of which is the Fink truss (after Mr Fink) (see Figure 3.42). Trussed rafters are prefabricated off site and because of their engineered design, the timber sections used in the truss, typically 70–80mm deep × 30–47mm wide, are considerably smaller than traditional pitched roofs. Joints are formed using toothed metal plates, sometimes known as 'gang-nails' pressed into the timber. Trussed rafters are able to offer the following advantages over traditional roof construction:

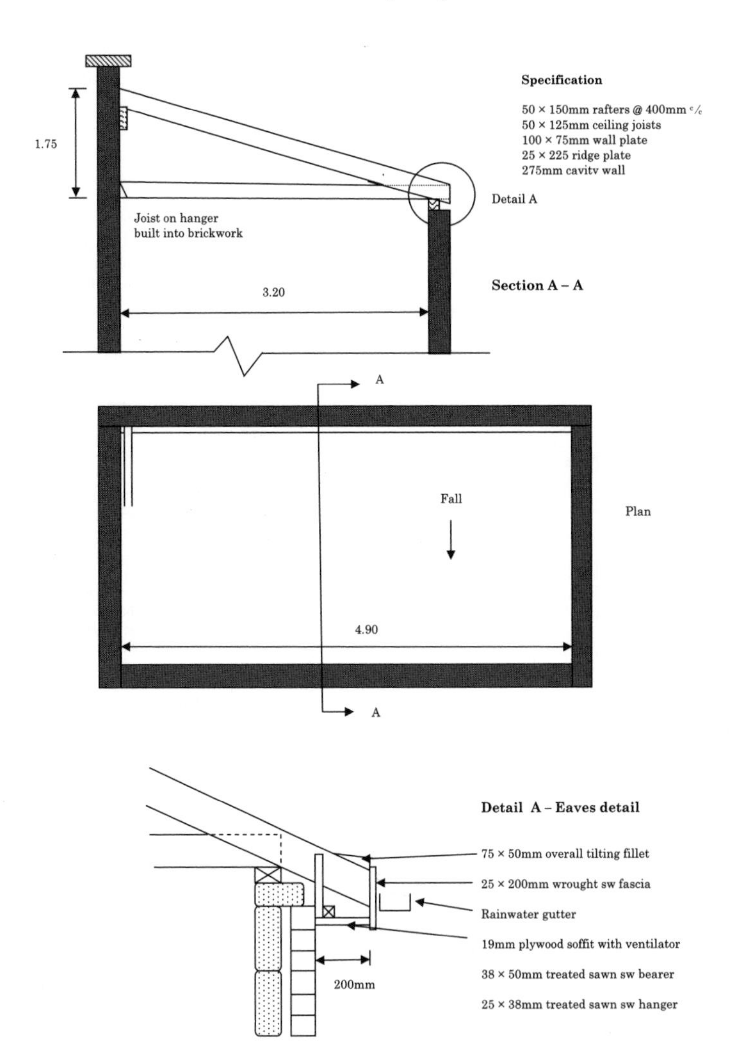


**Figure 3.40** Mono-pitched roof

- span up to 12m
- offer speed of construction
- do not require skilled labour
- eliminate load-bearing internal partitions.

| | | | | |
|---|---|---|---|---|
| | | | **Example 4 – Roofs** | Note: This example is in accordance with SMM7. |
| | 4.90 | | 100 × 75mm treated sawn softwood plate. | |
| | | | Rafters<br>4.900<br>Less 2/0.050 0.100<br>2/½/ 0.50 0.050 0.150<br>÷ 0.400 4.750<br>12 + 1 = 13<br>Length<br>3.200<br>Add External wall 0.275<br>3.475<br>$1.750^2 = 3.063$<br>$3.475^2 = 12.076$<br>$\sqrt{15.139}$<br>Length of rafter 3.891 | The rafters are spaced at 400mm centres that gives almost exactly 12 spaces + 1 = 13 rafters.<br><br>Length of rafters<br><br>Length ?<br>1.75<br>3.475 |
| 13/ | 3.89 | | 50 × 150mm treated sawn softwood pitched roof member | If the drawing was to scale then it could be quite possible to measure the length of the rafters off the drawing. However, given that figured dimension take preference over scaled dimensions it is preferable to calculate the length of the rafters using basic trigonometry. |
| | | | Joists | |
| 13/ | 3.48 | | 50 × 125mm treated sawn softwood pitched roof member | |
| | 4.90 | | 25 × 225mm treated sawn softwood plate plugged and screwed to brickwork | |
| | 13 | | 2mm galvanised ms joist hanger for 50 ×125mm joist built into brickwork | |
| | 4.90<br>3.20 | | Insulation<br>150mm thick insulation laid between joists at 400mm centres | |
| | | | 1 | |

Figure 3.41

| | | | | |
|---|---|---|---|---|
| | | | Eaves detail A | In this example a traditional form of eaves construction is detailed. The end of the eaves must be sealed with a single piece of plywood or timber referred to as a Spandril boxed end in the following shape. |
| | | | Tilting fillet | |
| | 4.90 | | 75 × 50mm overall treated sawn softwood triangular individual support | |
| | | | 4.900<br>Add Brick walls 2/0.275 0.550<br>5.450<br>Less Spandrils 2/0.025 0.050<br>5.400 | In order to prevent the spread of rot or other timber defects in the roof timbers, the eaves should contain permanent ventilation. |
| | 5.40 | | 25 × 200mm wrot softwood fascia board grooved | Sundry items of softwood support are grouped together and measured in accordance with G20.13. |
| | | | &<br>Eaves board 0.200<br>Rebate 0.010<br>0.210 | |
| | | | 19 × 210 mm external quality plywood eaves soffit board | Joint of fascia and eaves soffit |
| | 2 | | 19mm external quality plywood spandril boxed end size 225 × 250mm overall | |
| 13/ | 0.20 | | 25 × 38mm treated sawn softwood individual support | |
| | 5.40 | | 38 × 25mm ditto plugged and screwed to blockwork | |
| | 4.90 | | Manthorpe crossflow eaves ventilator fixed to plywood | |
| | | | 2 | |

Figure 3.41 (continued)

| Timesing | Dimension | Squaring | Description |
|---|---|---|---|
| 2/ | 5.40 | | Prime only before fixing wood general surfaces n.e. 300mm girth |
| | | | Spandrils |
| 2/ | 1 | | Ditto isolated areas n.e. $0.50m^2$ |
| | | | Fascia & soffits |
| | | | Fascia 0.200 |
| | | | Add 2/0.025 0.050 |
| | | | Soffit 0.200 |
| | | | 0.450 |
| | 5.40<br>0.45 | | Knot, prime, stop, two undercoats and one coat gloss to general surfaces of wood exceeding 300mm externally |
| | | | Rainwater goods |
| | | | 5.450 |
| | | | Add overhang 2/0.050 0.100 |
| | | | 5.550 |
| | 5.50 | | 114mm Hunter plastics Squareflo black rainwater gutter (ref R114) with and including support bracket (ref R395) |
| | 1 | | Extra over for running outlet (ref R376) |
| 2/ | 1 | | Ditto stopend (Ref R380) |
| | 3.00 | | 65mm Hunter plastics Squareflo rainwater pipe (ref R300) with push fit socket joints and stand off brackets (ref R388) at 2m centres plugged to masonry |
| | 1 | | Extra over for adjustable offset (ref R399)<br>&<br>Ditto rainwater shoe (ref R391) |
| | | | 3 |

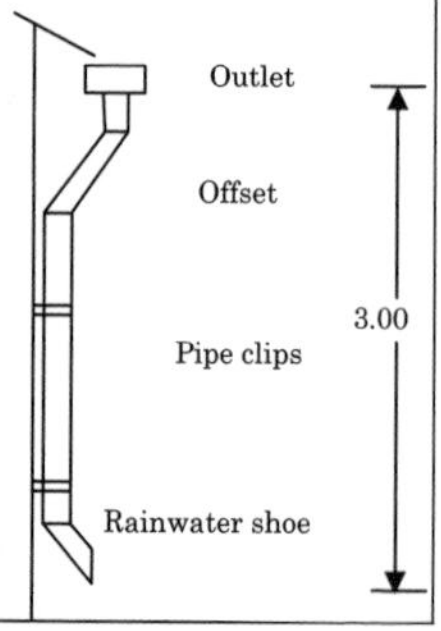

In this example timber and plywood have been used for boxing out the eaves and therefore it will need to be painted for protection against the elements. The traditional specification is:

Knot, prime, stop, two coats of undercoat and one coat gloss.

Knot refers to painting knots in the wood with shellac knotting to prevent sap leaking from the knot and spoiling the finish. Prime refers to painting with a coat of primer before any paint is applied.
Stopping refers to filling in any imperfections in the surface of the timber with linseed oil putty/premixed wood filler. After all this preparation the timber is sanded down and two undercoats and one coat of gloss is applied.

This convention relates to the time when most paint was oil-based and three coats were required in order to provide adequate protection. However with modern painting systems the specification should be studied to discover the correct approach to decoration.

Items such as skirtings and architraves have the backs of the timber sections primed before they are fixed.

Rainwater goods include rainwater gutters and pipes and associated fittings such as stop ends and offsets.

Outlet
Offset
Pipe clips
3.00
Rainwater shoe

Figure 3.41 (continued)

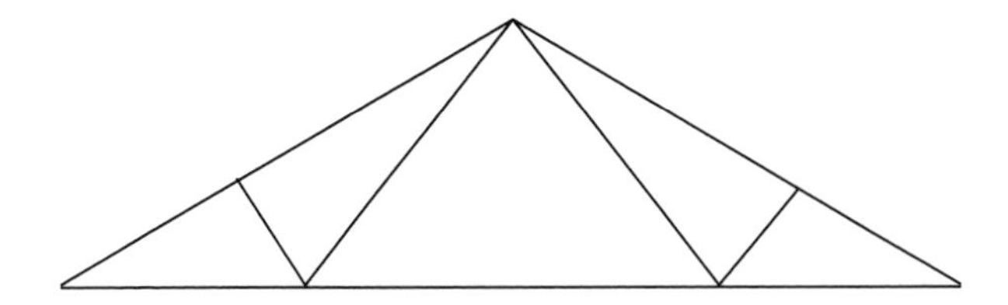

**Figure 3.42** Typical Fink and W type truss

As with traditional roof construction, a trussed rafter sits on a wall plate and is fixed by either nailing or with the use of special clips. It is good practice to strap the wall plate down to external walls. Once the trusses are erected, stability is achieved with longitudinal binders and diagonal braces that are required to prevent deformation of the roof from wind, etc. Overhanging verges at gables can be formed with the use of prefabricated gable ladders. It may be necessary to locate water tanks in the roof space, in which case the load should be distributed evenly over at least three trusses.

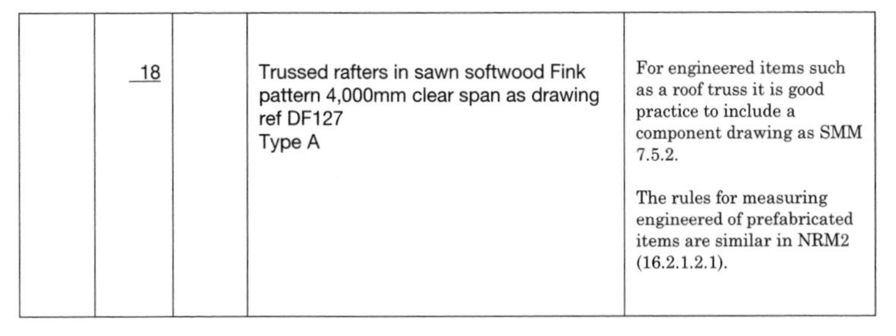

| | | | | |
|---|---|---|---|---|
| | 18 | | Trussed rafters in sawn softwood Fink pattern 4,000mm clear span as drawing ref DF127<br>Type A | For engineered items such as a roof truss it is good practice to include a component drawing as SMM 7.5.2.<br><br>The rules for measuring engineered of prefabricated items are similar in NRM2 (16.2.1.2.1). |

**Figure 3.43**

The calculation of roof members is slightly different between hipped and gable ends as roof measurement requires the quantity surveyor to carry out a number of calculations depending on the amount of information available in order to calculate lengths of rafters and areas of roof coverings; once again figured dimensions take preference over scaled dimensions. It is common to have the span and the rise of a pitched roof marked on the drawings.

The length of rafter and the area of sloping roof coverings can usually be calculated using simple trigonometry, for example: the length of the rafter C can be determined using Pythagoras' theorem: $C = \sqrt{B^2 + A^2} = 2.83$.

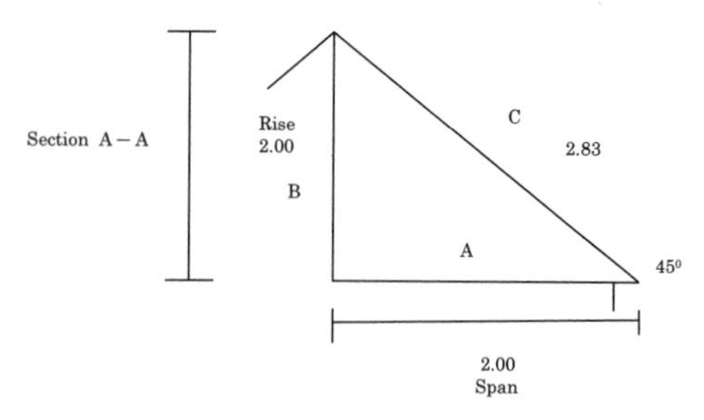


The span is taken to the extreme projection of the roof and includes eaves projections etc.

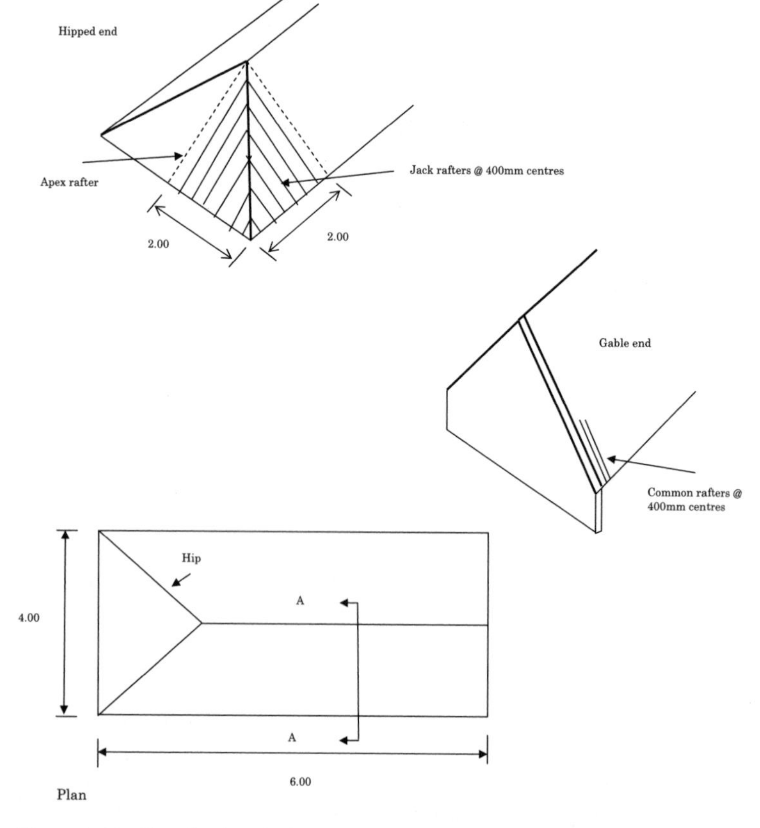


**Figure 3.44** Details of pitched roof with one hipped and one gable end

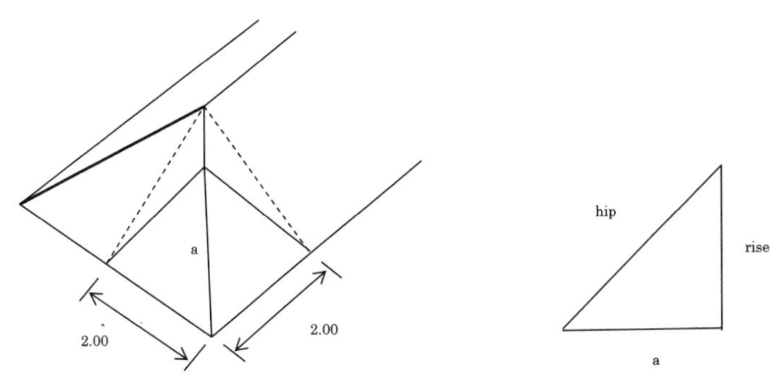


**Figure 3.45** Details of pitched roof with one hipped and one gable end

The length of the hip also has to be calculated as follows:

To calculate a: $a = \sqrt{2.00^2 + 2.00^2} = 2.83$

Therefore length of hip $= \sqrt{2.83^2 + 2.00^2} = 3.47$

As can be seen, the formation of the hipped end requires a number of rafters to be cut in order to fit against the hip rafter. These shorter rafters are referred to as jack rafters and present no real problems to the measurer as the shorter rafters off-set each other lengthwise in cases where the roof pitch at the hip equals the pitch of the main roof. However, an additional rafter, known sometimes as an apex rafter, must be added as indicated in Figure 3.46, as this is a complete addition to the structure.

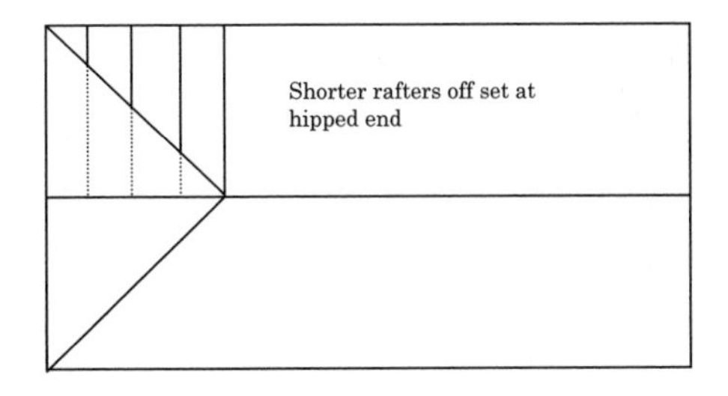


**Figure 3.46**

Therefore the number of rafters required for the hipped roof in Figure 3.46 is as follows:

| | | | | |
|---|---|---|---|---|
| | | | Rafters<br>6.000<br>Gable end<br>Less<br>External wall 0.275<br>Clearance 0.050<br>CL of rafter 0.025 0.350<br>5.650<br>÷ 0.400<br>14.13 spaces = 15 rafters | NRM2 rules have been applied here.<br><br>The length of the roof 6.00 is adjusted to take account of the gable end where the first rafter is on the inside face of the external wall. As previously, a clearance of 50mm is allowed between the first rafter and the wall and the length is further adjusted to the centre line of the first rafter. |
| 2/15/ | 2.83<br>2.83 | | 50 × 100mm treated rafters and assoc roof timbers fixed at 400mm centres. | |
| | | | Hip rafters | The number of rafters is multiplied by two for both slopes and an additional Apex rafter is included. |
| 2/ | 3.47 | | 50 ×150mm ditto | Note: All timber is deemed to be sawn, unless otherwise stated. |

Figure 3.47

**Figure 3.48** A trussed roof with diagonal wind bracing and showing a gable ladder (left of picture)

## *Pitched roof coverings*

Coverings to pitched roofs in the UK are usually tiles or slates both of which are laid on underlay and battens. In addition to the general roof area the eaves, verges, ridges, hips and valleys require separate items to be measured in linear metres in accordance with SMM7. Although not strictly required for measurement purposes, the differences between the various types of slates and tiles is explained in Chapter 5.

Roof tiling is covered in SMM7 starting at clause H60 and clause 18.1 in NRM2.

| | | | | |
|---|---|---|---|---|
| | | | **Example 5 – Pitched roof coverings** | This section is measured in accordance with NRM2. |
| | | | The following in: Marley Eternit Hawkins Staffordshire Blue plain clay roofing tiles and fittings | The following section is devoted to a proprietary form of tiling and therefore to avoid constant repetition it is common practice to include a heading 'The following in:' The end of the section should also be signposted. |
| 2/ | 6.00<br>2.83 | | Roof coverings 265 × 165mm tiles with 75 mm head lap $40^0$ pitch each tile nailed with 2 no 44mm composition nails to 38 × 25mm treated sawn softwood battens to one layer of felt to BS Type 1F with 100mm minimum laps fixed with galvanised clout nails<br><br>Ridge<br>6.000<br>Less 2.000<br>4.000 | This example relates to Figure 3.45.<br>The dimensions have been multiplied by 2 for both roof slopes. The area of the roof coverings does not need to be adjusted for the hipped end provided the hip and the main roof have the same pitch. |
| | 4.00 | | 250mm half round horizontal ridge | |
| 2/ | 3.47 | | 125mm angle arris raking hip | Boundary work is deemed to include undercloaks, rough and fair cutting, bedding, pointing, ends, angles and intersections. |
| 2/ | 6.00<br>4.00 | | Double course of horizontal tiling at eaves including 165 × 215mm eaves tile | |
| 2/ | 2.83 | | Verge raking including tile and half | |
| | | | End of Marley Eternit Hawkins Staffordshire Blue plain clay roofing tiles and fittings | |

Figure 3.49

**Timber flat roof: taking-off list (SMM7 and NRM2)**

- Construction — Wallplate/joists/strutting; Joist hangers/straps
- Coverings — Decking/firings/fillets; Coverings/upstands/flashings
- Eaves & rainwater goods — Fascia/eaves soffit/support/spandrel boxed ends/tilting fillet/painting; Rainwater gutters/pipes/fittings

### *Metal flashings*

When roof coverings abut vertical brickwork some type of flashing is required to prevent the ingress of water. The traditional material for flashings was lead, but today zinc is a more common material due to the high cost of lead. Where the joint to be flashed is raking then a soaker is also used and SMM7 states that soakers must be measured in two items: supply and handing to others for fixing. This arcane procedure is because traditionally soakers were supplied by plumbers but fixed by roofers. The number of soakers required is calculated by dividing the length of the abutment by the gauge of the tiling or slating. The size of the soakers will vary according to circumstances and are calculated as follows:

Length = Gauge + Lap + 25mm

Width = 100mm + 50mm (upstand)

In NRM2 soakers are measured once under Fittings to Sheet Roof Coverings (17.9.5).

### *Flat-roof coverings*

Flat roofs are generally cheaper to construct than pitched roofs however from a whole-life cost point of view they have a shorter life span depending on the degree of exposure and climate as follows:

- 3 layer built-up felt: 15–20 years
- asphalt: 20–60 years
- copper: 100–300 years
- zinc: 20–40 years
- uPVC: 30–40 years
- lead: 60–100 years

compared with natural slate and clay tiles that have an expected life of 100 years plus.

### *Solar protection*

Continued exposure to sunlight causes asphalt and felt roofing to deteriorate more quickly and therefore some form of solar protection is required. Traditionally, solar protection has been provided with a layer of white stone chippings to reflect the sunlight and protect the coverings. Unfortunately, chippings, even if bedded into the covering, tend to become dislodged and block up rainwater systems, therefore, as an alternative, solar reflecting paints are now widely used.

### *Insulation*

One of the problems with flat roofs is the build up of condensation within the roof. The position of the insulation in a flat roof will determine the category as either cold roof or warm roof.

When a cold roof is constructed, as shown in Figure 3.50, the insulation is placed above the plasterboard finish and therefore the roof space remains cold. Importantly, cold roofs need to be well ventilated to dissipate condensation.

Warm roofs can be subdivided into two categories: sandwich and inverted.

**Cold roof**

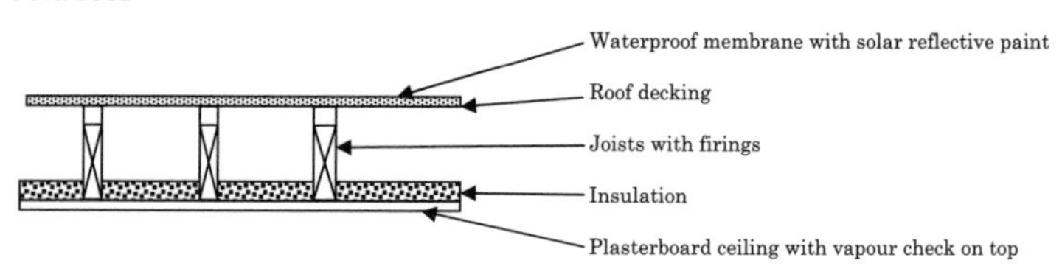


**Warm roof – sandwich type**

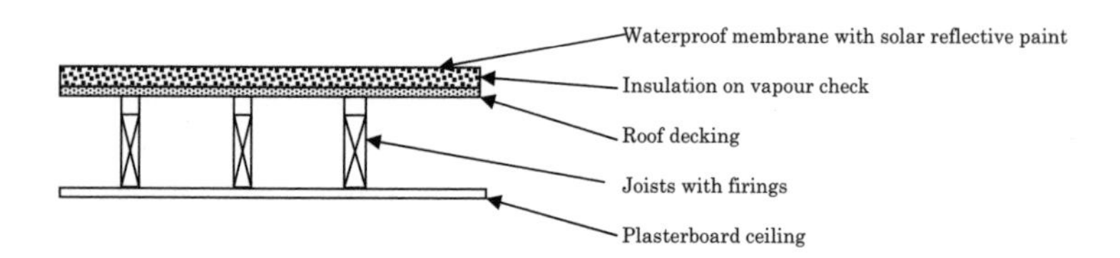


**Warm roof – inverted type**

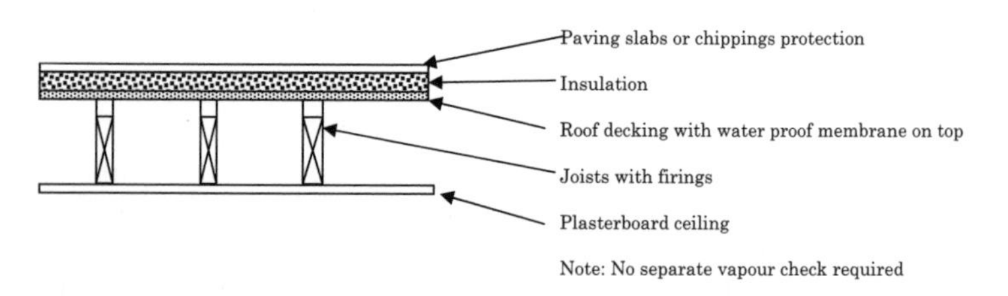


**Concrete flat roof**

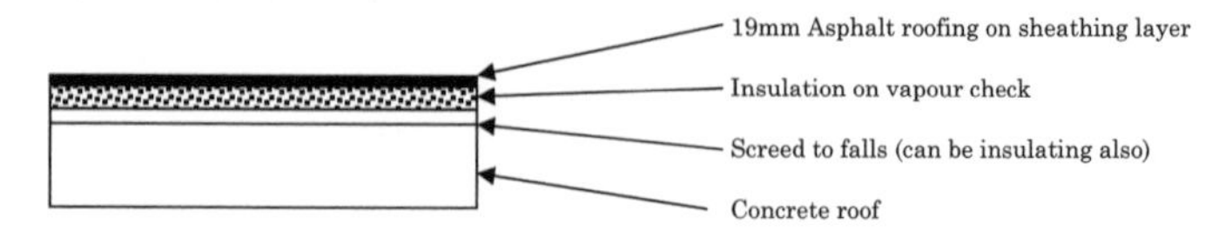


**Figure 3.50** Types of flat roof

## *Flat roofing example*

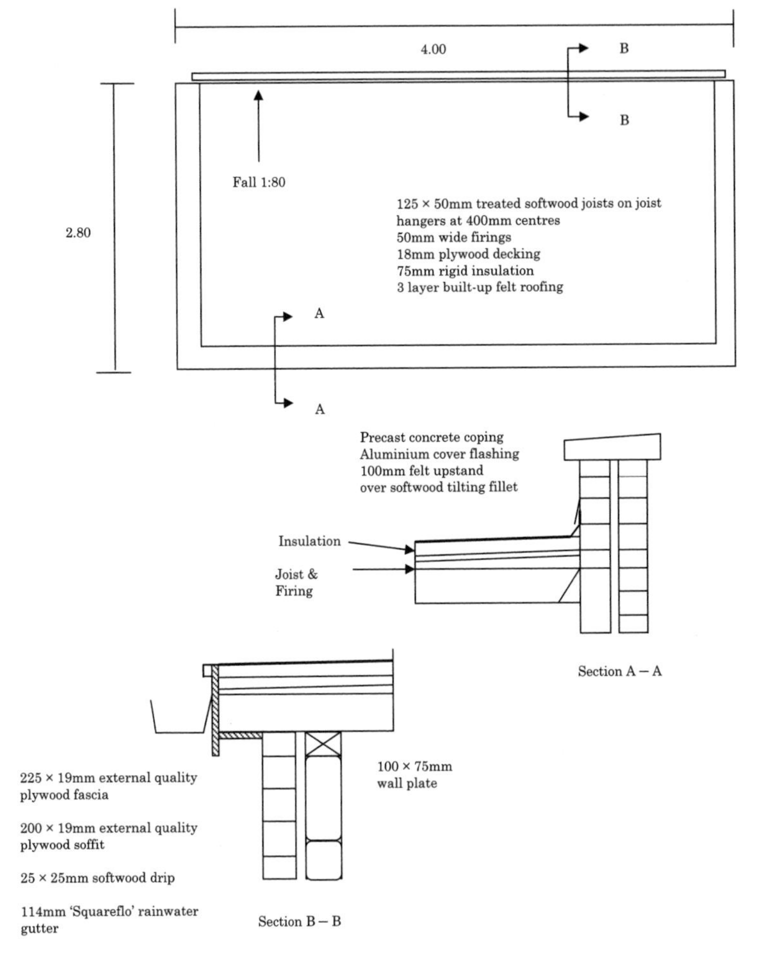


**Figure 3.51**

## Internal finishes

### *Generally*

The majority of finishes are either applied in situ, e.g. plaster and other forms of rendering, or tile, slab and sheet finishes. There can be a considerable amount of repetition when measuring finishes, therefore it is important to ensure that dimensions that relate to one type of finish are grouped together on the dimension sheets. Most finishes are dealt with by Section M in SMM7 although items such as skirtings and other unframed isolated trims are dealt with in Section P. Even for fairly simple projects the use of a schedule can greatly simplify the process of specifying materials and measurement (see Table 3.3 on p. 161).

Most modern in situ plasterwork is carried out in lightweight gypsum plaster, known by its trade names Carlite or Thistle, although Carlite has now been merged into the Thistle brand. Thistle plasters come in a variety of types depending on where they are to be used and the background to which the plaster is applied. According to SMM7, work over 300mm wide is measured in square metres while work under 300mm wide is measured in linear metres. NRM2 requires work over 600mm wide to be measured in square metres and work under 600mm to be measured in linear metres.

In many cases the area of floor and ceiling finishes are identical and can be anded-on.

It is very important that the correct type of plaster is used; the five types (four undercoat and one topcoat) available are as follows.

### *Undercoats*

**Browning plaster** is an undercoat plaster for moderate suction solid backgrounds that have a good mechanical key, such as brickwork or blockwork. A slow-setting variety is available that gives greater time for application.

**Bonding plaster** is an undercoat plaster for low suction backgrounds, for example concrete or plasterboard or surfaces sealed with pva (a universal water-based adhesive).

**Toughcoat** is an undercoat plaster for solid backgrounds of high suction with an adequate mechanical key.

**Hardwall** is an undercoat plaster that provides a much harder and more durable finish and is also quick drying.

### *Finishing coats*

**Finishing plaster** is an ideal choice over sand and cement bases and can be used on still damp backgrounds.

| Timesing | Dimension | Squaring | Description | Side notes |
|---|---|---|---|---|
| | | | **Example 6 – Flat-roof construction and coverings** | Note: This example is measured in accordance with NRM2. |
| | | | **Roof plate**<br>4.000<br>Less 2/ 0.275 0.550<br>Length 3.450 | |
| | 3.45 | | 100 × 75mm treated softwood wall plate fixed to blockwork | |
| | | | 3.450<br>Less<br>Clearance 2/0.50 0.100<br>Centre line 2/0.025 0.050 0.150<br>3.300<br>÷ 0.400<br>= 8 spaces +1 = 9 | In this case dividing the roof length by 400mm centres gives 8 spaces and 100mm remainder. Therefore it was decided to include 8 + 1 = 9 joists. Had the remainder been say 200mm then an extra joist would be added to equal 8 + 1 + 1 = 10 joists. |
| 9/ | 2.73 | | 50 × 125mm treated softwood roof joists @ 400mm centres | |
| | | | Length<br>2.800<br>Less Bkwk 0.275<br>2.525<br>Add overhang 0.200<br>2.725 | |
| | 3.45 | | 38mm thick treated softwood strutting to joists 125mm deep | |
| 9/ | 9 | | 2mm galvanised ms joist hanger for 50 × 125mm joist built into blockwork | |
| | | | Firings<br>Max 0.080<br>Min 0.030<br>0.110<br>Avg 0.055 | |
| 9/ | 2.73 | | 50 × 55mm (avg) treated softwood firing | |
| | | | 1 | |

Figure 3.52

| | | | | |
|---|---|---|---|---|
| 2/ | 3.45<br>2.53 | | 50 × 50mm (avg) triangular ditto | |
| | 3.45<br>2.73 | | 19mm plywood external grade flat-roof boarding over 600mm wide<br><br>&<br><br>75mm Rocksilk horizontal board insulation in plain areas<br><br>&<br>**Coverings**<br><br>3 layer ruberoid built-up felt flat roofing over 500mm wide to plywood sheathing laid in accordance with manufacturer's instructions | |
| | | | Upstands | |
| | 3.45 | | Ruberoid felt roofing ab horizontal upstand ne 500mm high<br><br>&<br>22 SWG aluminium horizontal flashing 100mm girth<br><br>& Eaves<br>Ruberoid roofing felt ab at eaves 200mm girth | |
| 2/ | 2.53 | | Ditto at verges 200mm girth | |
| | | | 2 | |

Figure 3.52 (continued)

| | | | | |
|---|---|---|---|---|
| | | | **Eaves and rainwater goods** | |
| | 3.45 | | 25 × 25mm treated sawn softwood individual support | |
| | | | & | |
| | | | 225 × 18mm external quality horizontal plywood fascia | |
| | | | & | |
| | | | 200 ×18mm ditto as soffit board | |
| | 2 | | 18mm external quality plywood boxed end size 200 × 225mm | |
| 2/ | 3.45 | | Prime only before fixing wood general surfaces n.e 300mm girth external | |
| | | | Spandrils | |
| 2/ | 1 | | Ditto isolated areas n.e 1.0m² ditto | |
| | | | Fascia & soffits | |
| | | | Fascia 0.225<br>Add 2/0.025 0.050<br>Soffit 0.200<br>0.475 | |
| | 3.45<br>0.48 | | Knot, prime, stop, two undercoats and one coat gloss to general surfaces of wood exceeding 300mm externally | |
| | | | 3 | |

Figure 3.52 (continued)

| | | | | |
|---|---|---|---|---|
| | | | Rainwater goods 3.450<br>Add overhang 2/0.050 0.100<br>4.550 | Alternative 1 used – fitting 38.4 and 5. |
| | 4.55 | | 114mm Hunter plastics Squareflo black straight rainwater gutter (ref R114) with and including support bracket (ref R395) fixed to softwood. | |
| | 1 | | Running outlet (ref R376) | |
| 2/ | 1 | | Ditto stopend (Ref R380) | |
| | 2.30 | | 65mm Hunter plastics Squareflo rainwater pipe (ref R300) with push fit socket joints and stand-off brackets (ref R388) at 2m centres plugged to masonry. | Note that in NRM2 fittings are not required to be measured as extra over. |
| | 1 | | Adjustable offset (ref R399) | |
| | | | 4 | |

Figure 3.52 (continued)

**Board finishing plaster** is a one-coat plaster for skim coats to plasterboard.

**Multi-finish** is used where both undercoat and skim coat are needed on one job. Suitable for all suction backgrounds and ideal for amateur plasterers.

In addition, **universal one coat** is a one-coat plaster for a variety of backgrounds, suitable for application by hand or machine.

*The White Book* published by British Gypsum is a good source of reference for all types of plaster finishes. It can be viewed at the following web page:

http://www.british-gypsum.bpb.co.uk/literature/white_book.aspx

### *What is used in practice?*

For plasterboard ceilings: board finish plaster.

For blockwork walls: an 11mm-thick undercoat of browning plaster followed by 2mm skim coat of finishing plaster. The undercoat is lightly scratched to form a key.

Metal lathing: two undercoats are often required followed by a finishing coat.

Galvanised or stainless steel angle beads are used to form external angles in in situ plaster.

In situations where damp walls are plastered following the installation of an injection dpc, most gypsum plasters are not suitable as they absorb water and fail.

**Table 3.3** Example of an internal finishes schedule

| **Internal finishes schedule** | | | | | | **Job No. ABC/09** |
|---|---|---|---|---|---|---|
| **Room** | **Floor** | **Ceiling** | | **Walls** | | **Skirting** |
| | | **Finish** | **Decoration** | **Finish** | **Decoration** | |
| G1 | 35mm cement & sand (1:3) screed.<br>Carpet PC £25/m$^2$ | 9.5mm Plasterboard & 5mm skim | Two coats of emulsion paint | 13mm two coats of plaster | Wall paper<br>PC £7.00 per piece | 19 × 75mm softwood skirting.<br>KPS & 3 coats |
| G2 | 35mm cement & sand (1:3) screed<br>305 × 305mm Vinyl floor tiles | 9.5mm Plasterboard & 5mm skim | Two coats of emulsion paint | 13mm two coats of plaster | Two coats of emulsion paint | 80mm Vinyl skirting |
| G3 | Etc. | | | | | |

### *Sand/cement screeds*

The function of a floor screed is to provide a smooth and even surface for finishes such as tiling and is usually a mixture of cement and coarse sand, typically in the ratio of 1:3. It is usually a dry mix with the minimum of water added and has a thickness of between 38 and 50mm and is laid on top of the structural floor slab.

### *Board finishes*

Plasterboard is available in a variety of thicknesses, the most common being 9.5mm and 12mm, and is fixed to either metal or timber studs with screws. Standard sheets are 900 × 1800mm and 1200 × 2400mm. It is also possible to fix plasterboard to block work with plaster dabs. Boards can either be pre-finished and require no further work or finished with a skim coat of plaster. Boards with tapered edges and foil backing are also available.

### *Floor and wall tiling*

Most wall tiles are 100 × 100mm with spacers to ensure ease of fixing. Tiles require adhesive for fixing and grout for the joints. Thickness can vary and floor tiles can be found in a wider variety of sizes.

## Windows, doors and joinery

In days gone by, items such as doors and windows would have arrived on site separately from their associated linings, frames and ironmongery, to be assembled on site. The increased use of standard units, dimensional co-ordination, AutoCAD, etc. means that for most projects, with the exception of so-called signature schemes, items of joinery are available in standard sizes. In the case of doors, for example, these are regularly specified as door sets that include: the door leaf or leaves, frame or lining and ironmongery that arrive on site packaged together.

Many windows and doors are UPVC and pre-glazed and require only to be built in and then protected until the works are complete.

Irrespective of whether standard units or individual units are used the measurement of doors and windows will require adjustment for door or window opening. As has been previously mentioned, masonry, finishes and painting and decoration are measured gross; these must now be adjusted. Work is also required to be measured to cover:

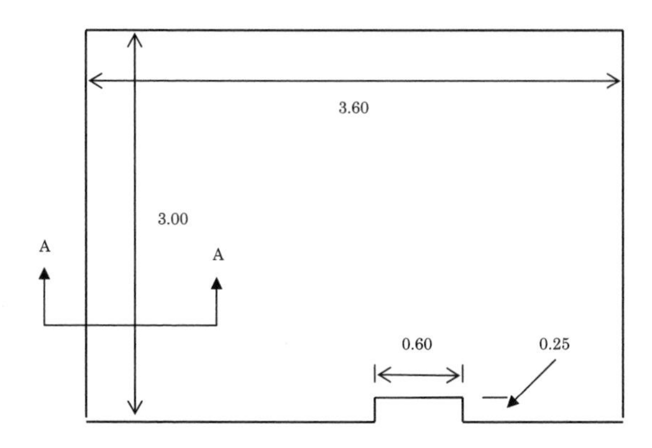


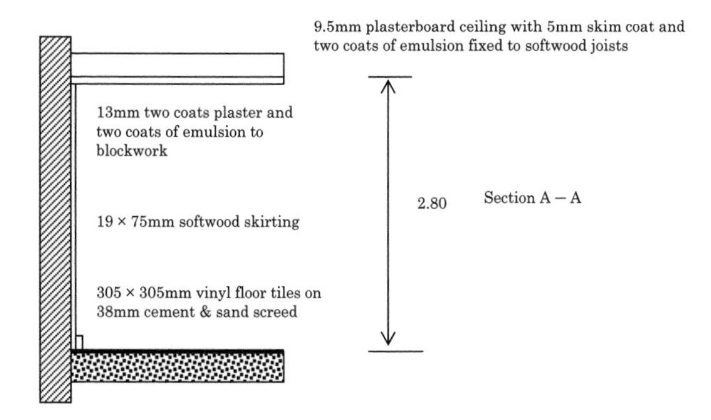


**Figure 3.53**

- lintels at the window head, these can be steel or precast reinforced concrete
- damp-proof courses
- work to window reveals and sills both inside and out.

External window and door frames are usually fixed with galvanised fixing cramps that are screwed, in the case of softwood, to the door or window frame and then built into the masonry as work proceeds (see Figure 3.55).

Modern timber staircases are also usually prefabricated and brought to site for final assembly.

| | | | | |
|---|---|---|---|---|
| | | | **Example 7 – Internal finishes** | Note: This example is measured in accordance with NRM2. |
| | 3.60<br>3.00 | | 9.5mm gypsum plasterboard with 5mm skim coat in Thistle board finishing plaster ceiling (14.5mm thickness overall) fixed with plasterboard nails exceeding 600mm wide | Note: Work to ceilings over 3.5m high have to be kept separately and described as such in both SMM7 and NRM2. |
| | | | & | Note: painting must be classified as internal or external (NRM2). |
| | | | Two coats of emulsion general surfaces of plaster over 300mm girth internal | Note that the room has a small pier size 600 × 250mm. It is not necessary to deduct this from the area of floor and ceilings because it is less than 1.0$m^2$ NRM2 or 0.50$m^2$ SMM7. |
| | | | & | |
| | | | 305 × 305 × 2.5mm thick vinyl tiles fixed with approved adhesive ad to cement sand screeded bed | |
| | | | & | Note: Wall finishes are measured gross and deductions for windows, doors, etc. are deducted with door and window measurement; see next example. |
| | | | 38mm thick cement and sand (1:4) level and to falls only $15^0$ from horizontal screeded bed on concrete over 600mm wide. | Work not exceeding 600mm wide has to be measured in linear metres. Therefore the sides and face of the pier are measured in linear metres. |
| | | | **Walls**<br>2/3.000 6.000<br>2/3.600 7.200<br>13.200 | Skirtings are fixed to blockwork by either plugging them directly or, on softwood grounds. |
| | 13.20<br>2.80 | | 13mm thick two coat Thistle plaster ad exceeding 600mm wide to blockwork. | Softwood skirting fixed to softwood ground, plugged to blockwork |
| | 0.60<br>2.80 | | Ddt Pier<br>Last | |
| 3/ | 2.80 | | Pier<br>Ditto not exceeding 600mm wide | If grounds are used then the plaster will not extend to ground level and the appropriate deduction will have to be made. |
| | | | 1 | |

Figure 3.54

| | | | | |
|---|---|---|---|---|
| 2/ | 2.80 | | Expamet galvanised mild steel angle bead ref 595 to blockwork with plaster dabs | The external angles of the pier will have metal angle beads to give a hard finish to the detail as well as providing a guide for the plasterer. The angle beads are fixed with plaster dabs or nails. |
| | | | Height<br>2.800<br>Less<br>Screed 0.038<br>Sktg 0.075 0.113<br>2.687<br>Girth<br>3.600<br>3.000<br>2/6.600<br>13.200<br>Add 2/0.250 0.500<br>13.700 | Note that the height of the emulsion has been adjusted to take account of the timber skirting. |
| | 13.70<br>2.69 | | Two coats emulsion ab | The rules for painting and plasterwork are slightly different when it comes to narrow widths. Plasterwork in narrow widths is kept separate and measured in linear metres, but painting is only measured in linear metres if it is less than 300mm wide in isolated areas. In other words if the narrow width is bordered by different finishes or materials. In this example although the side of the pier is 250mm wide, because it is not in an isolated area, it is included with the main wall. |
| | 13.70 | | 19 × 25mm wrot softwood chamfered skirting plugged to blockwork<br>&<br>Prime only general surfaces of woodwork before fixing n.e. 300mm wide internal<br>& | |
| | | | KPS one coat undercoat one coat gloss on general surfaces of woodwork n.e. 300mm girth internal | Classified as isolated in this case. |
| | | | 2 | |

Figure 3.54 (continued)

Figure 3.55 Frame cramp

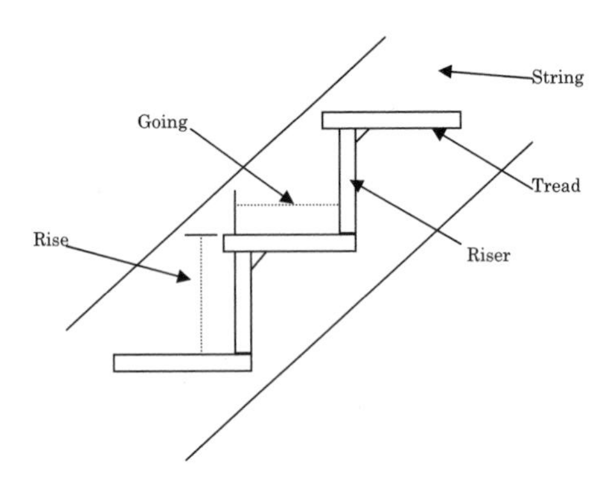


**Figure 3.56** Section through staircase

## Windows

### Softwood windows: taking-off list (NRM2)

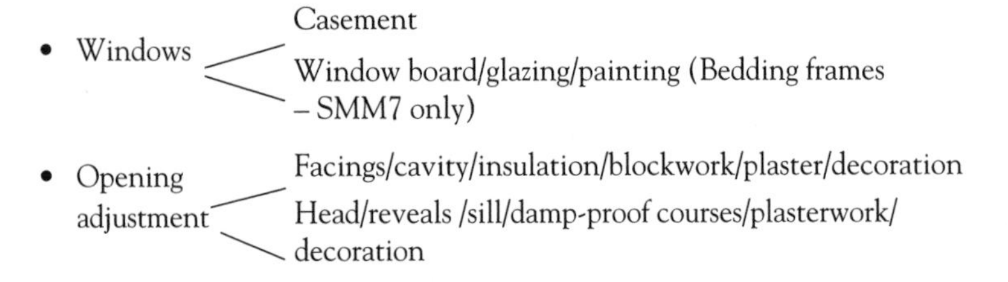


### Softwood door: taking-off list (SMM7)

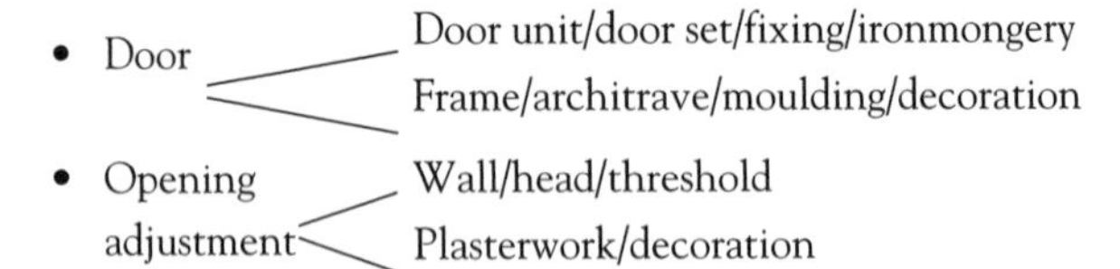


### Internal doors

For the example shown in Figure 3.60 it is assumed that the door lining is not a prefabricated unit but made up on site. As this is not a fire door the door stop is nailed to the lining and there is no intumescent strip. Fire doors are used to slow down and/or limit the spread of fire and smoke and are designated according to the time period that they are designed to resist fire, while maintaining their integrity. Intumescent strips are fitted into grooves in the door lining or door and expand when exposed to heat to fill the gap between the door and frame.

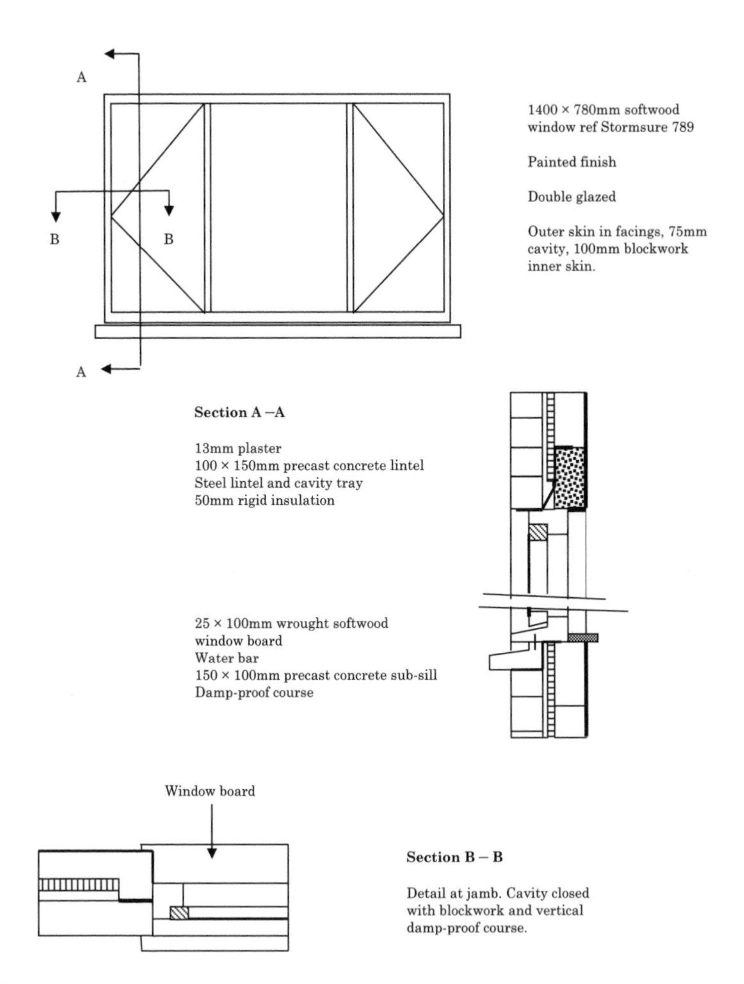


Figure 3.57

Fire doors are generally fitted with automatic door closers to ensure that they are always in place.

## Plumbing installations and drainage

The measurement of plumbing is often characterised by a lack of detailed information and drainage is seldom designed at bills of quantities stage, being treated

| | | | | |
|---|---|---|---|---|
| | | | **Example 8 – Softwood window** | Note: This example is measured in accordance with NRM2. |
| | 1 | | 1400 × 780mm Stormsure softwood window ref 789 and fixing with 4 No galvanised frame cramps, screwed to softwood and built into blockwork as work proceeds as drwg ref 1234 | Window/glazing and ironmongery schedules must accompany this section. This is a standard window unit and comes already fitted with standard ironmongery. Note that hinges or butts are normally specified in pairs. Therefore 3 butts or hinges would be measured as 1½ PAIR.<br><br>Glazing and ironmongery measurement rules are included in the windows section of NRM2.<br><br>Bedding the frame is deemed to be included. |
| | | | 1.400<br>Returns 2/ 0.038 0.076<br>1.476 | Note that window boards continue past the opening. |
| | 1.48 | | 25 × 100mm wrot softwood window board plugged and screwed to masonry | Window boards can be plugged or plugged, screwed and pelleted in work requiring a high-class finish. Traditionally, a pellet is a piece of timber dowel that fits into the screw hole to disguise the head. Today filler is most likely to be used in lieu of timber dowel. |
| 2/ | 1 | | 4mm clear sheet glass size 400 × 700mm | Windows are supplied either glazed or unglazed. If glazing is not included then it must be measured separately. |
| | 1 | | Ditto size 600 × 700mm | Work to the head and sill will be measured with the open adjustment. |
| | | | 1 | |

Figure 3.58

| | | | | |
|---|---|---|---|---|
| | 1.40<br>0.78 | | Prime only wood glazed surfaces before fixing over 300mm girth internal<br><br>&<br><br>Ditto external<br><br>&<br><br>One coat undercoat one finishing coat on glazed wood surfaces over 300mm girth internal<br><br>&<br><br>Ditto external | NRM2 – internal work and external work is kept separate. |
| | 1.48 | | **Window board**<br>Prime only general surfaces of woodwork before fixing not exceeding 300mm girth internal<br><br>&<br><br>One coat undercoat one finishing coat on general surfaces exceeding 300mm girth internal<br><br>× 0.16 = m$^2$ | The window board is not isolated as it will be painted together with the window frame. |
| | | | 2 | |

Figure 3.58 (continued)

| | | | | |
|---|---|---|---|---|
| | | | Opening adjustment | Firstly the opening is adjusted to take account of the window, followed by work and adjustments to the head, reveals and sill. |
| | 1.40<br>0.78 | | Ddt<br>Half bk skin hollow wall in facgs ab | |
| | | | & | |
| | | | Ddt<br>Formg cavity 75mm wide ab | |
| | | | & | |
| | | | Ddt<br>50mm cavity insulation ab | |
| | | | & | |
| | | | Ddt<br>100mm blockwork skin hollow wall ab | |
| | | | & | |
| | | | Ddt<br>13mm plaster to block walls exceeding 600mm ab | |
| | | | & | |
| | | | Ddt<br>Two coats of emulsion to plaster walls internal ab | |
| | | | 3 | |

Figure 3.58 (continued)

| | | | | |
|---|---|---|---|---|
| | | | **Head**<br>Lintel<br>Opening 1.400<br>Building in 2/ 0/150 0.300<br>1.700 | Lintels are built in at either end by 150mm minimum. |
| | 1 | | 100 x 150mm × 1700mm long precast concrete (mix A) reinforced with and including 2 No 12mm diameter mild steel bars built into blockwork in gauged mortar (1:1:6)<br>&<br>Hy-ten galvainsed mild steel lintel and cavity tray 1700mm long built into brickwork in gauged mortar (1:1:6) | |
| | 1.40 | | 13mm plasterwork ab to concrete walls not exceeding 600mm wide<br>&<br>Expamet galvanised mild steel angle bead ref 595 to concrete with plaster dabs<br>&<br>Two coats of emulsion to plaster walls intl ab<br>× 0.10 = m² | Plasterwork to underside of lintel. |
| | | | **Reveals** | |
| 2/ | 0.78 | | Extra over half bk skin of hollow wall in fcgs for closing cavity at reveals 75mm wide<br>4 | |

Figure 3.58 (continued)

| | | | | |
|---|---|---|---|---|
| | | | **Reveals (Contd.)** | |
| 2/ | 0.78<br>0.10 | | Vertical pitch polymer dpc exceeding 300mm wide bedded in gauged mortar (1:1:6) | |
| 2/ | 0.78 | | 13mm plasterwork ab to blockwork walls not exceeding 600mm wide.<br>&<br>Expamet galvanised mild steel angle bead ref 595 to blockwork with plaster dabs<br>&<br>Two coat emulsion ab to plaster walls intl.<br>× 0.10 = m$^2$ | |
| | | | **Sill** | |
| | 1 | | 150 × 100 × 1700mm long Precast concrete (mix 123) sill, sunk, weathered, grooved and throated, bedded on brickwork in gauged mortar (1:1:6) | |
| | 1.70<br>0.10 | | Ddt<br>Half-brick skin hollow wall in facings ab | |
| | 1.70 | | 5 × 30mm galvanised mild steel water bar bedded in concrete in gauged mortar (1:1:6) | |
| | | | 5 | |

Figure 3.58 (continued)

| | | | | |
|---|---|---|---|---|
| | | | To take Dpc | When taking a break from measuring, for whatever reason, it is good practice to leave yourself a note as to what is to be done next! |
| | 1.70<br>0.33 | | 1 layer horizontal pitch polymer stepped dpc width exceeding 300mm bedded in gauged mortar (1:1:6)<br><br>6 | It should be scored through once measured. |

Figure 3.58 (continued)

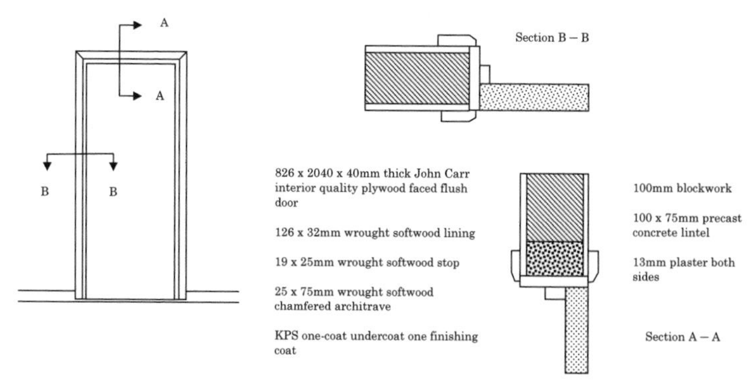


Figure 3.59

as 'Provisional' for measurement purposes to be remeasured when actually carried out on site. Drainage therefore is often the first item in the final account and its measurement is usually reserved for trainees or year out students.

### *Plumbing*

Traditionally, the approach to dealing with services, that is to say, plumbing, heating, electrical installations, is to include a provisional sum for specialist named subcontractors to tender for a later date. However, over the years clients have become less tolerant to this approach, as the quantity surveyor would still receive a fee for the services sections of the bills of quantities, as if they had been measured in detail. The demand to have services measured is even more understandable when it is considered that for some projects the value of the services elements can be in excess of 50%. That is not to say, as service installations become ever more complicated, that the quantity surveyor should have a detailed knowledge of every system on the market but it is useful for a quantity surveyor to have a basic knowledge of how services are measured.

| | | | **Example 9 – Internal door and lining** | Note: This example is measured in accordance with SMM7. |
|---|---|---|---|---|
| | 1 | | 826 × 2040 × 40mm Thick interior quality flush door faced both sides with plywood as John Carr ref KL56 | A dimensioned diagram or some other source of John Carr joinery information should be referenced. |
| | Item | | Include the Provisional Sum of £25.00 for the supply of ironmongery from a named supplier | In this example the ironmongery has been covered by the inclusion of a Provisional Sum as the decision has not yet been taken to a supplier. |
| | | | Fix only the following ironmongery to hardwood and softwood | The hardwood refers to the hardwood lipping to the edge of the door. |
| | 1 | | Pair 75mm pressed steel butts<br>&<br>Mortice latch<br>&<br>Set lever handle furniture<br>End of fixing the following | |
| | | | 1 | |

Figure 3.60

| | | | | |
|---|---|---|---|---|
| | | | Lining | Note: SMM7 L20.7 requires the number of lining sets to be stated. In addition repeats of identical sets are required to be stated. |
| | | | The following in 1 No wrot softwood lining set(s) tongued at angles | The traditional approach is to include all items relating linings under an appropriate heading as shown. |
| | | | Door 0.826<br>Add Lining 2/ 0.032 0.064<br>Head 0.890 | Linings are constructed by being tongued at the angles (the joint of the jamb and head as shown below). |
| | 0.89 | | 32 × 126mm head | 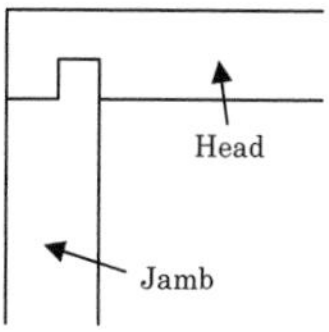 Head<br>Jamb |
| | | | Door 2.040<br>Tongue 0.012<br>Jamb 2.052 | |
| 2/ | 2.05 | | 32 × 126mm jamb plugged to blockwork | |
| | | | End of 1 No lining set | |
| 2/ | 2.04<br>0.79 | | 19 × 25mm wrot sw stop<br>0.826<br>Ddt 2/ 0.019 0.038<br>0.788 | |
| | | | Architraves | The joint between the plaster and the lining is covered by an architrave, mitred at the angles as shown below: |
| | | | 0.826<br>Add<br>Lining 2/0.010 0.020<br>2/0.075 0.150<br>2.040<br>0.010<br>0.075<br>2/ 2.125 4.250<br>5.246 | |
| | | | 2 | |

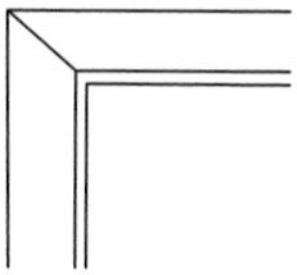

Figure 3.60 (continued)

| Timesing | Dimensions | Squaring | Description | Notes |
|---|---|---|---|---|
| | 5.25 | | 25 x 75mm wrot sw chamfered architrave | |
| | | | Painting<br>Door<br>0.826<br>2/½/0.040 0.040<br>0.866<br>2.040<br>½ /0.040 0.020<br>2.060 | Doors are painted on both sides as well as the vertical edges. However, it is the convention to paint either the top or bottom edge, not both! Check how many flush doors have a painted top edge. |
| 2/ | 0.87<br>2.06 | | KPS one undercoat and one finishing coat gloss paint to general wood surfaces exceeding 300mm wide | |
| 2/<br>2/ | 0.89<br>2.05<br>2.04<br>0.79<br>5.25 | | Prime only general wood surfaces before fixing n.e. 300mm girth | |
| | 4.91<br>0.40 | | KPS one undercoat and one finishing coat gloss paint to general wood surfaces exceeding 300mm girth<br>2/ 0.075 0.150<br>2/0.025 0.050<br>0.126<br>2/0.019 0.038<br>2/0.020 0.040<br>0.404<br>2/ 2.040 4.080<br>0.826<br>4.906 | The individual parts of the lining/stop/architrave are collected together and a girth calculated for painting purposes. |
| | | | 3 | |

Figure 3.60 (continued)

| Timesing | Dimensions | Squaring | Description | |
|---|---|---|---|---|
| | | | Opening Adjustment<br>0.826<br>Add 2/ 0.032 0.064<br>0.890<br>2.040<br>Add 0.032<br>2.072 | |
| | 0.89<br>2.07 | | Ddt<br>100mm thick blockwork wall ab<br>&<br>Ddt<br>13mm plaster to wall exceeding 300mm ab<br>× 2= m$^2$ | |
| | | | 0.890<br>Add Architrave 2/0.070 0.140<br>1.030<br>2.072<br>Add Ditto 0.070<br>2.142<br>Less skirting 0.075 | |
| 2/ | 1.03<br>2.07 | | Ddt<br>Two coats of emulsion on plaster ab<br>2.067 | |
| | | | 0.890<br>Building in 2/0.150 0.300<br>1.190<br>Say 1200 | |
| | 1 | | 100 × 75mm × 1200mm long Precast concrete (mix A) reinforced with and including 1 No 12mm diameter mild steel bars built into blockwork in gauged mortar (1:1:6) | |
| | 1.20<br>0.08 | | Ddt<br>100mm thick blockwork wall ab | |
| 2/ | 1.03 | | Ddt<br>Softwood skirting ab<br>&<br>Ddt<br>Prime only before fixing not exceeding 300mm girth ab<br>&<br>Ddt<br>KPS one undercoat and one finishing coat gloss finishing not exceeding 300mm girth | |
| | | | To take<br>Adjustment for flooring | |
| | | | 4 | |

Figure 3.60 (continued)

### *Incoming supply*

The incoming supply from the water main should be run in a duct at a minimum depth of 750mm below ground or finished level, in order to avoid damage from frost. The detail of how the incoming main is dealt with varies according to the water supplier and some customers are metered. At the point that the rising main enters the building there should be a stop valve and drain tap.

### *Distribution pipework*

The pipework from the incoming main to the fittings and storage is referred to as the distribution pipework and is often not shown on the drawings. It is left to the quantity surveyor to determine the size and length of pipes as well as fittings such as bends, elbows, stop valves, etc.

The rising main is usually 15mm copper and the distribution pipes from the cylinder are 22–28mm diameter, reducing to 15mm for sinks and WCs and 22mm for baths. In recent years plastic has become a popular

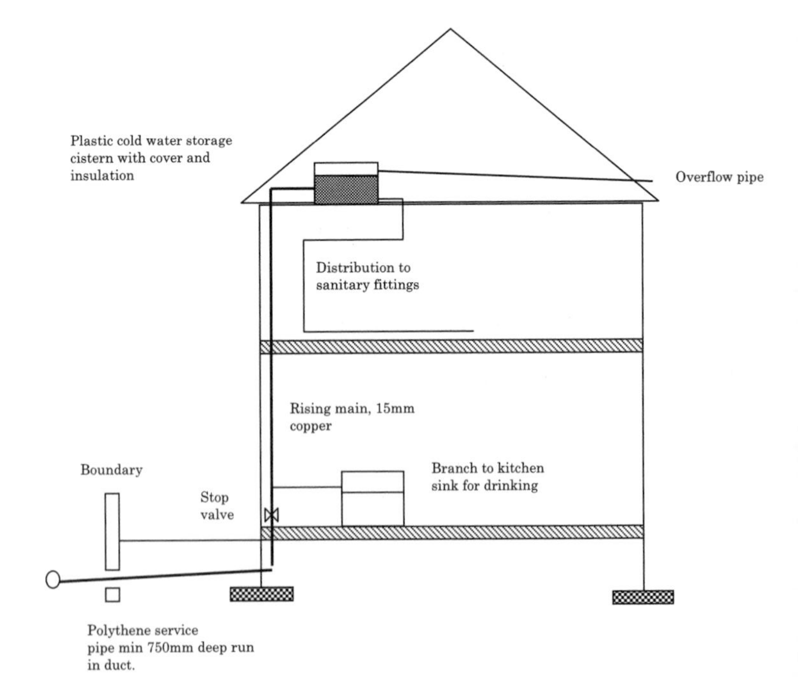


Figure 3.61

alternative to copper, it should be remembered when positioning pipework to make allowances for ease of access in the case of maintenance. When measuring pipework it is a good idea to have a copy of the manufacturer's catalogue to hand so that the appropriate fitting can be chosen.

### *Hot water supply*

As well as cold water, most buildings also require a hot water supply and there are a number of alternatives available to provide this depending on the nature of the project and the amount required. The broad choice is between local instantaneous heaters, or 'combi' boilers that can vary between a single appliance over a sink to a larger unit that feeds several appliances and centrally heated and stored hot water. If a centrally stored system is selected there are two types of cylinder; direct and indirect. With a direct system the central heating hot water is the same water that is used for the taps of the appliances and therefore has to be continuously replaced and is the less common form. With the indirect system the hot water is passed through the hot water cylinder by means of a heat exchanger in the form of a copper coil. The hot water to the taps therefore is heated indirectly by the heat exchanger and is the most commonly used.

### *Central heating*

Most wet central heating systems are based on either a one or two pipe system. With a one pipe system the water is fed to and flows from the radiators by a single pipe. The disadvantage of this approach is that the water tends to cool more quickly than the two pipe system detailed in Figure 3.62. The two pipe system has a separate flow and return pipe therefore the water to the radiators is more or less the same temperature.

### *Drainage*

Drainage is another trade that uses schedules to simplify the measurement process and contains many of the trades previously discussed, such as excavation, masonry, etc. Traditionally, manholes or inspection chambers were built from semi-engineering bricks on an in situ concrete base, but it is more common now to use pre-formed units made from plastic or concrete; these have considerable advantages as they are quicker and cheaper to use. Inspection chambers are essentially the same manholes but they are shallower.

Drainage pipes, according to their position and type require to be bedded, or bedded and surrounded with suitable granular material (see Figure 3.65).

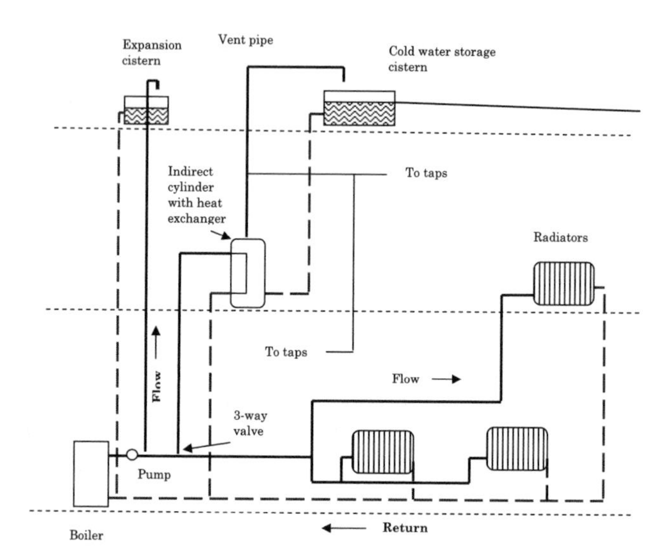


**Figure 3.62**

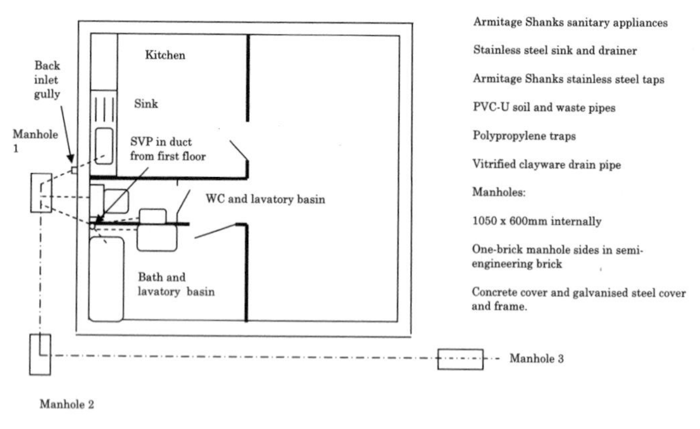


**Figure 3.63** Ground floor sanitary appliances and drainage

## Alterations: spot items

The final section in the measurement section is the drafting of spot items for alteration works and is covered in SMM7 clause C20 or NRM 4.1. The art of writing good spot items is firstly, a knowledge of the construction technology

| | | | | |
|---|---|---|---|---|
| | | | **Example 10 – Plumbing installations** | This example is measured in accordance with NRM2 Alternative 1 used 38.4–5. |
| | | | **Sanitary appliances** | |
| | | | Accolade Sanitary fittings (as manufactured by Armitage Shanks), including all assembly and jointing together of components and fittings. | As all the sanitary appliances are from the same manufacturer, it is good practice to use a heading as shown. |
| | 1 | | Bathroom<br>600 × 490mm white vitreous china wash basin and pedestal with one tap hole, plug and chain and 12mm Accolade chromium plated basin mixer tap including plugging and screwing brackets to masonry and bedding pedestal to floor and basin in mastic | The fittings in this example are measured on a supply and fix basis. It is also possible to include a provisional sum for supply only fittings. In this case an item for fix only would be included in the bills of quantities. |
| | | | & | |
| | | | 1685 × 800mm overall white vitreous china bath with one tap hole, plug and chain and 19mm Accolade chromium plated bath mixer tap including setting and levelling on feet plugging and screwing brackets to masonry and feet to concrete | |
| | | | & | |
| | | | WC<br>Close-coupled vitreous china WC unit with white plastic seat and cover, 9 litre cistern with dual flush and refill unit and P trap connector, including plugging and screwing to masonry and screwing pan to floor. | |
| | | | 1 | |

Figure 3.64

| | | | | |
|---|---|---|---|---|
| | 1 | | 450mm white vitreous china cloakroom basin and pedestal with tap holes, plug and chain and pair 12mm Accolade chromium plated basin pillar taps including plugging and screwing brackets to masonry and bedding pedestal to floor and basin in mastic | |
| | | | **End of sanitary fittings** | |
| | | | Kitchen | |
| | 1 | | Leisure stainless steel single bowl and drainer with pair Armitage Shanks 12mm Supataps, chromium plated waste and fixing in position on timber base unit (measured elsewhere) | Sanitary appliances need to be fitted with a trap to prevent obnoxious smells and vermin from entering the building from the sewer. The trap is formed from a water seal and there are basically 3 types: bottle, P traps and S traps and are generally made from plastic and have ring-sealed screwed joints. There is a chance that the water in the trap can be siphoned out and the trap made ineffective and so the soil stack, into which the appliance discharges is ventilated, by leaving the stack open to the atmosphere at the top. A WC does not require a separate trap as a seal is designed into the pan. |
| | | | Traps and fittings<br>LB | |
| 2/ | 1<br>1 | | 32mm Osma polypropylene 'P' trap with 76mm seal including screwed joint to fitting and solvent weld joint to PVC-U pipe<br>Kitchen sink | |
| | | | & | |
| | | | Bath<br>40mm ditto with 32mm overflow | |
| | | | 2 | |

Figure 3.64 (continued)

| | | | | |
|---|---|---|---|---|
| | | | WC | |
| | 1 | | 110mm diameter Osma PVC-U WC connector | The waste pipework for measurement purposes in separated into above and below ground. |
| | | | & | In this example the above ground drainage is carried out in plastic, whereas the below ground drainage pipework and fittings are in vitrified clayware. The advantage of clayware is that it is very durable. Plastic pipes are also popular for underground drainage as they are light in weight and easy to lay although undue loading can cause plastic pipes to deform and cause blockages to form. |
| | | | Ancillaries: rest bend (ref 4589B) | |
| | | | Above ground drainage | |
| | 5.00 | | 110mm Straight pipes, OsmaSoil solvent weld PCV-U (ref 4SO3B) with and including pipe bracket (ref 4SO82B) plugged and screwed to masonry. | Length has been assumed. |
| | 1 | | Ancillaries: Osma drain coupler (ref 4S 107G) | |
| | | | Sink 0.600<br>LBs 2/2.50 5.000<br>5.600 | |
| | 5.60 | | 32mm straight OsmaWeld pipes with welded joints with and including pipe bracket (ref 4ZO81G) plugged and screwed to masonry | |
| | | | Bath | |
| | 1.00 | | 40mm ditto | |
| | | | 3 | |

Figure 3.64 (continued)

| | | | | |
|---|---|---|---|---|
| 2/ | 1 | | Ancillaries: 110mm OsmaSoil pipe for bossed access pipe for 32mm OsmaWeld pipe | When bills of quantities are being prepared the following additional items have to be added where appropriate:<br><br>Marking the position of and leaving or forming all holes, mortices, chases and the like required in the structure. |
| | 1 | | Ditto for 40mm ditto | Testing and commissioning.<br><br>Preparing drawings.<br><br>Operating manuals and instructions. |
| | | | Builder's work is dealt with by SMM7 clause P31 and includes the measurement of all of the hole, chases and other work required for the installation of pipework, etc. through various parts of the structure. The approach of NRM2 is somewhat different as there is a separate section (41) dealing with builder's work in connection Mechanical, Electrical and Transportation installations. | |
| | | | Cold water installation | |
| | | | British Standard Table X light gauge copper pipes with capillary joints and fixed with and including pipe clips to masonry | NRM2 gives two alternatives for measuring pipe work fittings. Pipework can be measured inclusive of fittings or measured separately as here. |
| 2/ | 2.50<br>3.00<br>2.50 | | 15mm pipe straight | |
| | | | Bath | |
| | 3.00 | | 22mm pipe ditto | |
| | | | Fittings | For rising main |
| | 1 | | Ancillaries<br>Combined high pressure screw down stop valve and drain tap with compression joints to copper tube | |
| | | | 4 | |

Figure 3.64 (continued)

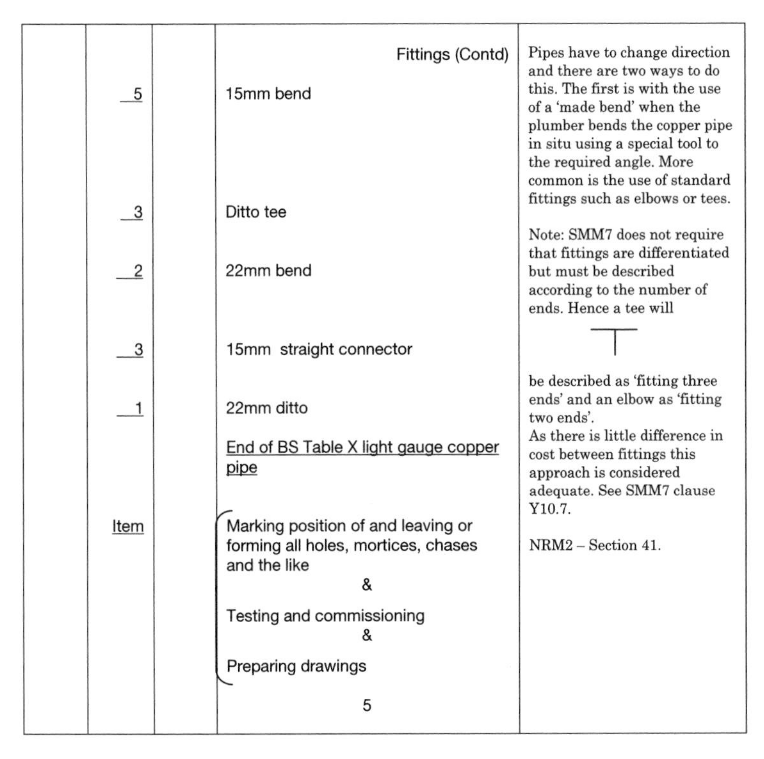

| | | | | |
|---|---|---|---|---|
| | | | Fittings (Contd) | Pipes have to change direction and there are two ways to do this. The first is with the use of a 'made bend' when the plumber bends the copper pipe in situ using a special tool to the required angle. More common is the use of standard fittings such as elbows or tees. |
| | 5 | | 15mm bend | |
| | 3 | | Ditto tee | Note: SMM7 does not require that fittings are differentiated but must be described according to the number of ends. Hence a tee will |
| | 2 | | 22mm bend | |
| | 3 | | 15mm straight connector | |
| | 1 | | 22mm ditto | be described as 'fitting three ends' and an elbow as 'fitting two ends'. As there is little difference in cost between fittings this approach is considered adequate. See SMM7 clause Y10.7. |
| | | | End of BS Table X light gauge copper pipe | |
| | Item | | Marking position of and leaving or forming all holes, mortices, chases and the like & Testing and commissioning & Preparing drawings | NRM2 – Section 41. |
| | | | 5 | |

Figure 3.64 (continued)

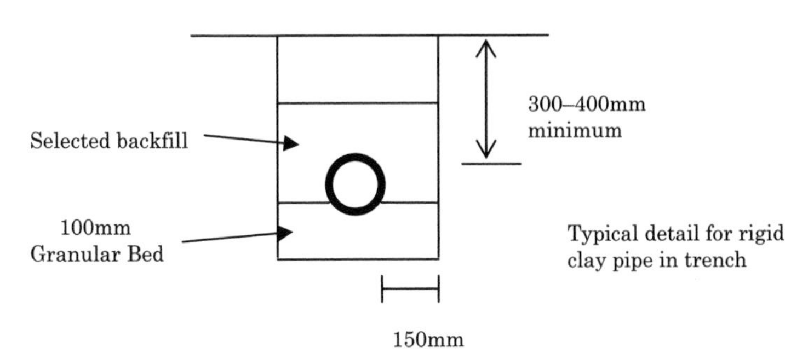


Figure 3.65

and methodology involved, as this is not normally marked on the drawings, combined with a logical approach, that makes the item straightforward for the contractor to visualise and price.

Alteration work is composed from a number of items such as cutting openings, where the following items need to be considered:

- cutting opening – size and purpose of opening, for example new door, window or blank opening. Note if the opening is for a new door or window these will be measured separately in the new work;
- finishes to existing wall, plastered on one or both sides;
- treatment at threshold/sill;
- treatment at jambs/closing cavities/damp-proof courses;
- treatment at head lintel/damp-proof courses/wedging and pinning;
- clearing away and making good all work disturbed.

## SPECIFICATIONS

The specification is an important part of the contract documentation, it lays down the quality and types of materials that must be priced for and used by the contractor, as well as setting out the expected standards of workmanship. It is also a vital point of reference for a number of members of the design team during both pre- and post-contract stages. In its traditional format a specification is divided and presented in the following sections:

- materials
- workmanship
- construction – these clauses are only used when the contract does not have a bill of quantities.

| | | | | |
|---|---|---|---|---|
| | | | **Example 11 – Drainage** | Note: This example is measured in accordance with SMM7. |
| | | | Manholes | For this example a traditional brick built manhole is used. |
| | | | 1.050 | |
| | | | Add Bkwk 2/0.225 0.450 | |
| | | | 1.500 | |
| | | | 0.600 | |
| | | | 0.450 | |
| | | | 1.050 | Invert level of pipe |
| | | | Depth |  |
| | | | Ground level 45.000 | |
| | | | Invert level 44.600 | |
| | | | 0.400 | |
| | | | Pipe and bedding 0.040 | |
| | | | Manhole base 0.150 | |
| | | | 0.590 | |
| | | | 1 | |

**Figure 3.66**

| | | | | |
|---|---|---|---|---|
| 3/ | 1.50<br>1.05<br>0.59 | | Excavate pit n.e. 1m deep ( 3 No.)<br><br>&<br><br>Remove excavated material from site | There is no spread on the foundations, therefore no backfilling is required. |
| | | | 1.500<br>1.050<br>2/2.550<br><br>5.100 | There are major changes in the way below ground drainage and manholes are dealt with in NRM2.<br><br>Manholes are enumerated instead of being measured out as shown here. |
| 3/ | 5.10<br>0.59 | | Earthwork support n.e. 1m deep and n.e. 2m between opposing faces | Drain runs include the excavation, pipes, beds, etc. in one description. |
| 3/ | 1.50<br>1.05 | | Surface treatment, compact bottom of excavation<br><br>&<br><br>Plain in situ concrete (Mix A) bed n.e. 150mm thick poured against earth.<br><br>X 0.15 = $m^3$ | |
| | | | 5.100<br>Ddt 4/2/½/0.225 0.900<br>MEAN GIRTH 4.200<br><br>0.590<br>Less base 0.150<br>bedding 0.040 0.190<br>0.400<br><br>2 | |

Figure 3.66 (continued)

| | | | | |
|---|---|---|---|---|
| 3/ | 4.20<br>0.40 | | One brick wall, vertical in semi-engineering bricks built English bond in cement mortar (1:3) built fair face and flush pointed one side. | |
| | 3 | | Benching 1050 × 600mm × 200mm average high in plain concrete (Mix A) and finished with cement & sand (1:3) with trowelled finish | When using a traditional brick manhole the bottom of the manhole is finished with cement and sand benching. The benching is haunched up from the sides of the channels at the bottom of the chamber to direct any spillages or overflow back into the main channel. |
| | 1 | | 100mm vitrified clay half round curved main channel 1400mm girth bedded in cement mortar (1:3)<br>&<br>Ditto 900 mm girth ditto<br>&<br>100mm vitrified clay half round straight main channel, 1050mm long ditto | |
| | 2 | | 100mm ditto three quarter section branch channel bends ditto | |
| | 7 | | Building in end of 100mm pipe to one brick wall | |
| | | | 3 | |

Figure 3.66 (continued)

| | | | | |
|---|---|---|---|---|
| | 3 | | 1500 × 1050mm × 100mm thick precast reinforced concrete (Mix B) manhole cover slab with rebated opening size 600 × 400mm bedded in cement mortar (1:3) on masonry | Deep manhole will need galvanised mild steel step irons built into the sides of the manhole. |
| | | | & | |
| | | | 600 × 400mm galvanised mild steel light duty manhole cover and frame bedding in cement mortar (1:3) | |
| | | | Drain runs | |
| 3/ | 4.80<br>7.90<br>1.00 | | Exc trenches for pipes n.e. 200mm diameter average depth 500–750mm deep | The pipes are longer than the trenches as they have to pass through the manhole walls whereas the bed under the pipe will be the same as the trench. |
| | | | 4.800<br>0.450<br>5.250<br>7.900<br>0.450<br>8.350<br>1.000<br>0.225<br>1.225 | |
| 3/ | 5.25<br>8.35<br>1.23 | | 100mm vitrified clay pipe with push fit flexible joints | |
| 3/ | 1 | | E.O. 100mm pipe for bend | 2 bends per branch |
| | | | 4 | |

Figure 3.66 (continued)

| | | | | |
|---|---|---|---|---|
| 3/ | 4.80<br>7.90<br>1.00 | | In situ concrete (Mix A) bed to 100mm pipe, 425mm wide × 100mm deep | |
| | 1 | | Clay trapped gully with 100 outlet and horizontal back inlet for waste pipe<br>To take<br>Testing drains | |
| | | | 5 | |

**Figure 3.66 (continued)**

| | | | | |
|---|---|---|---|---|
| | Item | | Cut or form blank opening in rear of existing stage size 0.95 × 2.10m high in existing one brick wall plastered one side including preparing a level bed at threshold quoining up jambs and providing and building in 225 × 225mm precast concrete lintel (Mix LC30) lintel reinforced with and including 4 No 12mm diameter mild steel bars bedded in gauged mortar (Group 3) on brickwork, extend finishes and make good all work disturbed. | SMM7/NRM2<br><br>Note that the location of the alteration item should be included in the description to facilitate location. |
| | Item | | Alter and adapt door opening between Room A and toilet to form new blank opening size 1.20m wide ×2.50m high including taking down half-brick wall for a length of 1.60m and a height of 2.50m including single door and frame, extending one jamb for a width of 0.40 and a height of 2.50m in common brickwork one brick thick in gauged mortar (Group 3) cut, toothed and bonded to existing, preparing level bed at threshold, extend finishes and make good all work disturbed. | |

**Figure 3.67**

The JCT(05) Standard Form of Contract has an edition specifically drafted for instances where the specification and not the bill of quantities is a contract document. This is usually the case when the project is reasonably small, say up to £250,000.00, as there can be considerable savings in both time and cost for a client compared to preparing a detailed bill of quantities. However, the decision to base the contract on a specification and drawings does not come without risk, as this strategy is far less prescriptive than a bill of quantities and therefore can carry a higher degree of uncertainty.

Uses of a specification:

- As previously stated, for small projects when the cost of a bill of quantities cannot be justified, the specification together with drawings, forms the basis on which the contractor calculates the tender price. Under these circumstances the contractor when preparing a bid, usually takes their own dimensions using builders' quantities (see Chapter 2) and prices them accordingly.
- Used by the quantity surveyor during the preparation of the bills of quantities. In this case the specification is incorporated into the contract documents as the preamble to the bill of quantities in much the same format with only the materials and workmanship sections being used.
- Used during the post-contract stages by the clerk of works, site agent, etc. as a point of reference for quality and standards, but *never* for the ordering of materials.
- At final account stage when using a specification and drawings, the specification forms the basis of valuing variations and the final account.

There are two main types for specifications:

1. Traditional descriptive specifications, as noted previously.
2. Performance specifications – this format is being increasingly used, especially in Private Finance Initiative (PFI) contracts. Unlike the traditional descriptive format, a performance specification is not prescriptive, but rather it lays down the standards and the operational requirements of the project or element and allows the contract to, in the case of a PFI contract, design and build the appropriate solution to meet the performance criteria. A performance specification need not be as wide ranging as this though and is a more conventional contract used for mechanical and electrical installations for example. The advantages of using this less prescriptive approach is that it enables the contractor to use their expertise to design and deliver the best possible solution and value for money as well as allowing the development of alternative solutions.

Examples of these formats are given below.

### Traditional (prescriptive) format

The rules for preparing a descriptive specification are as follows. This format is used when the contract is based on a specification and drawings. Unlike bills of quantities, prepared in accordance with a standard method of measurement, there are no rules of measurement. Nevertheless, the order of the specification generally follows trade order, for example:

- preliminaries
- ground work
- concrete work
- masonry, etc.
- through to external works and drainage.

When using this approach, each trade is a self-contained item which contains clauses relating to:

- materials
- workmanship
- construction.

Another approach is to present the specification in an elemental format, where the project is divided into elements (see Chapter 2 for definition of element) for example:

- preliminaries
- substructure
- frame
- etc.

Unlike the trade order approach, when using an elemental format the materials and workmanship clauses that apply to the job as a whole are grouped together in the front of the specification, the construction clauses are then included for each element.

### *Example*

The completed specification is presented with a cover sheet similar to a bill of quantities, as shown in Figure 3.68.

There are a number of sources of information available to the specification writer and these include the following.

### *For materials*

- British Standards (BS). British Standards Institution is the UK national standards organisation that produces standards and information that share best practice.

**SPECIFICATION OF WORKS**

for

**THE REINSTATEMENT OF FIRE DAMAGE AND REMEDIAL WORK**

at

**WESTERN AVENUE, ASHFORD, KENT**

Duncan Cartlidge FRICS
Chartered Surveyors
36 Castle Hill Road
Folkestone
Kent

Daniel Hunt & Partners
Chartered Building Surveyors
7 Station Road
Ashford
Kent

April 2012

**Figure 3.68** Sample specification cover sheet

**Section E**

**MASONRY**

**Item**

**Materials**

A. Portland cement

Cement shall be normal setting Portland Cement to comply with BS12.

B. Sand

Sand for use in mortar shall be clean, sharp sand obtained from an approved source and complying with Table 10 of B.S. 1200.

C. Lime

Lime for use in gauged mortar shall be hydrated lime to BSEN 459-1 : 2001

D. Water

Water shall be clean, fresh drinking water.

E. Mixes

Gauged mortar for brickwork and blockwork shall be composed of a 1:2:9 mix (by volume) of cement/lime/mortar.

F. Concrete blocks

Aerated concrete blocks shall be to BS 6073 strength 3.5N/mm$^2$

G. Common bricks

Common bricks shall comply with BS 3921 Class 3; 20.5N/mm$^2$

H. Facing bricks

Facing bricks shall comply with BS 3921 and shall be Westbrick Multi-facings

| | £ | |
|---|---|---|
| 1 Masonry Carried to Collection £ | | |

Figure 3.68 (continued)

| Section E | £ | |
|---|---|---|
| **MASONRY (Contd.)** | | |
| **Item** | | |
| **Materials (Contd).** | | |
| A. Air bricks | | |
| Air bricks shall comply with BS 493 Part 2 to match main facings, size 225 x 75mm. | | |
| **Workmanship** | | |
| B. Mixing mortar | | |
| 1. Mix mortar by mechanical mixer or by hand on a clean, level banker board.<br>2. Mix thoroughly so that all individual constituents are incorporated evenly but do not over-mix mortars containing plasticisers.<br>3. Use mortar within two hours of mixing at normal temperatures.<br>4. Do not use after initial set has taken place and do not re-temper.<br>5. Keep plant and banker boards clean at all times. | | |
| C. Laying and jointing | | |
| 1. All bricks shall be wetted before being laid.<br>2. All courses shall be kept level and perpends shall be plumb throughout.<br>3. No part of the brickwork shall be built up higher than 915mm above any other.<br>4. The contractor shall be deemed to have included for all labours including forming chases, pointing, toothing and bonding new work to existing, wedging and pinning new work to existing.<br>5. No brickwork shall be built in frosty weather and in seasons liable to Frost, all brickwork laid during the day shall be properly protected against frost at night.<br>6. Facing brickwork shall be built fair to match existing and care shall be taken to ensure that all face work is kept free from mortar droppings. | | |
| D. Cavities | | |
| Keep cavity and ties completely free from mortar and debris with laths or other suitable means. | | |
| 2 — Masonry Carried to collection £ | | |

Figure 3.68 (continued)

| Section E | £ | |
|---|---|---|
| MASONRY (Contd.) | | |
| Item | | |
| Construction | | |
| A. Build the 275mm external cavity walls for the lengths, widths and heights shown of Drawing No. 123/b. | | |
| B. Build the 100mm thick internal partitions to the lengths, widths and heights as shown on Drawing Nos 123/b and 124/c. | | |
| Masonry<br>Carried to collection £ | | |
| | COLLECTION<br>£ | |
| Page 1<br>Page 2<br>Page 3 | | |
| Masonry<br>Carried to summary £ | | |
| 3 | | |

Figure 3.68 (continued)

- European Standards – Comite Europeen de Normalisation (CEN). CEN develops European Standards and promotes voluntary technical harmonisation in Europe in conjunction with worldwide bodies and its partners in Europe. BSI is a leading member of CEN in the development of European Standards.
- Manufacturers' literature: now widely available online.
- Trade associations such as the Copper Development Association, the Brick Development Association, etc.

### *Co-ordinated Project Information (CPI)*

Lack of co-ordination in the preparation of project information can manifest itself in the following ways:

- impractical design, resulting in difficulties during construction;
- conflicting information;
- inadequate information;

- a high volume of variations during the contract;
- a high volume of contractor claims;
- a high degree of defects in completed building.

In 1979 the Co-ordinating Committee for Project Information was established to investigate the ways by which construction documents are produced and presented. It finally presented its recommendations in 1987 and included proposals for ways in which problems of unco-ordinated project information could be prevented. The recommendations include:

- better procedures for producing documents, including the use of standard libraries of specification clauses;
- checking the specification against the information contained on the drawings;
- better arrangement of contract documents.

If the CPI approach is used then the specification is arranged in the same order as SMM7, for example:

A Preliminaries and general conditions
C Demolition/Alterations
D Ground work
E Concrete
F Masonry
Etc.

### Standard library of descriptions

In practice, very few specifications are written from scratch, they are usually prepared using a cut and paste approach. To help specification writers a number of standard libraries of specification clauses have been developed that can be adapted as necessary for each individual project. The National Building Specification (NBS) is one such provider of a number of solutions to cover new work, reburbishment work and small domestic projects. The library allows the specification writer to use his/her own clause instead of the library option. The NBS is available by subscription in electronic format. Standard libraries can cope with standard items well enough, however in the case of alteration work a more flexible approach is required to take account of the bespoke nature of the works.

### *For workmanship*

- **Codes of practice.** Codes of practice are also published by the BSI and lay down standards for good practice, although these are not mandatory.
- **Building Research Establishment (BRE) Digests.** These digests cover a wide range of topics relating to construction design and defects in existing buildings.

Note: when writing a specification the imperative tense should be used; that is to say in the form of an unambiguous direct command. The reason for this is, that it leaves no doubt about what needs to be done. For example: 'The contractor shall …' or 'Portland Cement shall comply with BS EN 197-1: CEM I 52,5N …'

### *Cover-all clauses*

It is common practice to include a clause in a specification such as:

> The contractor is to include in his price everything necessary to complete the works, whether or not included in this specification.

Putting aside whether it is either professional or ethical to include such clauses, the effect is to transfer large amounts of risk to the contractor, who realising this will either price the works accordingly, or use the vagueness of such a clause as the basis for a claim against the client should there be substantial amounts of extras not included in the specification.

### *Performance specifications*

Perhaps the ultimate performance specification is the Output Specification that is used as the basis of obtaining bids for Private Finance Initiative projects. The output specification is included in the outline and full business cases and is regarded as one of the most difficult and important parts of the PFI bidding process and concentrates on outputs, that is to say what is actually consumed by the user of the services. For example, in the case of a new PFI prison, instead of asking for bids on the basis of drawings and bills of quantities a statement of outputs will be given, for example: 'Full custodial services for 800 category B prisoners including: secure management and control of prisoners, visiting facilities, healthcare, training and employment, record keeping, etc.'

The point about using a statement of outputs is that it gives the maximum freedom to the consortia to develop alternative solutions and innovative forms of construction, etc., whilst still conforming with all statutory and legal regulation relating to prisons and prisoners.

### *Specifications for works of repair and alteration*

As is the case when preparing a bill of quantities, works of alteration and repair present the quantity surveyor with particular challenges. There are no set rules of measurement and all embracing omnibus items must be drafted that include all the items that require to be carried out and priced by the contractor. It therefore follows that the person preparing the specification needs a sound and wide ranging knowledge of construction technology. The following example is typical of alteration work items.

| **Item** | **Zone E – Existing caretaker/porch** |
|---|---|
| A | Form new door opening size 1.45 × 2.10m in one-brick thick wall, east of existing porch, including preparing level bed at threshold, quoining up at jambs, cutting away and building in 225 × 225 precast reinforced concrete lintel (Mix LC30), reinforced with and including 4 No 12mm diameter mild steel bars bedded in gauged mortar (Group 3) on brickwork and make good all work disturbed and remove materials from site. |

Note that the location of the item is referred to in the description. When commencing a specification for alteration works it is good practice to adopt a logical sequence for the section, for example:

- all internal works starting on the lowest floor and moving through the building room by room in a clockwise direction, then progressing to the upper floors, if present;
- roof and external walls;
- external works.

In this way the builder trying to price the specification can easily trace and follow various items. In the above example, the new door to complete the opening will be included in the woodwork section of the specification.

## Measurement for Energy Performance Certificates (EPC)

From the start of 2009 most new and existing public, commercial and domestic properties have to have an Energy Performance Certificate. In order to do this several specialised forms of measurement have been developed as follows for:

- new build domestic EPCs;
- existing domestic EPCs;
- non-domestic (commercial) EPCs;
- public buildings – Display Energy Certificate (DEC).

### *New build domestic EPCs*

The assessment methodology for new build houses is Standard Assessment Procedure (SAP), a system devised in the early 1990s. A SAP-derived $CO_2$ emissions rate is now required by the Building Regulation for all new houses. A SAP rating needs data that can only be obtained from plans and the specification of the house. The data is fed into software to produce the rating as follows:

- 1–10 Very poor
- 11–30 Poor
- 31–50 Standard
- 51–60 Above average
- 61–80 Good
- 81–100 Very good
- 101–120 Outstanding

The UK average SAP Rating is 48, new build properties tend to be above 80 points.

### *Existing domestic buildings' EPCs*

Reduced data Standard Assessment Procedure (RdSAP) is the system used to calculate energy ratings and produce EPCs for existing domestic buildings. RdSAP is based on SAP which requires the input of many measurements and details that cannot be readily seen or accessed in an existing dwelling and therefore the RdSAP was developed based on a much reduced requirement for data collection.

Irrespective of which accreditation scheme the energy assessor belongs to, the same RdSAP data set is used to calculate the energy rating and $CO_2$ emissions. The assessor has to visit the site and record details using a standard pro forma on the following, based on the information required by NHER.

Property type:

- The assessor should note:
  - whether the property is a house, bungalow, flat or maisonette. The property type is required as different types of property have differing heat loss walls;
  - the procedure for dealing with flats is different to houses as flats very often open onto unheated corridors and have unexposed upper floors.

Built form:

- The assessor should note:
  - whether the property is detached, end-terrace, semi-detached, mid-terrace, enclosed mid-terrace or enclosed end-terrace;
  - the number if any of rooms in the roof;
  - the number of habitable rooms;
  - the number, age and form of any extensions;
  - the number of open fireplaces;
  - in addition to the above the energy assessor is required to calculate the floor area, room height and heat loss perimeter for each floor.

Property age:

- There are eleven age bands ranging from pre-1900 to 2007 onwards. The property age is required for the software to select the default heat-loss values (U values) for the wall, roof and floor and to calculate the window area. Houses of different ages have different window-to-wall and window-to-floor ratios. The age bands for newer properties correspond to changes in the Building Regulations, when regulations stipulated maximum window areas in order to reduce heat loss.

Construction:

- The materials in the construction of the house have to be recorded, for example granite, cavity wall, filled cavity, etc.
- If more than 10% of the external walls are built from an alternative type of construction, e.g. tile hanging on battens, then this has to be measured and recorded.
- Other items that concern the assessor are rooms in the roof (loft conversions), etc., external porches and conservatories.

Windows:

- The assessor should note:
  - the area of glazing compared to the average for its type;
  - the proportion of double glazing; and
  - the age of the double glazing.

Heating systems:

- The following details are required of the heating system(s):
  - type of electricity meter;
  - type of heating system and controls together with the date of manufacture;
  - the type of secondary heating system, bottled gas, coal, portable heater should also be recorded;
  - the type of water heating system;
  - hot water cylinder size, type and insulation;
  - whether there is a hot water cylinder thermostat.

Other details:

- The assessor should record any solar or photovoltaic panels.

### *Non-domestic (commercial) EPCs*

As explained earlier, the process of carrying out a domestic EPC is generally a one-person job involving fairly straightforward processes. Non-domestic EPCs can involve several people and take several days depending on the size of the building. Basically the process is as follows:

- Assessors visit the site and make a note of the building characteristics with regards to:
  - building fabric – materials, insulation, windows, etc.;
  - heating, hot water and lighting;
  - ventilation and air conditioning;
  - fuels used.
- At the same time the assessor will begin the process of zoning the building into various activities. Large buildings have several activities in progress, from office zones, rest and restaurant areas to car parks, and each of these zones will have different heating and cooling load from an energy assessment point of view.
- After the site visit the assessor returns to the office and for each zone, using a set of scaled floor plans, enters the details for each zone, together with the dimensions: floor areas and room heights for each zone.

Using an approved data set, described earlier, the assessor then inputs the information into the software and the EPC is produced.

### *Display Energy Certificates (DECs)*

Display Energy Certificates are required for public buildings greater than 1,000m$^2$ floor area from 1 October 2008. Public buildings include:

- public libraries
- schools and education centres
- hospitals (private hospitals are excluded)
- care homes under local authority control
- offices
- courts
- NHS trusts
- universities and colleges
- police
- prisons
- MOD
- army
- executive agencies of the government
- statutory regulatory bodies
- leisure centres (but not private clubs)
- public golf club houses
- museums and art galleries which are sponsored by public authorities.

The DEC must be accompanied by an Advisory Report. DECs are valid for 12 months, Advisory Reports seven years. This report is a listing of recommendations provided by the energy assessor after completion of the energy survey. It will detail improvements that would improve the energy rating of the building. Such recommendations may include improved glazing or lighting, for example heating systems, etc.

The penalty is £500 for failing to display a DEC at all times in a prominent place clearly visible to the public and £1,000 for failing to have possession of a valid advisory report. In addition to these penalties, it is still necessary to commission the documents.

A DEC must contain, by law, the following information:

- the operational rating and the asset rating (if available) as determined by the government approved method;
- the operational ratings for the building expressed in any certificates displayed by the occupier during the last two years before the nominated date;
- a reference value such as a current legal standard or benchmark;
- the unique certificate reference number, the address of the building, the total useful floor area of the building, the name of the energy assessor, their employer (or trading name if self-employed), the name of their accreditation scheme and the date when the DEC was issued.

### *Building information modelling (BIM)*

Building information modelling (BIM) is the process of creating a computer model of a building project that can be used to design, analyse, build, manage and even demolish a building. The key to BIM is that it is not just a model but a database of information that enables different parties to a project to store and retrieve information in a consistent and shareable format. From 2015 it has been proposed that the use of BIM will be compulsory on all government projects in an attempt to reduce costs in the public sector. The advantages that are claimed for BIM include:

- improved design process;
- better communication;
- environmental assessments;
- quicker and easier design revisions;
- consistency of standards;

- more accurate scheduling of information;
- quantity take-offs;
- more accurate tendering processes;
- project planning and resource allocation;
- more efficient construction phasing;
- links to facilities management systems.

There are a number of software publishers currently marketing BIM as an add-on to existing model packages.

# 4

# Procurement

## INTRODUCTION

The dictionary definition of procurement is: 'the obtaining of goods and services'. Put in a construction context this can be taken to mean obtaining the whole spectrum of goods, materials, plant and services in order to design, build and commission a building that delivers the best possible value for money for the client over its life-cycle.

Traditionally, the criteria for the selection of goods, materials, plant and services has been the cheapest or lowest priced bid; however, recently the emphasis has switched away from cheapest cost to best.

The approach to construction procurement and the choice of the appropriate construction strategy will vary between clients (public or private sector) and projects (new or refurbishment). Traditionally, the public and the private sector have had very different approaches towards construction procurement, with the public sector being heavily regulated, with an emphasis on transparency and accountability. For the private sector, the emphasis has been less regulated and accountable with the end result being more important than the process. During the last twenty years or so there has been increasing co-operation between the two sectors, for example the increasing use of Public–Private Partnerships, explained later in this chapter. For the public sector another layer of procurement legislation in the form of the EU public procurement Directives must also be taken into account.

## GUIDELINES TO PROCUREMENT

There are a number of standard guides to construction procurement such as those published by The National Joint Consultative Committee for Building (NJCC). Although the NJCC was disbanded in 1996 the documentation is still used as a reference. In 1969, the NJCC, in conjunction with the Ministry of Public Building and Works, published the *Code of Procedure for Selective*

*Tendering*, which was the first of a long line of practice guides published for the guidance of clients using the industry as well as their professional advisors administering their projects. This document was republished as the *Code of Procedure for Single Stage Selective Tendering* 1977. By this time the Ministry of Public Building and Works had been replaced by the Department of the Environment and they collaborated in its production, together with the JCCs in Scotland and Northern Ireland. The Code of Practice for Single Stage Selective Tendering was superseded by JCT Practice Note 6: Main Contract Tendering for Works Executed under JCT Contracts. There followed codes of procedure for:

- *Tendering for Industrialised Building Projects* (1974).
- *Two Stage Selective Tendering* (1983).
- *Selective Tendering for Design and Build* (1985).
- *Letting and Management of Domestic Sub-Contract Works* (1989).
- *Selection of a Management Contractor and Works Contractors* (1991).

These publications were augmented with nine guidance notes, twenty procedure notes and two tendering questionnaires. Interestingly, the NJCC Code of Practice under the heading, 'Assessing Tenders and Notifying Results' places emphasis for the selection of bids on the lowest cost as it recommends that: 'The lowest tenderer should be asked to submit his priced bill(s) of quantities as soon as possible ....'

In addition to NJCC guidance the UNCITRAL model law provides a useful starting point and sets out the general principals for successful construction procurement as follows:

- maximising economy and efficiency in procurement;
- fostering and encouraging participation in procurement proceedings by suppliers and contractors, especially where appropriate, participation by suppliers and contractors, regardless of nationality, and thereby promoting international trade;
- promoting competition among suppliers and contractors for the supply of goods, construction or services procured;
- providing for the fair and equitable treatment of all suppliers and contractors;
- promoting the integrity of, and fairness and public confidence in, the procurement process;
- achieving transparency in the procedures relating to procurement.

Although primarily for use in the public sector, the UNCITRAL model law's reference to competition, equity, fairness, confidence and transparency can said to be the essential requirements for any procurement strategy whether public or private sector.

To an extent the procurement strategy that is dominant at any particular time is influenced by a number of external factors as illustrated in Table 4.1.

## A GENEALOGY OF PROCUREMENT

### 1834–1945

- 1834 – the birth of the quantity surveyor, a profession still synonymous with construction procurement.
- The Joint Contracts Tribunal in their publication *The Use of Standard Forms of Building Contract* advises that from about 1870 the possibility of a standard form was discussed among various trade bodies and the RIBA.
- Agreed forms were subsequently issued in 1909 and 1931. New editions followed to take account of change in industry practice.

### 1946–1969

- The Second World War left a massive demand for new buildings and infrastructure.
- The election of a Labour government in 1964 attempted to shackle developers from making what were considered to be excessive profits, with the introduction of a Betterment Levy, that was included in the Land Commission Act of 1967.
- The predominant forms of procurement/contract during this period were single-stage lump sum contracts based on bills of quantities. A Code of Practice for Single Stage Selective Tendering was developed as a highly prescriptive guide to this form of procurement.
- The emergence of cost reimbursement contracts which allowed a contractor to be reimbursed for the costs of a project on the basis of actual cost of labour, materials and plant plus a previously agreed percentage, to cover profit and overheads. The theory was that work could commence on site much more quickly than via traditional procurement routes although the higher the cost the greater the contractor's profit.

### 1970–1979

A highly volatile period characterised by the following:

- The election of a Conservative government in 1970 and the start of a property boom. The previous Labour government restrained property development with legislation as well as controlling house building with mortgage lending restrictions.

**Table 4.1** A genealogy of procurement

| Dates | Economic milestones | Procurement trends | Construction activity |
|---|---|---|---|
| 1934–1945 | Few corporate clients | Sequential, fragmented process<br>Bills of quantities, competitive tendering | Traditional approach |
| 1946–1969 | Post-war regeneration | High value = Low cost<br>Lump sum competitive tendering<br>Cost reimbursement | Rebuilding post-war Britain. |
| 1970–1979 | Rampant inflation 25%+ pa<br>Historically high interest rates | Management contracting<br>Two-stage tendering | Property boom 1970–74 |
| 1980–1989 | 1989 Base rate reaches 15%<br>Financial deregulation<br>Privatisation<br>1987 Inflation reaches 7.7% pa<br>1987 Stock market crash | Construction management<br>Management contracting<br>CCT<br>Bespoke contracts to load risk onto contractors | Property slump 1980–84<br>Property boom 1985–90 |
| 1990–2000 | Globalisation<br>Low interest rates & inflation<br>World economic slump | Partnering<br>PPP/PFI | Property slump 1991–97<br>Property boom 1997–2000 |
| 2001–2008 | Globalisation<br>Sustained economic growth<br>Low interest rates & inflation | e-procurement<br>Prime contracting<br>Relationship contracting | Property boom |
| 2009– | Sub-prime market collapses<br>End of ten years of economic growth | Design and construct | Slump in housing sector |

- Between 1970 and 1974 the Conservative government decided to increase the money supply and much of this money found its way into property development. Between 1970 and 1973 bank lending increased from £71bn to £1332bn, with most of the increase going to property companies.
- Rents doubled within two years.
- Developers became increasingly impatient with existing procurement techniques and they demanded short lead-in times and fast completion.
- Inflation rose to 25% per annum and historically high interest rates. The faster that buildings could be procured the less profits were eaten up by interest charges and inflated material and labour prices.
- In mid-1972 the continuing Arab/Israeli war caused oil prices to soar and there was a loss of confidence in the UK economy.
- There was a period of comparative political instability with the election of a minority Labour government in February 1974 followed by another general election in October 1974. Labour was to remain in power until 1979 when a new Conservative government was elected.
- In the private sector fast-track procurement strategies were developed. Although faster than traditional methods they had little cost certainty. However, lack of cost certainty was to some extent tolerated in times when property values were rising so quickly. Many of the problems that were to be associated with the so-called fast-track methods of Construction Management and Management Contracting were due to lack of pre-contract planning. These new methods, however, sought to collapse the construction programme by using a concurrent approach with, in some cases, design and construction being carried on in parallel.
- In the public sector traditional procurement, that is single-stage selective tendering, still predominated in an attempt to ensure accountability.

**1980–1989**

- High interest rates contributed to the recession of 1979–1982.
- The emergence of the developer-trader, often an individual, who had as many projects as money could be found for. The completed projects were sold for a profit at a time of rising rental values, the bank repaid and the next project would be undertaken.
- By the end of the decade UK banks were exposed to £500 billion of property-related debt.
- Fee scales under threat – during the 1980s the Conservative government introduced compulsory competitive tendering for public sector projects

which by the end of the decade had spread to the private sector and consultants faced with a reduction of fee income of up to 60% had to adopt a more pragmatic approach to procurement.
- The later part of the 1980s was characterised by the increasing use of modified standard forms of contract. Private clients/developers became increasingly frustrated with what was perceived as unfair risk allocation and as a result lawyers were increasingly asked to amend, in some cases substantially, standard forms of contract to redress the balance.
- Design and build procurement began to see a rise in popularity as clients perceived it as a strategy that transfers more risk to the contractor.

## 1990–2008

- The 1990s started with the worst recession the construction industry had known in living memory. Almost 500,000 people left the industry.
- The construction industry was not the only sector experiencing difficult trading conditions. In 1992 the oil and gas industries came to the conclusion that the gravy train had finally hit the buffers. It launched an initiative known as CRINE or Cost Reduction Initiative for the New Era, with the stated objective of reducing development and production costs by 30%. The chief weapon in this initiative was partnering and collaborative working in place of silo working and confrontation and the pooling of information for the common good.
- Latham and Egan Reports recommended fundamental changes in procurement practice.
- In 1999 the construction industry started to come to terms with the new order. A new breed of client was in the ascendancy demanding value for money and consideration of factors such as life-cycle costs. Predominant forms of procurement/contract during this period were partnering, PPP/PFI and design and build.
- 2008 saw the end of ten years of economic growth and the collapse of the housing and retail sectors.

## 2009–2012

- Began with predictions of economic meltdown with the housing and retail sectors being particularly badly hit.
- PFI was withdrawn as a procurement path in Scotland by the SNP and in England and Wales several large-scale PFI models were withdrawn by the coalition government in 2010.
- Coalition government demanded better value for money from procurement.
- BIM to be compulsory by 2016 for some larger public sector projects.

## RISK AND PROCUREMENT

Construction, in whatever form, is a process that involves various amounts of risk and various procurement strategies and forms of contract have different mechanisms for the allocation of risk. From a client's perspective risk can manifest itself in the following forms:

- Cost risk – the risk that the final cost may exceed the initial estimate.
- Time risk – the risk that the project will be delivered later than planned.
- Design or quality risk – the risk that the client is unable to easily influence and evaluate the quality of the project.

A widely accepted definition of risk is: an uncertain event or set of circumstances, that could, should it occur, have an effect on the achievement of the project objectives. From a procurement perspective the questions that should be addressed are:

- What are the risks?
- What will their impact be?
- What is the likelihood of the risks occurring?

And most importantly:

- Who will be responsible for the management of risk?

These procurement drivers are sometimes illustrated in the somewhat simplistic model as shown in Figure 4.1. The quantity surveyor, when determining the appropriate procurement strategy asks the client: which of the drivers

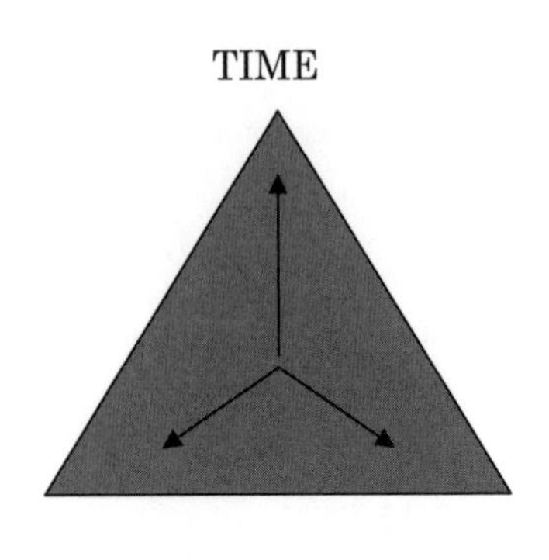


**Figure 4.1**

is important? Depending on the response a strategy is chosen that matches most closely the client's requirements. For example, for a project where it is considered that design, quality and cost are high priorities but where completion is not critical, traditional lump sum procurement based on drawings and a bill of quantities may be considered. Other publications are also available from the Royal Institution of Chartered Surveyors and the Cabinet Office, but in practice there is no substitute for a thorough understanding of the client's requirements and experience of the characteristics of the various available procurement strategies. There follows therefore details of the most common procurement paths that have been grouped:

- traditional
- design and build
- management.

## TRADITIONAL PROCUREMENT

During the 1960s in the UK, the traditional strategy described below was the most commonly used form of construction procurement with approximately 60% plus of all contracts being let on this basis in both the public and private sectors. In recent times client pressure has seen its popularity decrease to around 40%. Figure 4.2 illustrates the traditional model for procurement. The client has, in this illustration, appointed a project manager, usually a quantity surveyor to co-ordinate the project on their behalf. From a client perspective this has the advantage that the project manager is a single point of contact for the answering of day-to-day queries, instead of having to approach the individual members of the design team. This approach to procurement is also commonly referred to as 'architect led' procurement as traditionally the client has chosen and approached an architect in the first instance and it is then the architect who assembles the rest of the design team: structural engineer, services engineer, quantity surveyor, etc.

The chief characteristics of traditional single-stage competitive tendering are:

- It is based on a linear process with little or no parallel working, resulting in a sometimes lengthy and costly procedure.
- Competition or tendering cannot be commenced until the design is completed.
- The tender is based on fully detailed bills of quantities.
- The lack of contractor involvement in the design process, with the design and technical development being carried out by the clients' consultants, unlike some other strategies described later.

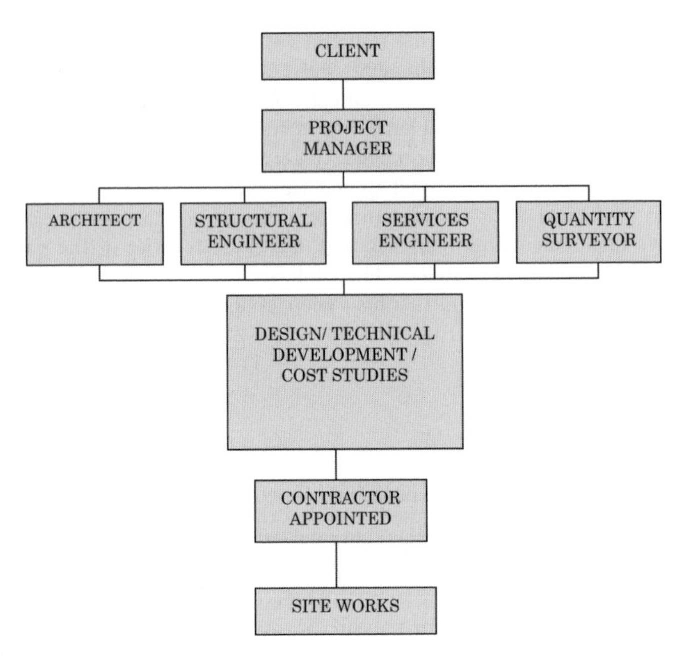


**Figure 4.2**

Other procurement paths have attempted to shorten the procurement process with the introduction of parallel working between the stages of client brief, design, competition and construction. The following is based on the RIBA Outline Plan of Work 2007 as included as Table 2.1.

## Preparation, Stages A & B: Appraisal and Design Brief

Having taken the decision to build, the client, whether a corporate body or an individual, finds an architect and discusses the proposals. Depending on the scale and type of project, the design team (see Figure 4.2) is now brought together.

Following on from the formation of the design team the next stage in the process is for the client to brief the design team on their requirements. For the quantity surveyor there is usually a requirement from the client to provide an estimate of cost based on the preliminary information that may, or may not, be available. It can of course be the case that the client has a cost limit to which the design team has to produce its solution, either way the quantity surveyor's function is to monitor/advise on costs. It can be a problem as the information is so provisional and the techniques that are available for the quantity surveyor are described in Chapter 2. At this stage the decision will also be taken as to which procurement strategy is to be used.

### Design, Stages C, D & E: Concept, Design Development and Technical Design

During this period the architect and engineer will develop the design and specification. The basis of single-stage competitive tendering is detailed bills of quantities and the quantity surveyor must decide at what point the information is of sufficient detail to let measurement commence, without having to become involved with re-measurement. The usual practice is to draw up a taking-off schedule indicating when information is required and when sections will be completed. It is a feature of this type of procurement that all the work must be measured and included in the bills of quantities prior to obtaining bids. During this period the quantity surveying should draw up a list of tenderers. The list should comprise three contractors who are to be approached to carry out the work. The list may be extended to six in the public sector, although it is increasingly difficult to have this number of competent contractors that are available at the same time. During the preparation of the bills of quantities the quantity surveyor should contact prospective firms that have the approval of the client and the architect, to determine whether they are available to bid for the project. In the first instance this is done by telephoning the chief estimator of a prospective contractor and giving brief details of the proposed project including the approximate value, date for dispatch of documents and the starting date. The decision on whether or not to tender for a project will be influenced by:

- workload
- future commitments
- market conditions
- capital
- risk
- prestige
- estimating workload
- timing – one of the decisions that will have to be taken by the design team is the length of the contract period, a critical calculation, as successful contractors will be liable to pay damages in the event of delays and non-completion.

If the contractor is interested and available for the project the enquiry is followed up with a letter giving the following information:

- name of client, architect and other lead consultants;
- name and type of project;
- location;
- approximate value;
- brief description;
- date for dispatch and return of tenders;
- start on site and contract duration;
- form of contract;
- particular conditions applying to the contract.

In an attempt to make the selection of contractors easier, key performance indicators or KPIs were introduced in 1999 by the government body the Centre for Constructing Excellence. Benchmarking asks who performs better, why they are performing better and how a company can improve. The broad categories that are benchmarked and measured are:

- building performance – e.g. environmental performance, design quality;
- project performance – e.g. time, cost, defects;
- organisational performance – e.g. health and safety, respect for people;
- relationship quality – e.g. customer satisfaction service.

Therefore, it is claimed that the KPIs can be used by clients or consultants when selecting contracting organisations that perform highly across these categories. The information can be accessed on line.

## Pre-construction, Stages F, G & H: Production Information, Tender Documentation and Tender Action

This is the stage that the bills of quantities are finalised, as in reality they will have been started as soon as information is available. The quantity surveyor during this stage begins to assemble the tender documents to be sent to the listed contractors. The tender documents are typically:

- Two copies of the bills of quantities, one bound and one unbound. The bound copy is for pricing and submission, the unbound copy is to allow the contractor to split the bills up into trades so that they can be sent to subcontractors for pricing.
- Indicative drawings on which the bills of quantities were prepared.
- The form of tender – a statement of the tenderer's bid.
- Instructions (precise time and place) and envelope for the return of the tender.

Only indicative drawings are sent out with the tender documents; however, details are also given where tenderers may inspect a full set of drawings, usually the architect's office. Details should also be included on access to the site.

- The usual time given to contractors for the return of tenders is four weeks although for particularly complex projects this may be six weeks. The tender documents contain precise details for the return of the tender, usually 12 noon at the selected date at the architect's office. Note that the system described above is the one used in England and Wales; in Scotland it is slightly different in so far as the tenderers submit a priced bill of quantities, not just a form of tender for consideration.
- It is common during the tender period for contractors to raise queries with the quantity surveyor on the tender documentation. If errors, omissions or other anomalies come to light, then the quantity surveyor, once the problem has been resolved, must communicate the result in writing, to all tenderers, so that they continue to work on the same basis.
- Completed tenders are submitted to the architect's office on or before the required deadline stated in the instructions. Any tenders that are submitted late should be discarded as it is thought they it may have been possible to gain an unfair advantage. In England and Wales a submitted tender may be withdrawn any time prior to acceptance.
- The tenders are opened and one selected; but what criteria should be used for selection? Traditionally, the lowest price was chosen on the basis that this provides the client with the best value for money, however, there is an increasing realisation that cheapest price may not provide clients with best value for money over the life-cycle of a building.
- One tender is chosen and another bid is selected as a reserve, both selected and reserve tenderers are informed with the first choice bidder being instructed to submit fully priced bills of quantities for checking as soon as possible, usually within forty-eight hours. The unsuccessful tenderers are told that they have not won the project. Bills of quantities are very comprehensive documents, containing thousands of individual prices and other data relating to a contract and it is now the responsibility of the quantity surveyor to check the bills and prepare a tender report for the client.

## TENDER EVALUATION

The tender evaluation should be treated as a confidential exercise. Once the quantity surveyor is handed the priced bills of quantities the due diligence and evaluation can begin. The checks that should be carried out are as follows:

- Arithmetical checks, confirm that prices have been extended correctly, that page totals are accurate and that page totals have been correctly carried forwarded to trade/element summaries and from there to the main summaries. It is not uncommon for extended bills rates to have errors, any errors should be noted. In the JCT Practice Note 6 there are two accepted ways to deal with errors in computation:
  - The tenderer should be given details of the errors and allowed the opportunity of confirming or withdrawing the offer. In other words the tenderer would proceed as if the error had not been made. If the tenderer withdraws the offer, the second reserve bills of quantities should be called for and examined. An amendment, signed by both parties to the contract should be added to the priced bills indicating that all measured rates are to be reduced or increased in the same proportion as the corrected total of priced items exceeds or falls short of such items.
  - In the second alternative the tenderer should be given the opportunity of confirming the offer or of amending it to correct genuine errors. Should the tenderer elect to amend the offer and the revised tender is no longer the lowest, the offer of the reserve tendered should be examined. If the errors are accepted by the tenderer, then the rates in question should be altered and initialled.
- Any items left unpriced should be queried with the tenderer.
- The general level of pricing should be examined and in particular:
  - Check that similar items that appear in different parts of the bills are priced consistently. For example, excavation items that appear in substructure, external works and drainage.
  - Items that are marked *provisional* and are therefore to be re-measured during the course of the works have been priced at rates that are consistent with the rest of the works.
  - Check for caveats inserted by the contractor. For example, 'Removal and disposal of all protective materials has not been included in the prices'.

Once these checks have been carried out and the tender is free of errors then the quantity surveyor can recommend acceptance.

The advantages of single-stage competitive tendering are:

- it is well known and trusted by the industry;
- it ensures competitive fairness;
- for the public sector, it allows audit and accountability to be carried out;
- it is a valuable post-contract tool that makes the valuation of variations and the preparation of interim payments easier.

The disadvantages are:

- a slow sequential process;
- no contractor or special involvement;
- pricing can be manipulated by tenderers;
- expensive.

### Risk allocation

When using single-stage tendering, risk is transferred for a large part to the client. The contractor has no responsibility for design and follows the design teams' instructions. Where tenders are based on incomplete design, the bids can only be considered as indicative of the final cost although the client is not often made aware of this and is vulnerable to additional works and costs.

### Bills of reduction

The situation can occur where the tenders exceed the amount of the client's budget. Under these circumstances there are two approaches: to abandon the scheme, or to negotiate a reduction with the preferred tenderer. This is best done with the preparation of a bill of reductions, that is, a bill of quantities detailing how the cost is to be reduced, perhaps with reduced quantities of the substitution of a revised specification.

A modern variation on the traditional single-stage selective tendering theme is described below.

- A fully developed business case is subjected to rigorous cost planning by consultants in conjunction with the client.
- Suitable contractors, say two or three, are pre-selected using appropriate selection criteria, such as:
  - commitment to supply chain management
  - the ability to guarantee life-cycle costs
  - capability to deliver the project
  - etc.
- The pre-selected contractors, are asked to fully cost the project proposals and submit their best and final offer (BFO).
- On the basis of the BFO a contractor is selected and enters into a contract guaranteeing the BFO price and in some cases whole-life costs for a predetermined period.
- Work starts on site.

In the first part of the twenty-first century, traditional lump sum procurement, based on bills of quantities, still accounts for approximately 40% of all

UK construction activity. The following section is a review of some traditional procurement strategies plus design and build:

- two-stage competitive tendering;
- design and build and variants.

## TWO-STAGE COMPETITIVE TENDERING

This was first used widely in the 1970s and is based on the traditional single-stage competitive tendering, i.e. bills of quantities and drawings being used to obtain a lump sum bid. Advantages include early contractor involvement, a fusion of the design/procurement/construction phases and a degree of parallel working that reduces the total procurement and delivery time. A further advantage is that documentation is based upon bills of quantities and therefore should be familiar to all concerned. Early price certainty is ruled out, as the client can be vulnerable to any changes in level in the contractor's pricing between the first and second stages.

### Stage 1

The first stage tender is usually based on approximate bills of quantities; however, this does not have to be the case. Other forms of first stage evaluation may be used, for example a schedule of rates, although there is perhaps a greater degree of risk associated with this approach. As drawn information, both architectural and structural, is very limited, the choice of bid documentation will be influenced by the perceived complexity and predictability of the proposed project. The two-stage approach places pressure on designers to take decisions concerning major elements of the project at an earlier stage than normal. A common feature with many non-traditional procurement systems, including 'design and constructer' and Public–Private Partnerships, is that the design development period is truncated and that as a result, the design that is eventually produced can lack architectural merit. Assuming that bills of approximate quantities are being used they should contain quantities that reflect:

- items measured from outline drawings;
- items that reflect the trades it is perceived will form part of the developed design;
- items that could be utilised during the pricing at second stage.

At an agreed point during the preparation of the first stage documentation the design for first stage is frozen thereby enabling the approximate bills of quantities to be prepared. Without this cut-off point the stage one documentation could not be prepared. Note that it is important to keep a register of which drawing revisions have been used to prepare the stage one documentation for later reference during the preparation of the firm stage 2 bills of quantities. However, while first stage documentation is prepared the second stage design development can continue. During the first stage tender period attention should be focused on the substructure, as it is advantageous if at the first tender stage this element is firm. If a contractor is selected as a result of the first stage tender then they may well be able to start on site to work on the substructure while the remainder of the project is detailed and the second stage bills of quantities are prepared and priced.

Once completed the first stage approximate bills of quantities, together with other documentation, are despatched to selected contractors with instructions for completion and return in accordance with normal competitive tender practice. On their return one contractor is selected to proceed to the next phase. It should be noted that selection at this stage does not automatically guarantee the successful first stage bidder award of the project; this is dependent on the stage two bidding process.

At this point the trust stakes are raised – assuming that a contractor is selected as a result of the stage one tender the following scenarios could apply:

| **Client** | **Contractor** |
|---|---|
| The client trusts the contractor to be fair and honest during the stage 2 negotiations. Failure could result in the client having to go back to the start of the process. | Although selected by stage one process – no guarantee of work. No knowledge as to accuracy of first stage documentation or client's commitment to continue. |
| Client could ask contractor to join design team to assure buildability. | Contractor could be asked to start on site on substructure while stage 2 is progressed. |
| Client relies on the design team to prepare documentation for stage two timeously. If stage 2 bills of quantities are not accurate the work will have to be remeasured for a third time! | Contractor rewarded on the basis of letters of intent, quantum meruit, etc. |
| The client could, under a separate contract, engage a contractor to carry out site clearance works while stage one bids are evaluated. | Contractor could exploit position during stage 2 pricing. |

### Stage 2

The purpose of stage 2 is to convert the outline information produced during stage 1 into the basis of a firm contract between the client and the contractor, as soon as possible. With a contractor selected as a result of the first stage process pressure is placed on the design team to progress and finalise the design. Between contractor selection and stage 2, usually a matter of weeks, the design team should prepare and price the second stage bills of quantities. During this phase it is usually the quantity surveyor who is in the driving seat and he/she should issue information production schedules to the rest of the design team. As design work on elements is completed it is passed to the quantity surveyor to prepare firm bills of quantities which are used to negotiate the second stage price with the contractor on a trade-by-trade or elemental basis. It is therefore quite possible that the contractor will be established on site before the stage 2 price is fully agreed. Unless the parameters of the project have altered greatly there should be no significant difference between the stage 1 and stage 2 prices. Once a price is agreed a contract can be signed and the project reverts to the normal single-stage lump sum contract based on firm bills of quantities; however, the adoption of parallel working during the procurement phase ensures that work can start on site much earlier than the traditional approach. Also the early inclusion of the main contractor in the design team ensures baked-in buildability and rapid progress on site.

### Critical success factors

- Trust between the parties.
- The appointment of a contractor that can innovate and is prepared to contribute to the design development.
- Information production keeping up with requirements; failure to achieve this can lead to a good deal of embarrassment.

The UNCITRAL Model Law referred to previously suggests the following approach to two-stage tendering:

- The first stage is used to ask suppliers or contractors to submit their technical proposals but without a tender price.
- The client enters into discussion with the supplier/contractor and those who appear to fall short of the criteria are discarded. At the same time the client, draws upon the proposals of the first stage tenderers to prepare the basis for the stage 2 documents.

The remaining contractors are then asked to submit tenders based on the revised documentation.

## DESIGN AND BUILD AND VARIANTS

Design and build (or design and construct) is a generic term for a number of procurement strategies where the contractor both designs and carries out the works. This approach is extensively used in France where both contractors and private practices are geared up to provide this service to clients. In the UK this approach has only become common during the last thirty years or so. The various forms of design and build are as follows.

### Traditional design and build (D&B)

The contractor is responsible for the complete design and construction of the project. Design and build is one of the procurement systems currently favoured by many public sector agencies. The use of D&B variants in all sectors has increased from 11% in 1990 to 40% in 2000, the reasons are:

- D&B gives a client the opportunity to integrate, from the outset, the design and the construction of the project.
- The client enters into a single contract with one company, usually a contractor who has the opportunity to design and plan the project in such a way as to ensure that buildablility is baked into the design.
- With specialist involvement from the start, this approach promises a shorter overall delivery time and better cost certainty than traditional approaches.
- The results of the studies indicated that D&B outperforms traditional forms of procurement in several respects; however, the differences are not that significant.
- One reason for this could be that, within the UK, for the organisations that provide D&B services, this is not their key competence and therefore when the opportunity comes to bid for a D&B project, temporary organisations of designers and constructors have to be formed specifically for a project. For the contractor and the designers the next project may be traditional contracting and therefore the temporary organisation is disbanded.
- Studies conclude that although delivery times are shorter when using D&B that improvements in cost certainty are only marginal.
- The total delivery speed of D&B compared with traditional approaches is 30–33% faster.
- The percentage of projects that exceeded the original estimate by more than 5% was 21% in B&D compared to 32% for traditional procurement.

- D&B is recommended by the Office of Government Commerce for procurement within a partnership arrangement.

The main criticisms of D&B procurement are centred around the lack of control over quality of design, with little time being allocated for design development and possible compromises over quality to provide cost savings by the contractor. It is possible for the client to employ independent professional advice to oversee a design and build contract.

Successful use of D&B relies on the contractor preparing proposals that include:

- a contract sum analysis that itemises the financial detail on an elemental basis;
- detailed proposals of how the requirements of the client's brief will satisfied.

Tender appraisal can be more difficult when using design and build as a means to decide which tender to accept as it does not only depend on the prices submitted but also the quality of the design and the delivery time. Design and build procurement is organised in exactly the same way as single-stage lump sum procurement, with the drawing up of a short list and bids being submitted to a strict timetable. Other variants of design and build are:

- Enhanced design and build – the contractor is responsible for the design development, working details as well as construction of the project.
- Novated design and build – the contractor is responsible for the design development, working details and supervising the subcontractors, with assignment/novation of the design consultants from the client. This means that the contractor uses the client's design as the basis for their bid.
- Package deal and turnkey – the contractor provides standard buildings or system buildings that are in some cases adapted to suit the client's space and functional requirements.

## MANAGEMENT PROCUREMENT

During the 1970s, and particularly the 1980s, commercial clients and property developers started to demand that projects were procured more quickly than had been the case with single-stage selective tendering. The three main management systems are:

1. management contracting
2. construction management
3. design and manage.

With fast-track methods the bidding and construction phases are able to commence before the design is completed and there is a degree of parallel working as the project progresses. This is obviously high risk as the whole picture is often unknown at the time the works commence on site. This risk is exacerbated when this strategy is used for particularly complex projects of reburbishment contracts.

### Management contracting

Management contracting is not only popular with developers, as projects are delivered more quickly, but also contractors, as the amount exposure to risk for them is substantially lower than other forms of procurement. This is because a management contractor only commits the management expertise to the project, leaving the actual construction works to others. Management contracting was first widely used in the 1970s and was one of the first so-called fast-track methods of procurement that attempted to shorten the time taken for the procurement process. This is achieved by running stages such as design and construction concurrently as illustrated in Figure 4.3. Procurement is as follows:

- Selection of a management contractor – as the management contractors' role is purely to manage, it is not appropriate to appoint a contractor using a bill of quantities. Selection therefore is based on the service

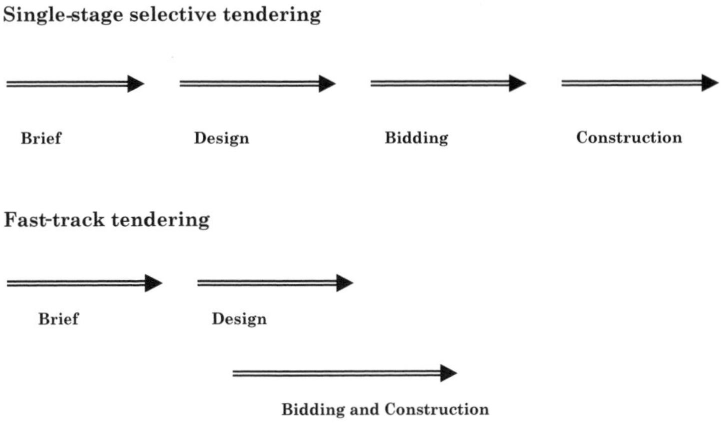

**Single-stage selective tendering**

**Brief** **Design** **Bidding** **Construction**

**Fast-track tendering**

**Brief** **Design**

**Bidding and Construction**

**Figure 4.3** Traditional and fast-track procurement compared

level to be provided, the submission of a method statement and the management fee, expressed as a percentage, of the contract sum. This can be done on a competitive basis. As the management contractor's fee is based on the final contract sum, there is little incentive to exert prudence.

- The management contractor only bids to supply management expertise, although sometimes they can also supply the labour to carry out the so-called 'builder's work in connection with services', as this work package can be difficult to organise.
- The project is divided into work packages – typically this is between twenty and thirty, for example, ground works, concrete work, windows and external doors, suspended ceilings, etc. The packages are entered into a schedule along with the following information:

| Package | To tender | Information from architect | Estimated cost | Start on site |
|---|---|---|---|---|
| Concrete work | 21/09/12 | 01/08/12 | £356,000 | 15/10/12 |

The work packages are in effect a series of mini bills of quantities, produced in accordance with NRM2 and therefore the time allowed for pricing an individual package can be reduced to around two weeks. The procedure for asking for bids is the same for single-stage selective tendering.

- At the same time the cost of the project must be determined and therefore an estimated prime cost is established for each package. For the quantity surveyor, this can be problematic as the design is incomplete and therefore estimates of costs tend to be detailed cost plans, agreed with the management contractor.
- Package by package the works are sent to tender: when a contractor is appointed based on the mini bills of quantities, the contractor enters into a contract directly with the client. Therefore, at the end of the process there are twenty or thirty separate contracts between the management contractor and the work package contractors; there is no contract between the client and the work package contractors.
- The management contractor role therefore is to co-ordinate the work packages on site and to integrate the expertise of the client's consultants.
- Payment is made to the management contractor on a monthly basis, who in turn pays the package contractors in accordance with the valuation.
- The advantages are: work can start on site before the design work is complete, earlier delivery of project and return on client's investment, the client has a direct link with package contractors.

- In order to provide a degree of protection for the client a series of collateral warranties can be put in place.
- The disadvantages are: high risk for client, firm price is not known until final package is let, difficult for the quantity surveyor to control costs and any delay in the production of information by the design team can have disastrous consequences on the overall project completion.
- A distinct JCT form of contract exists for management contracts.

### Construction or contract management

This approach is similar to management contracting in so far as the project is divided into packages, however the construction manager adopts a consultant's role with direct responsibility to the client for the overall management of the construction project, including liaising with other consultants. Construction managers are appointed at an early stage in the process and as with management contracting reimbursement is by way of a pre-agreed fee. Each work package contractor has a direct contract with the client, this being the main distinction between the above two strategies.

### Design and manage

When this strategy is adopted a single organisation is appointed to both design the project as well as managing the project using work packages. It is an attempt to combine the best of design and build and management systems. The characteristics are:

- A single organisation both designs and manages.
- The design and management organisation can be either a contractor or a consultant.
- Work is let in packages with contracts between the contractor or client, dependent on the model adopted.
- Reimbursement is by way of an agreed fee.

### Cost reimbursement contracts

#### *Cost plus*

Cost-plus contracts are best used for uncomplicated, repetitive projects such as roads contracts. The system works as follows:

- The contractor is reimbursed on the basis of the prime cost of carrying out the works, plus and an agreed cost to cover overhead and profit. This can be done by the contractor submitting detailed accounts for labour, materials and plant that are checked by the quantity surveyor. Once agreed the contractors costs are added.

- There is no tender sum or estimate.
- The greater the cost of the project, the greater the contractor's profit.

### *Target cost*

A variant of cost-plus contracts, this strategy incentivises the contractor by offering a bonus for completing the contract below the agreed target cost. Conversely damages may be applied if the target is exceeded.

### *Term contracts/schedule of rates*

Suitable for low-value repetitive works that occur on an irregular basis. Contractors are invited to submit prices for carrying out a range of items based on a schedule of rates. Contractors are required to quote a percentage addition on the schedule rates. Used extensively for maintenance and repair works.

### *Negotiated contracts*

This strategy involves negotiating a price with a chosen contractor or contractors, without the competition of the other methods. Generally regarded by some as a strategy of the last resort and an approach that will almost always result in a higher price than competitive tendering it has the following advantages:

- an earlier start on site than other strategies;
- the opportunity to get the contractor involved at an early stage.

The contractors selected for this approach should be reputable organisations with a proven track record and the appropriate management expertise.

## PARTNERING

Partnering is viewed to be a radical departure from the traditional approach not only to construction procurement, but also the management of supply chain relationships. The term partnering in connection with procurement emerged in the USA in the early 1980s as a technique for reducing the costs of commercial contract disputes by improving communications, process issues and relationships as the principal focus. Partnering relies on co-operation and teamwork, openness and honesty, trust, equity and equality between the various members of the supply chain.

Partnering is:

- a process whereby the parties to a traditional risk transfer form of contract, i.e. the client, the contractor and the supply chain, commit to work together with enhanced communications, in a spirit of mutual trust and respect towards the achievement of shared objectives;
- a structured management approach to facilitate teamwork across contractual boundaries that helps people to work together effectively in order to satisfy their organisation's (and perhaps their own) objectives;
- a means of avoiding risks and conflict. There isn't one model partnering arrangement; it is an approach that is essentially flexible, and needs to be tailored to suit specific circumstances;
- a model that enables organisations to develop collaborative relationships either for one-off projects (project-specific) or as long-term associations (strategic partnering);
- a process that is formalised within a relationship that might be defined within a charter or a contractual agreement.

Many major clients across all sectors have been adopting partnering in response to the proven long-term benefits that can be achieved through this approach and have reported to have reached savings of up to 40% on costs and 70% on time by using partnering approaches. There is, however, evidence that small, occasional clients have little to gain from the process.

For contractors the continuity of working repeatedly for the same clients is also thought to provide a number of benefits for contracting organisations, although whether a contractor's profit margin increases on a partnering project is unclear. Both supply and demand sides agreed that partnering provided a more rewarding environment in which to operate.

### Cultural issues

Cultural realignment for an industry is difficult to achieve. The approach generally adopted by organisations is to organise partnering workshops during which the approach is explained. This includes:

- the establishment of **trust** – a vital element of the partnering process;
- emphasising the importance of **good communications** between partnering parties;
- explaining that many activities are carried out by **teamwork** and how this is achieved in practice;
- the establishment of a **win–win** approach.

- breaking down traditional management structures and **empowering** individuals to work in new ways.

## Commercial issues

Organisations wishing to be involved in partnering arrangements should be able to demonstrate capabilities in the following:

- Establishing a **target (base) cost** that is based on open-book accounting and incorporating pain/gain schemes. Profit margins should be established and ring-fenced.
- A track record of using **value engineering** to produce best value solution.
- A track record of using **risk management** to identify and understand the possible impact of risk on the project.
- The ability to successfully utilise **benchmarking** to identify how to measure and improve performance.
- The willingness to take a radical look at the way they manage their business and to engage in **business process re-engineering** if necessary to eliminate waste and duplication.

## Key success factors

Some of the following are desirable for project partnering; all are essential for successful strategic partnering:

- There needs to be a commitment at all levels within an organisation to make the project or programme of work a success, which means a commitment to working together with others to ensure a successful outcome for all participants (win–win situation).
- Partners must have confidence in each other's organisations, and each organisation needs to have confidence in its own team, which means careful selection of the people involved.
- Partners should be chosen on the basis of the ability to offer best value for money and not on lowest price; their ability to innovate and offer effective solutions should also be considered.

  Clients should normally select their partners from competitive bids based on carefully set criteria aimed at getting best value for money. This initial competition should have an open and known prequalification system for bidders.
- Partners need collectively to agree the objectives of the arrangement/project/programme of work and ensure alignment/compatibility of goals.

This will require early involvement of the entire team to ensure a win–win situation for all. The agenda must be mutual interest with a focus on the customer; it must therefore be quality/value driven.
- To satisfy the relationship's agenda, there needs to be clarity from the client and continued client involvement. It is essential to define clearly the responsibilities of all participants within an integrated process.
- There needs to be a willingness to be flexible and adopt new ideas and different ways of doing things – e.g. different operational methodologies, different administrative procedures, different payment methods, different payment procedures, etc.
- All players should share in success in line with their contribution to the value added process (which will often be difficult to assess). There also needs to be a sharing of information, which requires open-book accounting and open, flexible communication between organisations/teams/people.
- Responsibility for risks must be allocated clearly and fairly, but there must be a collective responsibility for problems and an openness and willingness to accept and share mistakes. Adoption of such openness and sharing requires trust.
- It is important that all partnering arrangements incorporate effective methods of measuring performance. It has been identified that partnering should strive for continuous improvement, and this must be measurable to ascertain whether or not the process is effective.
- It is therefore important that agreed non-adversarial conflict resolution procedures are in place to resolve problems within the relationship. The principle of trying to resolve disputes at the lowest possible level should normally be adopted to save time and cost.
- Education and training is needed to ensure an understanding of partnering philosophy. It is important that, regardless of how well-versed participants are in the philosophy and procedures, teambuilding takes place at commencement of the relationship.

Trust is generally regarded as being crucial to the success of partnering; indeed it has been described as the cornerstone of a successful partnering relationship. Blois ('Trust in business to business relationships: an evaluation of its status', *Journal of Management Studies*, 36, 197–215, 1999) argues that only individuals can trust, and consequently trust between organisations (which are, after all, only collections of people) means a lot of people needing to trust a lot of other people. This makes relationships vulnerable to human fickleness. Careful selection of the correct individuals by each organisation is therefore crucial to success.

## ALLIANCING

The terms alliancing and partnering do not have the same legal connotations as partnership or joint venture and therefore there has been a tendency, particularly in the construction industry, to apply them rather loosely to a whole range of situations, many of which clearly have nothing to do with the true ethos of partnering or alliances, which is to be regretted. As with partnering, alliancing can be categorised as follows:

> Strategic alliances can be described as two or more firms that collaborate to pursue mutually compatible goals that would be difficult to achieve alone. The firms remain independent following the formation of the alliance. Alliancing should not be confused with mergers or acquisitions.

A project alliance is where a client forms an alliance with one or more service providers: designers, contractors, suppliers, etc. for a specific project and this section will continue to concentrate on this aspect of alliancing.

The principal features of a project alliance are as follows:

- The project is governed by a project alliance board that is composed from all parties to the alliance that have equal representation on the board. One outcome of this is that the client has to divulge to the other board members far more information than would, under other forms of procurement, be deemed to be prudent.
- The day-to-day management of the project is handled by an integrated project management team drawn from the expertise within the various parties on the basis of the best person for the job.
- There is a commitment to settle disputes without recourse to litigation except in the circumstance of wilful default.
- Reimbursement to the non-client parties is by way of 100% open-book accounting based on:

1. *100% of expenditure including project overheads*
   Each non-client participant is reimbursed the actual costs incurred on the project, including costs associated with reworks. However, reimbursement under this heading must not include any hidden contributions to corporate overheads or profit. All project transactions and costings are 100% open book and subject to audit.
2. *A fixed lump sum to cover corporate overheads and a fee to cover profit margin*
   This is the fee for providing services to the alliance, usually shown as a percentage based on 'business as usual'. The fee should represent the normal return for providing the particular service.

3. *Pain/gain mechanism with pre-agreed targets*
   The incentive to generate the best project results lies in the concept of reward, which is performance based.

A fundamental principle of alliances is the acceptance on the part of all the members of a share of losses, should they arise, as well as a share in rewards of the project. Risk–reward should be linked to project outcomes which add to or detract from, the value to the client. In practice there will be a limit to the losses that any of the alliance members, other than the client, will be willing to accept, if the project turns out badly. Unless there are good reasons to the contrary it may be expected that the alliance will take 50% of the risk and the owner/client the remaining 50%. The major differences between alliancing and project partnering are shown in Table 4.2.

For example, in project partnering one supplier may sink or swim without necessarily affecting the business position of the other suppliers. Therefore given the operational criteria of an alliance it is vitally important that members of the alliance are selected against rigorous criteria. These criteria, usually demonstrated by reference to previous projects undertaken by the prospective alliance members, vary from project to project but as a guide could be the ability to:

- complete the full scope of works being undertaken from the technical, financial and managerial perspectives;
- re-engineer project capital and operating costs without sacrificing quality;

**Table 4.2** Replacement expenditure profile

| | Partnering | Alliances |
|---|---|---|
| **The form of the undertaking** | Core group with no legal responsibilities. Binding/non-binding charters used in 65% of partnering arrangements. | Quasi joint venture operating at one level as a single company. |
| **The selection process** | Prime contractor responsible for choice of supply chain partners. | Rigorous selection process. |
| | Project can commence while selection continues. | Alliance agreement not concluded until all members appointed. |
| **The management structure** | By prime contractor. Partnering advisor. | Alliance board. |
| **Risk and reward mechanisms** | Partners losses not shared by other members of the supply chain. | Losses by one alliance member shared by other members. |

- achieve outstanding quality with an outstanding track record;
- innovate and deliver outstanding design and construction outcomes;
- demonstrate safety performance;
- demonstrate conversance with sustainability issues;
- work as a member of an alliance with a commitment to non-adversarial culture and change direction quickly if required;
- have a joint view on what the risks are and how they will be managed with reasonable provision made within the target cost for such items.

The circumstances under which variations arise are limited. Generally, normal changes due to design development are not considered. Changes that could give rise to variations are:

- significant increases or decreases in the scope of work, e.g. adding in new buildings, parts of buildings or facilities;
- fundamental changes in the performance parameters.

## PRIME CONTRACTING

Introduced in the 1990s, prime contracting is a long-term contracting relationship based on partnering principals and is currently being used by several large public sector agencies. A prime contractor is defined as an entity that has the complete responsibility for the delivery and, in some cases, the operation of a built asset and may be either a contractor, in the generally excepted meaning of the term, or a firm of consultants. The prime contractor needs to be an organisation with the ability to bring together all of the parties in the supply chain necessary to meet the client's requirements. There is nothing to prevent a designer, facilities manager, financier or other organisation from acting as a prime contractor. However, by their nature prime contractors tend to have access to an integrated supply chain with substantial resources and skills such as project management. To date, most prime contractors are in fact large firms of contractors, despite the concerted efforts of many agencies to emphasis the point that this role is not restricted to traditional perceptions of contracting. One of the chief advantages for public sector clients with a vast portfolio of built assets is that prime contracting offers one point of contact/responsibility, instead of a client having to engage separately with a range of different specialists.

As with other forms of procurement based on a long-term partnership, the objective of prime contracting is to achieve better long-term value for money through a number of initiatives such as supply chain management, incentivised payment mechanisms, continuous improvement, economies of scale and partnering. The approach to prime contracting differs from PFI because the prime contractors' obligations are usually limited to the design and construction of the built asset and the subsequent facilities management; there is no service delivery involved. Also, the funding aspects of this approach are much less significant for the prime contractor, in that finance is provided by the public sector client. In some models in current use, prime contractors take responsibility not only for the technical aspects of a project during the construction phase, including design and supply chain management but also for the day-to-day running and management of the project once completed. This may include a contractual liability for the prime contractor to guarantee the whole-life costs of a project over a pre-determined period for as much as thirty years.

The key features of prime contracting are:

- fewer larger long-term contracts
- shared risk
- partnering in the supply chain
- incentivisation of private sector contractors.

The key commercial issue surrounding prime contracting is setting up long-term relationships based on improving the value of what the supply chain delivers, improving quality and reducing underlying costs through taking out waste and inefficiency. It is claimed that the products and services provided by the companies in the supply chain typically account for 90% of the total cost of a construction project. The performance of the whole supply chain impacts on the way in which the completed building meets the client's expectations. By establishing long-term relationships with supply chain members it is believed that the performance of built assets will be improved through:

- the establishment of improved and more collaborative ways of working together to optimise the construction process;
- using the latest innovations and expertise.

The prime contractor's responsibilities might include the following:

- overall planning, programming and progressing of the work;
- overall management of the work, including risk management;
- design co-ordination, configuration control and overall system engineering and testing;
- the pricing, placing and administration of suitable subcontractors;
- systems integration and delivering the overall requirements.

## FRAMEWORKS

Framework agreements are being increasingly used to procure goods and services in both the private and public sectors. Frameworks have been used for some years on supplies contracts; however, in respect of works and services contracts, the key problem, particularly in the public sector, has been a lack of understanding as to how to use frameworks, whilst still complying with legislation, particularly the EU Directives and the need to include an 'economic test' as part of the process for selection and appointment to the framework. In the private sector, BAA was the first big player to use framework agreements which covered everything from quantity surveyors to architects and small works contractors. The EU public procurement Directives define a framework as: 'An agreement between one or more contracting authorities and one or more economic operators, the purpose of which is to establish the terms governing contracts to be awarded during a given period, in particular with regard to price and, where appropriate, the quality envisaged.'

A framework agreement therefore is a flexible procurement arrangement between parties, which states that works, services or supplies of a specific nature will be undertaken or provided in accordance with agreed terms and conditions, when selected or 'called-off' for a particular need. The maximum duration of a framework under current EC rules is four years and can be used for the procurement of services and works. An important characteristic of framework agreements is that inclusion in a framework is simply a promise and not a guarantee of work. Entering into such a framework, however rigorous and costly the selection process, is not entering into a contract, as contracts will only be offered to the framework contractors, supply chains, consultants or suppliers as and when a 'call-off' is awarded under the agreement. Interestingly, in November 2004, approximately one year after the roll-out of ProCure21 Framework Programme on a national basis, the first signs of discontent amongst the twelve framework consortia began to emerge. Contractors complained about lack of work and some even threatened to leave the framework.

The framework establishes the terms and conditions that will apply to subsequent contracts but does not create rights and obligations. The major advantages of framework agreements are that:

- they form a flexible procurement tool;
- they avoid repetition when procuring similar items;
- they establish long-term relationships and partnerships;
- whenever a specific contract call-off is to be awarded, the public body may simply go to the framework contractor that is offering the best value for money for their particular need;
- they reduce procurement time/costs for client and industry on specific schemes;

- they enable engagement with the supply chain early in the procurement process when most value can be added.

## PUBLIC–PRIVATE PARTNERSHIPS (PPP)

The procurement process for PPP projects varies according to the model that is used.

- Prior to 1989, British governments were not keen to allow private capital in the financing of public sector projects. In the UK the position was set out in the so-called Ryrie Rules.
- The Rules were revised in February 1988 to take account of the privatisation of the previously nationalised industries and the introduction of schemes such as contracting out, opting out, mixed funding and partnership schemes.
- The objective of the Ryrie Rules was to stop ministers from insulating private finance from risk so that it could be used to circumvent public expenditure constraints.
- The Ryrie Rules were formally retired in 1989. Subsequently, the Treasury promoted private finance as additional and not just substitutional; the Private Finance Initiative was launched in 1992 and revamped as Public–Private Partnerships (PPPs) by the Labour government in 1998.
- To many, PPPs were seen as a natural progression to the programme of privatisation that was undertaken in the UK during the 1980s and 1990s.

Public–Private Partnerships in the UK have developed and continue to be developed in many forms to suit the needs of particular sectors, e.g. education, health, etc., and in some cases subsectors, e.g. primary health care. The principal PPP models currently used in the UK construction sector are listed below and will be discussed in the section that follows:

- the Private Finance Initiative (PFI)
- Building Schools for the Future (BSF)
- NHS Local Improvement Finance Trust (LIFT)
- Frameworks
- Proure21+
- PRIME
- Public–Private Partnership Programme (4Ps)
- leasing
- concessions and franchises.

However, in most cases, in PPP arrangements private sector contractors become the long-term providers of services rather than simply upfront asset builders, combining some or all of the responsibilities for the:

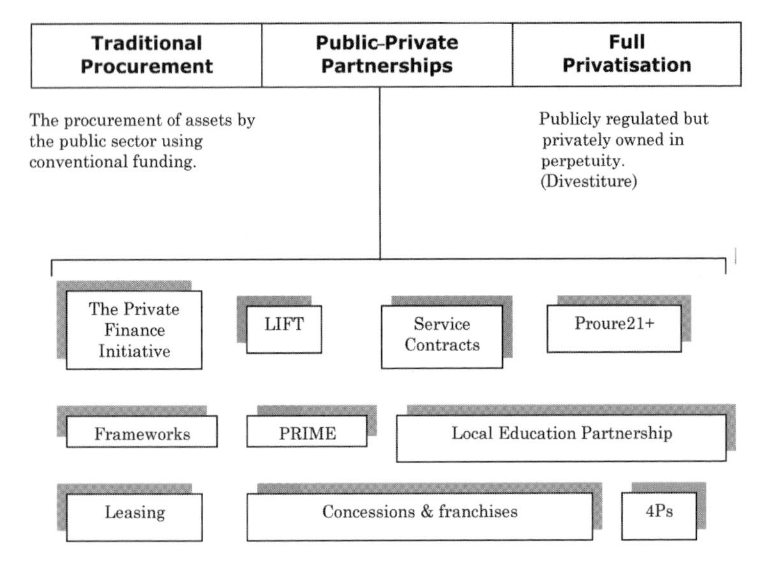


**Figure 4.4** PPP models

- design
- construction
- finance (which may be a mixture of public and private sources)
- facilities management, and
- service delivery

of a public service facility.

## The procurement process

In general a PPP project cycle can be broken down into stages as shown in Figure 4.5.

## The Private Finance Initiative (PFI)

In the UK during the last fifteen years or so, three main PFI procurement models have developed. These are:

1. joint ventures
2. financially free-standing projects
3. classic PFI.

### *Joint ventures*

Joint ventures are projects to which both the public and private sectors contribute, but where the private sector has overall control. The project as a whole must make economic sense and competing uses of the resources must be considered. The main requirements for joint venture projects are:

- private sector partners in a joint venture should be chosen through competition;
- control of the joint venture should rest with the private sector;
- the government's contribution should be clearly defined and limited. After taking this into account, costs will need to be recouped from users or customers; and
- the allocation of risk and reward will need to be clearly defined and agreed in advance, with private sector returns genuinely subject to risk.

The government's contribution can take a number of forms, such as concessionary loans, equity transfer of existing assets, ancillary or associated works, or some combination of these. If there is a government equity stake, it will not be a controlling one. The government may also contribute in terms of initial planning regulations or straight grants or subsidies.

### *Financially free-standing projects*

The private sector undertakes a project on the basis that costs will be recovered entirely through a charge for the services to the final user, for example the

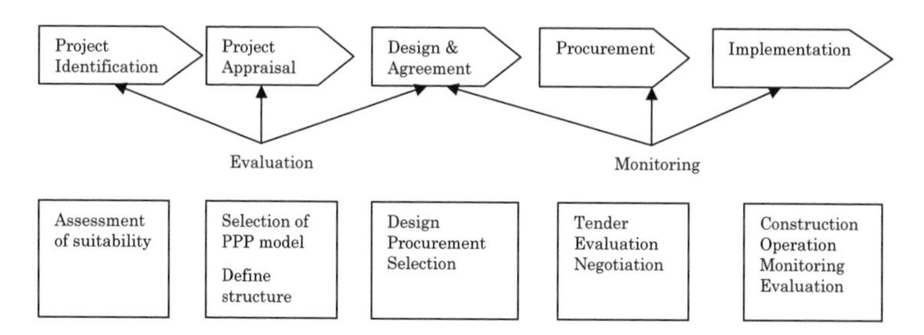


**Figure 4.5**

Queen Elizabeth II Bridge in Kent/Essex, where the toll charges go directly to the company that constructed and is running the bridge. The characteristics of this approach are:

- Government may contribute value to the project in terms of initial planning and statutory procedures, or determining the route of a linking road, etc.
- At the end of the concession period the ownership of the asset may be handed to the public sector.

### *Classic PFI*

The Private Finance Initiative (PFI) is the widest used, most controversial and best known form of PPP, accounting at its height for approximately 80% of all expenditure on PPPs in the UK construction sector.

Classic PFI is characterised as follows:

- bids are submitted by consortia to:
  - raise the finance
  - design
  - build
  - operate and maintain for a period of thirty years plus.

  During the currency of the contract the consortia will receive an annual payment, known as a unitary charge, providing that agreed performance standards are met. At the end of the contract the facility is handed back to the public sector in a good state of repair.

- Design Build Finance and Operate (DBFO) is the classic and perhaps most widely used PFI model, with a contract structure usually similar to the one illustrated in Figure 4.6.

In order to obtain approval early PFI projects were required to demonstrate two principal advantages over conventional public procurement strategies:

1. Value for money compared with traditional service provision. The mechanism used to demonstrate value for money is referred to as the Public Sector Comparator (PSC), a model still used that purports to show in black and white balance sheet terms, which of the two approaches (the PFI or traditional lump sum procurement) delivers the better value. Fortunately the PSC is no longer used as the sole litmus test for value for money

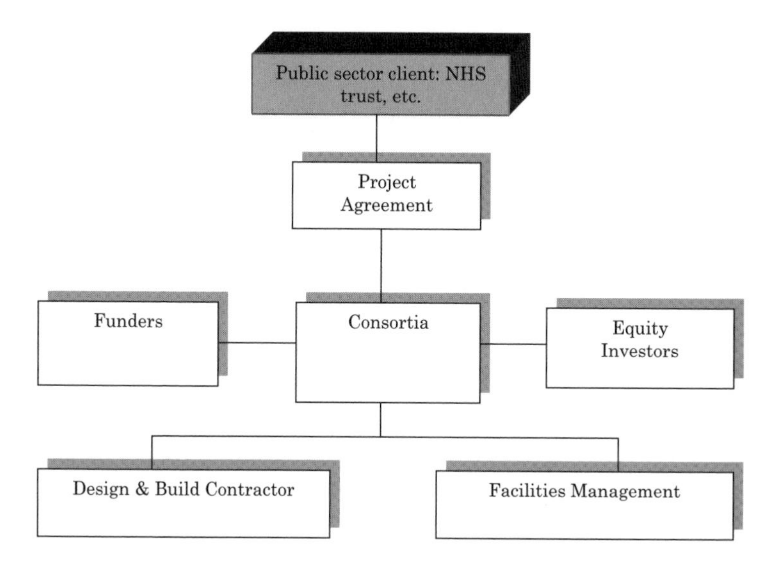


**Figure 4.6** PFI contract structure

and it is now recommended by the Office of Government Commerce that a more holistic approach is adopted when assessing which procurement path to select.

2. One of the basic premises of PPPs is that the private sector is better able to manage public sector facilities because of its superior management expertise and experience. The second important hurdle for PFI projects to clear is a demonstration that significant risk has been transferred from the public to the private sector operator (i.e. without a guarantee by the taxpayer against loss).

In addition to value for money and risk transfer other characteristics of PFI deals such as DBFO are:

- selection based on competition on the net present values of the unitary payment;
- an output-based specification rather than the traditional prescriptive model;
- a long-term contract, usually 30 years minimum;
- performance-related payments;
- task integration;
- operation of completed facility.

### *Unitary payment*

Unlike conventional procurement strategies where bids are calculated and assessed on the basis of the current capital cost of building a school or hospital, PFI projects are assessed on the proposed charge of a consortia to finance, build and operate a facility over the duration of the contract. Because of the long-term nature of the projects all costs likely to be incurred over the lifecycle of the PFI project have to be reduced to net present values in order that meaningful comparisons may be made.

### *Output-based specification*

Unlike conventional procurement where design and construction is based on a set of prescriptive documentation, in the case of a PFI deal it is the role of the consortia to interpret and comply with a series of outputs specified by the client. The consortia will have the responsibility for the design of the new facility therefore it is essential that the outputs are described in such a manner that they:

- comply with all existing regulations/sector benchmarks/etc.;
- allow the consortia the flexibility, within the constraints of the above, to develop solutions that are innovative. In particular the consortia should have the opportunity to identify sources of additional income stream from within the project, provided of course that the main service delivery is not compromised;
- safeguard the end users against suboptimal service delivery.

### *Long-term contracting*

Historically construction projects have been undertaken by what have been referred to as 'temporary multi-organisations'. That is to say organisations: contractors, clients, property professionals, etc. that come together for a specific project with a specified project period and them on completion of the project and hand over to the client the individual organisations go their own separate ways. With this approach it is entirely possible that the same team of organisations and individuals may never work together again. However, the length of the contractual relationship in PPP projects is such that during its currency there may well be several changes of government, wars, etc. This therefore marks a major cultural shift in the way in which built assets are procured and managed over their lifetime from short termism to long termism or to quote and paraphrase Sir Denys Linton, 'an approach that is based upon partnership rather than merely a hope of completing the contract unscathed'.

This approach has to be based on, among other things, mutual trust, transparency, alignment of aims and objectives and a desire to deliver high-quality public services in new and innovative ways.

### *Performance-related payments*

One of the main attractions of PPP/PFI procurement for the public sector is that it gives the procuring body the opportunity to incentivise the private sector contractor to deliver large and prestigious public projects to time and to budget as the public sector will not begin to pay for the assets until such time as they are completed and operational. What is more, once operational, continual performance monitoring ensures that performance remains at previously agreed and contracted levels.

### *Task integration*

Unlike conventional procurement the PFI requires tasks, which traditionally have been carried out in a discrete and fragmented way to be integrated.

Other strategies used are:

### *Build Operate and Transfer (BOT)*

Example – the Channel Tunnel, Spanish toll roads (autopistas), the Skye Bridge.

The process is as follows:

- **Build**: a private consortium agrees to invest in a public infrastructure project. The consortium then secures its own construction finance.
- **Operate**: the consortium then owns and maintains and manages the facility for an agreed concessionary period, say 25 years, and recoups its investment through charges or tolls.
- **Transfer**: after the concessionary period the consortium transfers ownership and operation of the facility to the government or relevant authority.

In its basic form a BOT project is one in which a public sector grants a concession to a private company for a fixed period of time. The private sector company constructs the project to the agreed specification and then operates and maintains it. This gives the private sector the opportunity to recoup the construction costs and to make a profit. At the end of the concession period the project is handed back to the private sector at no cost. The public sector is then free to run the project itself or appoint a contractor. Typically, the main parties to a BOT project are:

- the project company
- the government
- the government agency
- the investors, lenders
- the contractor
- the operator
- the suppliers.

The benefits of BOT schemes are:

- large infrastructure projects can be undertaken without risking public funds;
- the public sector benefits from private sector expertise;
- investment, construction and technology risks are shifted to the private sector;
- long-term income streams for the private sector consortia;
- combines responsibility for the disparate functions of design construction and maintenance under one single entity;
- project design can be tailored to the construction equipment and materials;
- tailored maintenance, attention to WLCs, incentivisation, smoother operations.

With a BOT approach the public sector relinquishes much of the control and day-to-day management they possess with more traditionally procured projects. Needs therefore must be identified up front – see Performance/Output specification.

The BOT scheme can also include a number of variations such as:

### *Build Rent Transfer (BRT)*

Similar to a BOT or BLT project except that the project site, buildings and equipment are rented to the private sector during the term of the project.

### *Build Lease Transfer (BLT)*

Similar to a BOT or BRT project except that a lease of the project site, buildings and equipment is granted to the private sector during the term of the project.

### *Build Own Operate (BOO)*

A method of financing projects and developing infrastructure, where a private company is required to finance and administer a project in its entirety and at its own risk. The government may provide some form of payment guarantee via long-term contracts, but any residual value of the project accrues to the private sector.

### *Design Construct Manage and Finance (DCMF)*

Adopted by the Prison Service when it embarked on its programme of modernisation in 1992, this model has produced the greatest efficiency savings to date. In most PFI models the public sector retains the responsibility for key staff, teachers in schools, clinicians in health schemes, etc.; however, in prison schemes the private sector consortia also provides the prison staff. As the consortia has also been responsible for the design of the prison the need to follow a heavily prescriptive specification has been dispensed with and innovative new ways to deal with questions such as high levels of security and surveillance were introduced. National Audit Office figures show that compared to traditionally procured and operated schemes, PFI prisons save about 10% or £40 million a year (CBI) over their life span. The public sector retains the demand risk as clearly the private sector has no control over the numbers admitted to prison.

## E-PROCUREMENT

There has been a growth in the number of tender documents being prepared in electronic format. The degree to which the entire procurement process is electronically based can vary widely. The main advantages claimed for e-procurement are:

- simplification of the process;
- reduction of costs and waste;
- avoidance of the need for duplication of effort when issuing documents;
- fairer assessment of bids.

The main methods available to distribute procurement documents are:

- disk
- an e-mail attachment
- web-based systems.

### *Disk*

Floppy disks are all but extinct now, so this method of data exchange will involve using CD, DVD or flash drive. The advantages are:

- familiarity with the technology;
- if encrypted, can be secure.

The disadvantages are:

- once transferred to a disk this then has to be posted or couriered to recipient;
- comparatively expensive;
- open to theft and physical damage/corruption.

### *E-mail*

When tendering information is exchanged using e-mail then it will be as an attachment. The advantages are:

- quick and cheap;
- familiarity with technology.

The disadvantages are:

- not secure – vulnerable to attack by virus, hackers, etc.;
- some files may prove too big to send;
- difficult to track revisions;
- may get lost in a sea of spam.

### *Web-based systems*

The most efficient and secure method of conducting e-procurement. An extranet (a closed/restricted) web-based portal is used to send information to the selected bidders only. The advantages are:

- the ability to securely communicate in real time;
- all bidders access a common pool of information.

The disadvantages are:

- compared to the other two options, it is relatively expensive;
- proprietary systems may require monthly subscription.

## EUROPEAN PUBLIC PROCUREMENT LAW

Procurement in the European public sector involves governments, utilities (i.e. entities operating in the water, energy and transport sectors) and local authorities purchasing goods, services and works over a wide range of market sectors, of which construction is a major part. For the purposes of legislation, public bodies are divided into three classes:

1. central government and related bodies, e.g. NHS Trusts;
2. other public bodies, e.g. local authorities, universities, etc.;
3. public utilities, e.g. water, electricity, gas, rail.

Public procurement is different from private business transactions in several areas; the procedures and practices are heavily regulated and, whilst private organisations can spend their own budgets more or less as they wish (with the agreement of their shareholders), public authorities receive their budgets from taxpayers and therefore have a responsibility to obtain value for money, traditionally based on lowest economic cost. However, in recent years the clear blue water between private and public sectors has disappeared rapidly with the widespread adoption of Public–Private Partnerships and the privatisation of what were once publicly owned utilities or entities.

### The Directives: theory and practice

EU Directives provide the legal framework for the matching of supply and demand in public procurement. A Directive is an instruction addressed to the EU member states to achieve a given legislative result by a given deadline. This is usually done by transposing the terms of the Directive into national legislation. The European public procurement regulatory framework was established by the public procurement Directives 93/36/EEC, 93/37/EEC and 92/50/EEC for supplies, works and services, and Directive 93/38/EEC for utilities, which, together with the general principles enshrined in the Treaty of Rome (1957), established the following principles for cross-border trading (references apply to the Treaty of Rome):

- a ban on any discrimination on the grounds of nationality (Article 6);
- a ban on quantitative restrictions on imports and all measures having equivalent effect (Articles 30 to 36);
- the freedom of nationals of one member state to establish themselves in another member state (Articles 52 *et seq.*) and to provide services in another member state (Articles 59 *et seq.*).

Enforcement Directives (89/665EEC and 92/13EEC) were added in 1991 in order to deal with breaches and infringements of the system by member states.

## The quantity surveyor and EU public procurement

How is the quantity surveyor likely to come into contact with the European public procurement system? The following scenarios are discussed:

- a surveyor working within a public body (contracting authority) and dealing with a works contract;
- a surveyor in private practice wishing to bid for work in Europe as a result of a service contract announcement.

### A *surveyor within a public body*

A quantity surveyor working within a body governed by public law (if in doubt, a list of European bodies and categories of bodies is listed in the Directives) should be familiar with procedures for compliance with European public procurement law. The Directives lay down thresholds above which it is mandatory to announce the contract particulars. *The Official Journal* is the required medium for contract announcements and is published five times each week, containing up to 1,000 notices covering every imaginable contract required by central and local government and the utilities – from binoculars in Barcelona to project management in Porto. Major private sector companies also increasingly use the *Official Journal* for market research. The current thresholds (effective from January 2010) for announcements in the *Official Journal* are:

1. For Works Contracts (i.e. construction) £3,927,260.00
2. For Supplies and Service Contracts (i.e. quantity surveying, project management)
   - Central government £101,323.00
   - Other public bodies £156,442.00
3. Utilities
   Water, Electricity, Urban
   Transport, Airports, Ports:
   Supplies and services £313,694.00
   Works £3,927,260.00

(NB: All figures exclude VAT).

Directorate General XV actively encourages contracting authorities and entities to announce contracts that are below threshold limits. Information on these impending tenders is published by the European Commission in the *Official Journal of the European Communities*, often otherwise known as the *OJEU* that is available free of charge, electronically, at europa.eu.int/.

The Directive also clarifies existing law in areas such as the selection of tenderers and the award of contracts, bringing the law as stated into line with judgements of the European Court of Justice.

## The EU procurement procedure

The OJEU announcement procedure involves three stages:

1. prior information notices (PINs) or indicative notices;
2. contract notices;
3. contract award notices (CANs).

Examples of these notices can be found in Annex IV of the Directive.

- **A prior information notice**, or PIN, which is not mandatory, is an indication of the essential characteristics of a works contract and the estimated value. It should be confined to a brief statement and posted as soon as planning permission has been granted. The aim is to enable contractors to schedule their work better and allow contractors from other member states the time to compete on an equal footing.
- **Contract notices** are mandatory and must include the award criteria, which can be based on either the lowest price or the most economically advantageous tender, specifying the factors that will be taken into consideration. Once drafted, the notices are published, five times a week, via the Publications Office of the European Commission in Luxembourg in the *Official Journal* via the Tenders Electronic Daily (TED) database, and translated into the official languages of the community, all costs being borne by the community. TED is updated twice weekly and may be accessed through the Commission's web site at http://simap.eu.int. Extracts from TED are also published weekly in the trade press. In order to give all potential contractors a chance to tender for a contract, the Directives lay down minimum periods of time to be allowed at various stages of the procedure – for example, in the case of the *open procedure* this ranges from 36 to 52 days from the date of dispatch of the notice for publication in the *Official Journal*. Restricted and negotiated procedures have their own time limits. These timescales should be greatly reduced with the wide-scale adoption of electronic procurement.

- **Contract award notices** inform contractors about the outcome of the procedure. If the lowest price was the standard criterion, this is not difficult to apply. If, however, the award was based on the 'most economically advantageous tender', then further clarification is required to explain the criteria – e.g. price, period for completion, running costs, profitability and technical merit, listed in descending order of importance. Once established, the criteria should be stated in the contract notices or contract documents.

### Award procedures

The surveyor must decide at an early stage which award procedure is to be adopted. The following general criteria apply:

- minimum number of bidders must be five for the restricted procedure and three for the negotiated and competitive dialogue procedures;
- contract award is made on the basis of lowest price or most economically advantageous tender (MEAT);
- contract notices or contract documents must provide the relative weighting given to each criterion used to judge the most economically advantageous tender and where this is not possible, award criteria must be stated in descending order of importance; and
- MEAT award criteria may now include environmental characteristics, e.g. energy savings, disposal costs, provided these are linked to the subject matter of the contract.

The choices are as follows:

**Open procedure** – which allows all interested parties to submit tenders.

**Restricted procedure** – which initially operates as the open procedure but then the contracting authority only invites certain contractors, based on their standing and technical competence, to submit a tender. Under certain circumstances, for example extreme urgency, this procedure may be accelerated.

**Negotiated procedure** – in which the contracting authority negotiates directly with the contractor of its choice. Used in cases where it is strictly necessary to cope with unforeseeable circumstances, such as earthquake or flood. Most commonly used in PPP models in the UK.

**Competitive dialogue procedure** – the introduction of this procedure addresses the need to grant, in the opinion of the European Commission, contracting authorities more flexibility to negotiate on Public–Private Partnership (PPP) projects. Some contracting authorities have complained that the existing procurement rules are too inflexible to allow a fully effective tendering process. Undoubtedly, the degree of concern has depended largely on how a contracting authority has interpreted the procurement rules as there are numerous examples of Public–Private Partnership projects which have been successfully tendered since the introduction of the public procurement rules using the negotiated procedure. However, the European Commission recognised the concerns being expressed, not only in the UK but also across Europe and it has sought to introduce a new procedure which will accommodate these concerns. In essence, the new competitive dialogue procedure permits a contracting authority to discuss bidders' proposed solutions with them before preparing revised specifications for the project and going out to bidders asking for modified or upgraded solutions. This process can be undertaken repeatedly until the authority is satisfied with the specifications that have been developed. Some contracting authorities are pleased that there is to be more flexibility to negotiations; however, for bidders this reform does undoubtedly mean that tendering processes could become longer and more complex. This in turn would lead to more expense for bidders and could pose a threat to new entrants to the PPP market as well as existing players. According to the Commission's DGXV department the introduction of this procedure will enable:

- dialogue with selected suppliers to identify and define solutions to meet the needs of the procuring body; and
- awards to be made only on the basis of the most economically advantageous basis.

In addition:

- all candidates and tenderers must be treated equally and commercial confidentiality must be maintained unless the candidate agrees that information may be passed onto others;
- dialogue may be conducted in successive stages. Those unable to meet the need or provide value for money, as measured against the published award criteria, may drop out or be dropped, although this must be conveyed to all tenderers at the outset;
- final tenders are invited from those remaining on the basis of the identified solution or solutions; and
- clarification of bids can occur pre- and post-assessment provided this does not distort competition.

To summarise, therefore, the competitive dialogue procedure is, according to the commission to be used in cases where it is difficult to access what would be the best technical, legal or financial solution because of the market for such a scheme or the project being particularly complex. However, the competitive dialogue procedure leaves many practical questions over its implementation, for example:

- the exceptional nature of the competitive dialogue and its hierarchy with other award procedures;
- the discretion of the contracting authorities to initiate the procedure; who is to determine the nature of a particular complex project?
- the response of the private sector, with particular reference to the high bid costs;
- the overall value for money;
- the degree of competition achieved, as there is great potential for post-contract negotiations.

## Electronic tendering

### *Electronic auctions*

The internet is making the use of electronic auctions increasingly attractive as a means of obtaining bids in both public and private sectors; indeed it can be one of the most transparent methods of procurement. At present, electronic auctions can be used in both open and restricted framework procedures. The system works as follows:

- The framework (i.e. of the selected bidders) is drawn up.
- The specification is prepared.
- The public entity then establishes the lowest price award criterion, for example with a benchmark price as a starting point for bidding.
- Reverse bidding on a price then takes place, with framework organisations agreeing to bid openly against the benchmark price.
- Prices/bids are posted up to a stated deadline.
- All bidders see the final price.

## Technical specifications

At the heart of all domestic procurement practice is compliance with the technical requirements of the contract documentation in order to produce a completed project that performs to the standards of the brief. The project must comply with national standards and be compatible with existing systems and technical performance. The task of achieving technical excellence becomes more difficult when there is the possibility of the works being carried out by a contractor who is unfamiliar with domestic conventions and is attempting to translate complex data into another language. It is therefore very important that standards and technical requirements are described in clear terms with regard to the levels of quality, performance, safety, dimensions, testing, marking or labelling, inspection, and methods or techniques of construction, etc. References should be made to:

- **A standard**: a technical specification approved by a recognised standardising body for repeated and continuous application.
- **A European Standard**: a standard approved by the European Committee for Standardisation (CEN).
- **European technical approval**: a favourable technical assessment of the fitness for use of a product, issued by an approval body designated for the purpose (sector-specific information regarding European technical approval for building products is provided in Directive 89/106/EEC).
- **Common technical specification**: a technical specification laid down to ensure uniform application in all member states, which has been published in the *Official Journal*.
- **Essential requirements**: requirements regarding safety, health and certain other aspects in the general interests that the construction works must meet.

Given the increased complexity of construction projects, the dissemination of accurate and comprehensive technical data is gaining in importance. It is therefore not surprising that the Commission is concerned that contracting authorities are, either deliberately or otherwise, including discriminatory requirements in contract documents. These include:

- lack of reference to European standards;
- application of technical specifications that give preference to domestic production;
- requirements of tests and certification by a domestic laboratory.

The result of this is in direct contravention of Article 30 of the Treaty of Rome, and effectively restricts competition to domestic contractors. In an attempt to reduce the potential problems outlined above, the EU has embarked on a campaign to encourage contracts to be based on an output or performance specification, which removes the need for detailed and prescriptive documentation.

# 5

# Pricing and tendering

The components of a unit rate, that is the rate used to price a bill of quantities or a specification, are composed from some or all of the following:

- Labour costs – the all-in labour rate. This is built up from operatives' wages plus statutory on-costs such as national insurance, etc.
- Material costs – the basic costs of materials plus the costs of delivery, unloading and storage and allowances for wastage.
- Plant costs – the hire cost of mechanical plant plus delivery to site, operating costs (drivers and fuel), etc. Can be included in the Preliminaries section, under the appropriate clause.
- Overheads and profit – overheads are such items as head office costs, etc. The profit margin will vary according to a number of external factors, including risk. Surprisingly, in the UK the profit margin for many general contractors is surprisingly low. The contractor may choose to include overheads and profit in the individual unit rates or may include suitable allowances elsewhere in the tender.

Given that contractors working within a given geographical area will almost certainly obtain their labour, materials and plant from a similar pool, it follows that the competitive edge between competing contractors must be within the overheads and profit.

## SUBCONTRACTORS

Increasingly, much of the work attributed to a general contractor is carried out by subcontractors. Subcontractors can be individuals or substantial firms who agree a contract with the main contractor to complete a section of a project, for example ground works and masonry. During the recent past the percentage of work being carried out by subcontractors has increased considerably. For the main contractor the advantages are:

- reduces the main contractor's liability to retain, on a full-time basis, all the specialists necessary for day-to-day operations;
- subcontractors are used as and when required, thereby reducing overheads.

The disadvantages to employing subcontractors are:

- programming – some types of subcontractors may, on occasions, be difficult to engage;
- control and co-ordination of subcontractors;
- quality of work.

Standard forms of contract vary in the amount of information that is required about subcontractors and the degree of authority vested in those responsible for giving permission for work to be subcontracted.

There are a number of different types of subcontractors as follows:

- domestic subcontractors
- named subcontractors
- nominated subcontractors.

**Named or nominated?**

Until the mid-1970s there were only two categories of subcontractor: nominated and domestic; but increasingly clients were using named subcontractors instead of nominated subcontractors. Contractually, a named subcontractor is the same as a domestic subcontractor except that the client states which contractors are acceptable without taking the responsibility for their final selection and appointment. The concept of naming rather than nominating started in the 1970s as relationships within the construction industry became more adversarial. It was first formalised in the JCT Intermediate Form of Contract in 1984. It is a procedure whereby a client selects a shortlist of preferred subcontractors, usually three, probably asks for quotations based on a common specification and then passes their names and quotations onto the main contractor. Neither the subcontractor nor the main contractor is obliged to use these quotations as a basis for a tender and they can be renegotiated.

Indeed, either party can decline to enter into a contract with the other party if they have reasonable grounds for not doing so. Named subcontractors are more popular with clients as:

- most forms of contract allow for the main contractor to claim an extension of time for delay caused by a nominated subcontractor, which the

main contractor has taken all reasonable measures to prevent. This is often quite an easy claim to make and a very difficult one to refute;
- clients pass on the risk of appointing subcontractors to the main contractor; and
- naming allows the client to keep some control on the subcontractor selection without having to take responsibility for performance.

A major change to the JCT(05) was to revise the process of sub-letting works with the demise of the provisions for nominating subcontractors in the industry flagship contract. In consequence all reference to nominated subcontractors has been removed to be replaced with simply 'subcontractor' clauses 3.7 to 3.9.

## DOMESTIC SUBCONTRACTORS

Domestic subcontractors are employed directly by the main contractor; it is a private arrangement between the two parties. Domestic subcontractors do not have a contract with the client but work on site as if main contractor's personnel and they are co-ordinated by the main contractor's site management team. Domestic subcontractors have a responsibility to deliver work that complies with the approval of the client and architect and are paid by the main contractor from monies received from the client in interim valuations. The terms of payment, discounts, etc., are negotiated between the domestic subcontractor and the main contractor. Although the various forms of contract vary slightly, so long as the main contractor informs the architect that certain parts of the works are to be carried out by domestic subcontractors, consent cannot be reasonably withheld. Domestic subcontractors are commonly used as follows:

- Subcontractors that carry out complete sections of the work, including all materials, labour and plant, for example.
- Labour-only subcontractors – as the title suggests, unlike nominated subcontractors, labour-only subcontractors supply only labour to the main contractor. Fears have been expressed over the use of this sort of labour that centre around:
  - lack of training and skills,
  - accountability, and
  - payment of taxation and other statutory obligations.
- The usual media image of labour-only subcontractors is of gangs of men on street corners, in the early hours, waiting to be collected by a white van. HM Revenue and Customs is constantly reviewing the process for employing this type of labour.

The contractual relationship in the case of the above two types of subcontracting is between the main contractor and the subcontractor only. The main contractor retains the responsibility for ensuring domestic subcontractors comply with all relevant statutory legislation and is answerable to the architect for the works done and materials supplied. The use of domestic subcontractors allows the main contractor the opportunity to transfer risk to the subcontractors.

Although not obligatory, there are a number of specialist contracts that have been developed for use with domestic subcontractors, for example DOM1 as developed by the Building Employers Confederation.

### Named subcontractors

Named subcontractors are often used in public sector projects and for projects based on the Form of Contract where there is no provision for nomination. If a named subcontractor has not been appointed at the time that the bills of quantities are ready for despatch, then a provisional sum is included for provision of the works to be carried out by the named subcontractor. The main features of named subcontracting are:

- the tender documents, usually the bills of quantities, include the names of potential named subcontractors. The main contractor has the opportunity to reasonably object to any firm on the list;
- the main contractor leads the tender process for each named subcontractor package, by assembling the tender documents, issuing and receiving tenders, selecting a named contractor; and
- after appointment, the named subcontractor, for all intents and purposes, is a domestic subcontractor. The main contractor is only paid the rates in the accepted subcontract tender.

Because the contractual relationship between a named and a domestic subcontractor is similar, named subcontractors are allowed for by the inclusion of a provisional sum.

### Nominated subcontractors

In cases when the architect or client wishes to restrict and control certain aspects of the project works then nominated subcontractors can be used. They may also be used in cases where, at the tender stage, parts of the project have not been fully detailed and therefore the use of nominated subcontractors allows the job to go to tender, with the nominated works being dealt with at a later date. The principal differences between domestic and nominated subcontractors are:

- the tender process is organised and run by the architect who invites suitable subcontractors to submit a tender;
- the architect selects the preferred tender and then instructs the main contractor to enter into a contract with the subcontractor.

In order to use nomination a prime cost sum is included in the bills of quantities at tender stage. A prime cost sum (or first cost sum) should be an accurate estimate of the likely cost of works when fully detailed. Unlike the domestic subcontractor, a nominated subcontractor does enter into a formal contract with the main contractor and has the benefit of stipulated payment procedures for which the nominated subcontractor has to agree to give the main contractor 2.5% discount on all sums due, to cover administration costs. In addition to the inclusion of a prime cost sum, the main contractor also has the opportunity to include for attendance on the subcontractors. These allowances vary according to the nature of the works being carried out by the subcontractor, but could include:

- unloading, storage and protection of the subcontractor's materials;
- provision of accommodation, plant, scaffolding and services required by the subcontractors;
- disposing of rubbish and packaging generated by the subcontractors; and
- protecting the finished work until the project is complete.

Attendances can be priced in a variety of ways by the main contractor, for example:

- as a lump sum;
- as a percentage addition;
- included within profit and overheads allowance.

From a contractor's point of view, the advantages of including a percentage is that this will increase pro rata, if the eventual cost of the work exceeds the prime cost sum allowance. To avoid duplication, a definition of 'general attendance' is included within the preliminaries section of the bill of quantities. The following is an example of how a prime cost sum should be included with a bill of quantities for the item of:

Include the prime cost sum of £**120,000.00** for vitreous enamelled steel cladding panels, complete with insulation and galvanised steel Z section framework, supplied and fixed by a nominated subcontractor. 120,000. 00

Add for profit

Add for general attendance as (refer back to preliminaries)

Add for special attendance as including:

Priced by contractor as a percentage or lump sum or included in overheads.

- mechanical offloading and distribution
- hoisting
- dry, covered and secure storage
- scaffolding and ladders
- provision of 110 volt AC supply
- protection of the completed work
- removal of all debris.

## THE NOMINATION PROCESS

The following process relates to contracts where the contract administrator is entitled to nominate:

- Suitable firms are identified by the architect and they are invited to submit tenders for the subcontract works carried out in accordance with strictly defined conditions. In the case of subcontracts where there is an element of design, then it is advisable to include a collateral warranty (see explanation later) that establishes a direct contract between the nominated subcontractor and the employer.
- When submitted the tenders are examined by the contract administrator and one is selected.
- Instructions are then given to the main contractor to enter into a subcontract with the selected nominated subcontractor and negotiate the items relating to programming, etc.
- Following agreement an employer/subcontractor agreement is put in place.

## COLLATERAL WARRANTIES

Collateral warranties are used in cases where the subcontract involves an element of design and are employed to give the client the right to pursue a subcontractor in the event that the design proves to be faulty. The JCT has developed standard forms of collateral warranty and it is a separate contract between the client and the subcontractors that operates alongside the main contract. Collateral warranties are dealt with in more detail in Chapter 6.

### Calculating bill rates using domestic subcontractors

If a contractor chooses to use domestic subcontractors, then the usual procedure is for the appropriate section of the tagged bill of quantities to be sent by the main contractor to subcontractors for pricing. When returned, one is selected by the main contractor, and then the main contractor adds on the domestic subcontractor's rates and the main contractor's profit and overheads before using the adjusted subcontractor's rates to price the bill of quantities as if they were their own.

## UNIT RATE AND OPERATIONAL ESTIMATING

There are two alternative approaches to estimating: operational and unit rate. Unit rate estimating, described below, is where prices for bill of quantities items are each calculated separately; in contrast, operational estimating considers a parcel of work as a package.

### Operational estimating

Civil engineering works have taken a broader brush approach to estimating, due in part to the nature of many civil engineering projects where large quantities are involved as well as extensive use of mechanical plant. In the case of civil engineering works, different approaches to carrying out the works can have a significant effect on prices. All the resources needed for parts of the construction are considered together, instead of in isolation. Examples where this approach is successfully used are:

- excavation and disposal
- concrete work
- drainage.

There is no hard-and-fast rule where operational-based estimating techniques stop and unit rate estimating begins. In fact it can be difficult to reconcile works priced on an operational basis with a bill of quantities. However, operational estimating is suitable for design and build tendering when the contractor can use their own approach and no bills of quantities have to be submitted.

### Unit rate estimating

A price is calculated for each item in the bill of quantities as if the item is to be carried out in isolation to the rest of the works. This is the traditional approach for pricing the majority of building work.

Unit rates will need to calculated by the estimator for directly employed labour and take into account the following factors.

### Labour costs

Labour costs are determined by the calculation of the so-called 'all-in' hourly rate and is the basic costs associated with labour with the addition of the costs that comply with a range of statutory requirements. These on costs may include all or some of the following items and can be found in the National Joint Council for the Construction Industry's Working Rule Agreement, which is published annually by the Building and Allied Trades Joint Industrial Council (BATJIC) and is available on line at: www.fmb.org.uk.

For example, the BATJIC Standard rate of wages (plain time rates) for a 39-hour week were determined as follows for September 2010/2011:

- craftsmen – between £356.07 and £414.18 per week or £9.13 and £10.62 per hour;
- operatives – £307.32 or £7.88 per hour.

Note: these rates change annually and for September 2011/2012 an increase of 1% has been agreed.

#### *On costs*

The principal on costs are as follows.

#### *Overtime payments – Working rule 7*

The first hour worked over 39 hours is at plain time rates and thereafter:

- Monday to Friday – for the next three hours, time and a half, and then double time after that;
- Saturday – time and a half up to 4.00pm and then double time;
- Sunday – double time.

### Holidays – Working rule 4

Annual (21 days) and public holidays (8 days).

### Guaranteed week – Working rule 9

Operatives are guaranteed 39 hours employment per week despite stoppages for inclement weather providing that they are available for work.

### Travelling, fares and lodgings – Working rule 11

There are a variety of daily allowances based on sliding scales to reimburse the costs of travelling to and from work as shown in Table 5.1.

### Sickness and injury benefit – Working rule 12

Payment to operatives for absence from work due to sickness or injury up to a maximum of 12 weeks per year. The first day of any absence is not paid.

**Table 5.1** Daily fare allowance

| Distance (km) | Fare £ | Distance (km) | Fare £ |
|---|---|---|---|
| 1–6 | Nil | 29 | 6.23 |
| 7 | 0.42 | 30 | 6.35 |
| 8 | 0.84 | 31 | 6.58 |
| 9 | 1.26 | 32 | 6.60 |
| 10 | 1.67 | 33 | 6.77 |
| 11 | 2.11 | 34 | 6.86 |
| 12 | 2.52 | 35 | 7.06 |
| 13 | 2.94 | 36 | 7.16 |
| 14 | 3.36 | 37 | 7.30 |
| 15 | 3.80 | 38 | 7.53 |
| 16 | 4.10 | 39 | 7.65 |
| 17 | 4.35 | 40 | 7.89 |
| 18 | 4.62 | 41 | 8.07 |
| 19 | 4.87 | 42 | 8.27 |
| 20 | 5.01 | 43 | 8.45 |
| 21 | 5.20 | 44 | 8.66 |
| 22 | 5.39 | 45 | 8.83 |
| 23 | 5.51 | 46 | 9.06 |
| 24 | 5.63 | 47 | 9.24 |
| 25 | 5.80 | 48 | 9.41 |
| 26 | 5.91 | 49 | 9.62 |
| 27 | 6.04 | 50 | 9.79 |
| 28 | 6.15 | | |

Current sick allowance is £118.04 per week or £23.61 per day and is in addition to statutory sick pay.

National insurance contributions are also payable at the appropriate rate and are calculated as follows. Note that the current threshold for NI payments is £102.00, although this does vary:

**Craftsman**

| | | |
|---|---|---|
| £390.00 per week × 46 weeks | = £17940.00 | |
| Threshold £102.00 × 46 weeks = | £4692.00 | |
| National Insurance 12.00% | £13248.00 | = £1589.76 |

**Operative**

| | | |
|---|---|---|
| £307.32 per week × 46 weeks | = £14136.72 | |
| Threshold £102.00 × 46 weeks = | £4692.00 | |
| National Insurance 12.00% | £9444.72 | = £1133.37 |

### *Retirement and death benefit – Working rule 13*

Payments made on the retirement or death of operatives.

### *Tool allowance – Working rule 18*

Now consolidated within the basic wage rates, the employer must provide secure storage and replacement if stolen or damaged.

### *Construction Industry Training Board (CITB) levy*

A levy paid to the training board to fund the training of new operatives and the development of new skills.

### *Other items*

Employer's liability insurances

Additional payments for intermittent responsibility – Working rule 1C as follows:

42 pence per hour for the use of air or percussion tools
63 pence per hour for operating a drag shovel, dumper up to 2,000kg, etc.
86 pence per hour for larger items of plant.

For continuous responsibility the following additions are payable as follows:

between £377.45 and £397.16 per week for tower cranes.

In addition the working rule agreements lay down standards for the following:

Health and welfare provisions – Working rule 15

Welfare benefit:

| | |
|---|---|
| Retirement contribution | £3.00 |
| Stakeholder pension scheme | £7.00 |
| Death benefit contribution | £0.90 |
| **Total** | **£10.90** |

Safety – Working rule 16
Trades union facilities – Working rule 17
Maternity/paternity leave – Working rule 24

### *All-in hourly rate calculation*

In order to calculate the all-in hourly rate, that is the hourly rate charged by a contractor for a construction operative, the estimator must first decide on the following:

- the number of hours worked per week;
- overtime rate – assumed to be time and a half – Working rule 7;
- summer working, in the following example taken as 30 weeks;
- winter working, in the flowing example taken as 20 weeks.

### *Hours worked*

**Summer period**

| | | |
|---|---|---|
| Number of weeks | 30 | |
| Weekly hours | 39 | |
| Total hours | | 1170 |
| Less holidays | | |
| Annual – 14 days | | (109.20) |
| Public – 5 days | | (39.00) |

**Winter period**

| | | |
|---|---|---|
| Number of weeks | 22 | |
| Weekly hours | 39 | |
| Total hours | | 858 |
| <u>Less</u> holidays | | |
| Annual – 7 days | | (54.6) |
| Public – 4 days | | (31.2) |
| Sickness | 8 | (62.00) |
| **<u>Total hours for payment</u>** | | 1731.60 |
| Allowance for bad weather | 2% | 34.63 |
| **<u>TOTAL PRODUCTIVE HOURS</u>** | | **<u>1766.23</u>** |

| **Annual earnings** | **Craftsman** | **Labourer** | **Craftsman** | **Labourer** |
|---|---|---|---|---|
| Basic weekly wages | £379.08 | £307.09 | | |
| Hourly rate (1/39) | £9.72 | £7.88 | | |
| Annual earnings | | | £16,831.15 | £12,911.14 |
| Add | | | | |
| Guaranteed minimum | | | | |
| Bonus per hour | £3.00 | £2.00 | £5,194.80 | £3,463.20 |
| **Add on costs** | | | | |
| Non-productive overtime* time and a half only | | | | |
| Hours per week summer | 10 × 26 weeks | | | |
| Hours per week winter | 5 × 20 weeks | | | |
| Hours per year summer | 260 | | £2527.20 | £2048.90 |
| Hours per year winter | 100 | | £972.00 | £788.00 |
| Sick pay per day | £20.22 | | | |
| Allowance per year – 8 days | | | £161.74 | £161.74 |

| | | |
|---|---|---|
| Working rule agreement says £0.25 per hour worked | £432.90 | £432.90 |
| **Sub-total** | **£26,119.79** | **£19,805.88** |
| **Add overheads** | | |
| National Insurance 12.00% above threshold | £1,589.76 | £1,133.37 |
| CITB Training levy | £123.63 | £94.11 |
| Holidays with pay allowance | | |
| Annual holidays allowance – 163.8 hours | £1,592.14 | £1,290.74 |
| Public holidays allowance – 70.2 hours | £682.34 | £553.18 |
| Welfare benefit 52 weeks @ £10.90 | £566.80 | £566.80 |
| **Sub-total** | **£30,745.59** | **£23,444.08** |
| Employer's liability & public liability insurance – 2% | £614.91 | £468.88 |
| **Annual cost of operative** | **£31,360.50** | **£23,912.16** |
| Hourly rate – divide by productive hours – 1766.23 | **£17.76** | **£13.54** |

* Non-productive overtime relates to the additional money paid to operatives working overtime. Whilst working overtime, half as much again is paid and yet the physical amount of work produced in not increased – this unproductive paid working is known as non-productive overtime.

### *Labour constants*

Labour output is the most uncertain part of a unit rate. It can vary considerably depending upon the skills and output of the operative, the site organisation, weather conditions and many other factors often outside the control of the contractor. Historical records of labour outputs are kept by most contractors and subcontractors based on a variety of sources including observing and benchmarking operations on site. These records which give the average unit time for each operation are called labour constants. Rates are usually expressed for an individual craftsman or labourer or in some cases, where more appropriate, for example brickwork, a gang rate may be used. In the case of brickwork this will be built up from calculating the output for two tradesmen and one labourer who will provide the bricklayers with the materials that they need.

#### GANG RATES

It is often the case that labour rates are calculated on the basis of gang rates rather than individual skilled and labourers' rates. This is thought to be a more realistic approach as often, for example in the case of bricklayers, roofers and plasterers, two or three skilled operatives will be furnished with materials by a labourer and the gang rate would be based on three skilled to one unskilled operatives per hour as follows:

Gang rate for bricklaying:

3 × £17.76 = £53.28
1 × £13.54 = £13.54
£66.82 per hour

Of course the output of the bricklaying gang will increase threefold plus compared to an individual bricklayer, economies of scale can also be achieved as a single labourer is used. A variety of approaches have been used in the following examples.

### Materials

The bills of quantities require the contractor to price items in terms of units that are in many cases unlike the units in which the materials are bought. For example, cement, sand and other constituents of concrete in foundations are bought in a variety of units, such as tonnes, whereas the bill of quantities requires concrete in foundations to be priced in $m^3$. Therefore, during the estimating process the builders' merchant rates and costs have to be converted to the unit required by the bills of quantities.

The all-in rate for materials comprises the following allowances:

- the price quoted by the supplier/builders' merchant. This will have to be adjusted for:
  - trade discounts, and
  - discounts for bulk orders;
- transport to the site;
- storage on site;
- an allowance for wastage, which will vary according to the type of material but on average is about 10%. This wastage can be attributed to items such as theft, breakages, misuse and mistakes during use; and
- bill of quantities items are measured net, in other words no allowances are made in the measurement for overlapping on roofing felt of damp-proof courses at joints, see NRM2. An addition has to be made to take account of this.

## Plant

For estimating purposes there are two scenarios relating to the pricing of plant: plant may be owned by the contractor or hired from a specialist plant hire firm.

Before deciding to own plant the contractor must consider the cost and the amount of use and work for the machine. Ideally, the plant should be used continuously. It is quite common for larger contracting organisations to have their own separate plant hire company so that the plant can be used in-house and also hired to other contractors. The principal advantages of hiring plant on an 'as and when' needed basis, is that it does not require large sums of capital to be tied up in plant as well as time devoted to maintenance and storage.

### *All-in rate for plant*

The factors that have to be considered when calculating an all-in rate for plant are:

- Fixed costs:
  - cost of plant and expected operating life
  - return on capital
  - maintenance costs, and
  - tax and insurance.

- Operating costs:
  - operator's wages
  - fuel
  - other consumables, including oil, etc.

***Fixed costs***

There are two principal approaches for the computation of fixed costs:

a) Straight line depreciation

| | |
|---|---|
| Cost of plant: | £20,000 |
| Less scrap value: | £3,000 |
| | £17,000 |

Assume a life of four years therefore annual cost

$$\frac{£17{,}000}{4} = £4{,}250$$

Assume a usage of 1500 hours per year:

Cost per hour $\frac{£4{,}250}{1500} = £\ 2.83$

b) Writing down depreciation

| | | Depreciation |
|---|---|---|
| Purchase price | £20,000 | |
| Initial writing down 60% during 1st year | £12,000 | £12,000 |
| | £8,000 | |
| Second year writing down allowance @ 25% | £2,000 | £2,000 |
| | £6,000 | |
| Third year writing down allowance @ 25% | £1,500 | £1,500 |
| | £4,500 | |

| | | |
|---|---|---|
| Fourth year writing down allowance @ 25% | £1,125 | £1,125 |
| Residual value | £3,375 | |
| Depreciation | | £16,625 |

### *Working life*

Construction plant comes in a variety of types and usually operates in very extreme conditions, however typical values are as follows:

| | |
|---|---|
| Concrete mixers | 6–7 years |
| Cranes | 8–10 years |
| Dumper | 3–4 years |
| Excavating plant | 5–7 years |
| Hoists | 5–7 years |
| Lorries | 3–5 years |

It is usual to assume that plant works for 1,500 hours per year. These figures should be used in the calculation of the cost of plant.

### *Tax and insurance*

A road fund licence is required for plant that uses public highways and generally is insured, for an annual premium, against theft.

### *Maintenance costs*

As with any mechanical item, construction plant will require regular maintenance to work reliably and effectively. Maintenance can be allowed for as a percentage of initial costs, based on manufacturers' recommendations as well as historical records. Allowances vary from 10–30% depending on type.

### *Operating costs*

OPERATOR'S WAGES

As mentioned earlier in the chapter, National Working Rule 1C allows operatives with continuous skills or responsibilities to be eligible for additional payments. In addition operators of plant may be eligible for an extra hour per day to fuel and maintain their plant.

FUEL

Typical fuel consumptions for items of plant are a matter of record, for example a rotary drum concrete mixer will use 1 litre of fuel per hour.

OTHER CONSUMABLES

This item includes oil, lubrication, etc. and can be allowed for by including 20–25% of fuel costs for the item of plant.

TRANSPORTATION AND MAINTENANCE

The cost of transporting mechanical plant to and from site and the setting up and erection of items such as tower cranes have to be allowed for.

### *Output*

The performance of construction plant will affect the cost and therefore it is essential to know the output or production of a particular item. Once again these statistics are based on historical records. For example:

> An excavator fitted with a shovel and having a 0.5 cubic metre bucket will be able to load 12 cubic metres of excavated material into lorries per hour. A 14/10 (280 litre) concrete mixer will be able to produce an average of 5.0 cubic metres of mixed concrete per hour.

The output/performance of mechanical plant will be affected by a number of factors, for example:

- site conditions, including time of year;
- the degree to which plant can be incorporated due to restriction on site, site organisation, etc.;
- skill of the operators.

These factors should be taken into account when pricing.

## Overheads

Overheads or establishment charges are very often difficult to define. For estimating purposes they include all items which are necessary to the efficient running of a building company and which are not normally charged to a specific contract. They include:

- head office costs – stationery, stamps, telephone, lighting, heating, depreciation of office furniture, head office salaries, etc.;
- head office building – rent, rates, depreciation, repairs, insurances, etc.;
- finance costs – interest on bank overdrafts, bank charges, etc.

Overheads are usually related to turnover and the required percentage addition is included in the unit rate build-up. For example, if the annual turnover of a company is £5 million and fixed costs are £250,000 then the required percentage is:

$$\frac{£250{,}000}{£5{,}000{,}000} \times 100 = 5\%$$

## Profit

The amount of profit required by a building contractor will vary considerably depending on the size of the company, the turnover, market conditions, the contract value and the perceived risks involved. Historically, UK contractors have operated on profit margins of around 4%; however, in time of work shortages contractors frequently tender for work on much less of a profit margin to ensure turnover is maintained. Profit and overheads may be included either as a percentage addition to measured work, or in the preliminaries section or as a combination of both.

The following examples comply with NRM2.

## Preliminaries

As described in Chapter 3, the preliminaries section of the bills of quantities contains items that are of a general nature that may affect the contractor's tender. NRM2 splits preliminaries down in to various sections as follows:

- Main contractor's preliminaries divided into:
  - Part 1: Information requirements, for example:
    - project particulars
    - the site
    - security.
  - Part 2: Pricing schedule, which is a simple list of cost centres divided into:
    - employer's requirements
    - main contractor's requirements.

In addition there is provision to compose preliminaries when procurement is based on work packages. The requirements are split into:

- Part 1: Information requirements
- Part 2: Pricing schedule
  - employer's requirements
  - work package contractors' cost items.

As with SMM7, NRM2 splits general cost items into:

- fixed charges, the cost of which, is considered to be independent of duration, that is to say charges that are not proportional either to the quantity of the work or its duration, for example:
  - temporary water supply connection
  - licences in connection with hoarding, scaffolding, gantries and the like, and
- time-related charges, the cost of which, is considered to be dependent on duration, that is to say charges that are directly proportional to either the quantity of the work or its duration, for example:
  - cleaning
  - general office furniture, including maintenance.

Although the preliminaries section of bills of quantities tends to contain many items, several of them are difficult to quantify and price and for this reason it is usually only the major items such as accommodation, staffing, mechanical plant that are priced. Appendices to NRM2 give templates for preliminaries pricing schedules.

### *Main contractor's cost items: site establishment*

#### EXAMPLE: SITE ACCOMMODATION

The contractor shall provide suitable accommodation for the clerk of works, with a minimum floor area of 20m$^2$ including desk, four chairs, filing cabinet and all necessary heating, lighting and cleaning.

**Data**

| | |
|---|---|
| Contract period: 144 weeks | Transport to and from site: £150 per trip |
| Hire of hut: £70 per week | Hire of furniture & heaters, etc.: £30 per week |

| **Fixed charges** | £ | £ |
|---|---|---|
| Transport to and from site 2 trips × £150 | 300.00 | |
| Erection 10 hours | | |
| Dismantle 5 hours | | |
| 15 hours labourer @ £13.54 | 203.10 | |

| | | |
|---|---|---|
| **Total of fixed charges** | 503.10 | 503.10 |
| **Time-related charges** | | |
| Hire of hut 144 weeks @ £70 per week | 10,080.00 | |
| Furniture & heaters 144 weeks @ £30 per week | 4,320.00 | |
| Energy costs for heating & lighting: 144 weeks @ £4.00 per week | 576.00 | |
| Cleaning – 2 hours per week × 144 weeks = 288 hours @ £13.54 | 3899.52 | |
| **Total of time-related charges** | 18,875.52 | 18,875.52 |
| **Rate per week = £131.08** | | |
| **Total (fixed & time-related charges)** | | **£19,378.52** |

### Example: hoardings, fences and gates

The contractor is to allow for enclosing all boundaries of the site. It shall be the contractor's responsibility to provide all necessary precautions, protection and security to safeguard the works using fences, hoardings, gates, etc. as considered necessary.

**Data**

Contract period: 144 weeks

Fencing £15 per m

| | £ | £ |
|---|---|---|
| Fixed charges: erection and taking down | | 1,000.00 |
| 50 metres of solid fencing; 1.8m metres high | 1250.00 | |
| @£25/m 1 pair gates and security | 200.00 | |
| | 1,450.00 | 1,450.00 |
| Maintenance and repairs | | |
| 1 hour per week labourer – 144 weeks @ £13.54 | | 1,949.76 |
| | **Total** | **£4,399.76** |

### Example: mechanical plant

The contractor shall provide all necessary plant, tools, vehicles, equipment and associated labour, etc. for the proper execution of the works.

Some items of plant are used by several trades and as such it is difficult to apportion the cost accurately to each trade and therefore the contractor has the opportunity to allow for them in the preliminaries section. Typical items included for are tower cranes, hoists, etc.

## Section D: groundwork

### *Excavation and filling*

An important point to remember with ground works is that when ground is excavated it increases in bulk. This is due to the fact that broken-up ground has a greater volume than compacted ground; a point that has to be taken into consideration when transporting or disposing of excavated materials. Volumes in the bills of quantities are measured net, without allowance for bulking; therefore the estimator must make an allowance. What is more, the degree to which excavated material bulk depends on the type of ground as follows:

- sand/gravel – 10%
- clay – 25%
- chalk – 33%
- rock – 50%.

Other points to take into account are:

- the time of year when the excavation is to be carried out and whether the excavated materials and the bottoms of the excavation are likely to be flooded;
- the nature of the ground will affect the type of earthwork support required and the labour constants for carrying out the works. Although earthwork support is required to be measured by SMM7 it is often regarded by contractors as a risk item and as such may not actually be used although priced in the bills of quantities. For this reason in NRM2 earthwork support is not measured unless specifically requested;
- the distance and availability of the nearest tip to which any excavated material has to be transported plus charges for tipping. Contaminated spoil must be dealt with and disposed of separated; higher tipping charges will be incurred for contaminated waste; and
- the sections of the works that are to be carried using hand digging or mechanical plant.

It is advisable to visit the site, study the results of test pits, soil samples, etc. and to check site access.

***Hand excavation or machine***

The majority of excavation is now done by machine; however, there may be instances where there is restricted access, or where there are small quantities involved, where hand excavation is used. The unit rate for hand excavation is generally more expensive than excavation by machine.

The following are average labour constants for hand excavation under normal conditions and in medium clay or heavy soil.

| **Item** | **Hours/m$^3$** |
|---|---|
| Bulk excavation commencing at reduced levels not exceeding 2m deep | 2.40 |
| Ditto over 2m not exceeding 4m deep | 2.70 |
| Foundation excavation commencing at reduced levels not exceeding 2m deep | 4.00 |
| Bulk excavation in trenches, commencing at reduced levels not exceeding 2m deep | 3.30 |
| Filling obtained from excavated materials final thickness exceeding 500mm deep | 1.00 |
| Retaining excavated material on site in temporary spoil heaps distance not exceeding 50 metres | 1.00 |

For excavation other than medium clay the following multipliers should be applied to the above:

| | |
|---|---|
| Loose sand | 0.75 |
| Stiff clay or rock | 1.50 |
| Soft rock | 3.00 |

### EXAMPLE: REDUCED LEVEL EXCAVATION

Bulk excavation commencing at reduced levels not exceeding 2m deep – cost per m$^3$.

Assume 10m$^3$

| | |
|---|---|
| 24 hours labourer @ £13.54 | £324.96 |
| Add Profit & Overhead 15% | £48.74 |
| Cost per 10m$^3$ | £373.70 |
| Cost per m$^3$ | **£37.37** |

### EXAMPLE: HAND EXCAVATION

Foundation excavation in trenches, commencing at reduced levels, not exceeding 2m deep – cost per $m^3$.

Assume $10m^3$

| | |
|---|---|
| 33 hours labourer @ £13.54 | £446.82 |
| Add Profit & Overheads 15% | £67.02 |
| Cost per $10m^3$ | £513.84 |
| Cost per $m^3$ | **£51.38** |

### EXAMPLE: MECHANICAL EXCAVATION

Foundation excavation in trenches, commencing at reduced levels, not exceeding 2m deep – cost per $m^3$.

**Data**
Assume $10m^3$

The use of $0.25m^3$ bucket excavator @ £17.00 per hour including labour and consumables. The output for a $0.25m^3$ bucket excavator is $5m^3$ per hour.
**Note:** when using mechanical plant for excavation a banksman is required who is paid the labourer rate plus a small addition for extra responsibility.

Plant

| | |
|---|---|
| 2 hours excavator @ £17.00 per hour | 34.00 |

Banksman

| | |
|---|---|
| 2 hours banksman @ £13.54 | 27.08 |
| | 61.08 |
| Add | |
| Profit & Overheads 15% | 9.16 |
| | 70.24 |
| Divide by 10 cost per $m^3$ | **£7.02** |

### EXAMPLE

Disposal of excavated materials from site cost per $m^3$.

Assume $6m^3$

Tipping charge: £10.00 per load
Site to tip: 4 kilometres round trip

$6m^3$ lorry @ £25 per hour including driver and fuel

Bulking of excavated material 33%

Load as dug $6m^3 \times \frac{133}{100} = 4.5m^3$

| | Time per load (minutes) |
|---|---|
| Assume average speed of lorry 30kph | |
| 1 kilometre in 2 minutes | |
| 4 kilometres travel = 4 × 2 = 8 minutes | 8.00 |
| Time to load lorry | |
| Excavator bucket size: $0.33m^3$ | |
| Cycle time per bucket: 0.5 minute | |
| $\frac{4.5}{0.33} \times 0.5 = 6.82$ | |
| Tipping time | 4.00 |
| **Total operation time** | **18.62** |
| 18.62 minutes × $\frac{25}{60}$ | 7.75 |
| Tipping charge | 10.00 |
| Cost per $4.5m^3$ | 17.75 |
| Cost per $m^3$ | 3.94 |
| Add Profit & Overheads 15% | 0.59 |
| Cost per $m^3$ | **£4.53** |

### *Hardcore filling*

Materials used as hardcore, such as brick, stone, etc. are bought either by volume or weight. If bought by volume an average of 20% should be added to the material to cover consolidation and packing. If bought by weight the following is the approximate amount required per cubic metre of compacted material:

- Brick ballast 1800kg
- Stone ballast 2400kg

On contracts where large volumes of hardcore are used it is usual to have the material tipped and then spread, levelled and compacted by mechanical means. However, below are typical labour constants for spreading and levelling hardcore by hand.

| **Item** | **Labourer hours** |
|---|---|
| Imported filling as bed over 50mm but not exceeding 500mm deep, 350mm finished thickness | 1.5/m$^3$ |
| Ditto exceeding 500mm deep | 1.0/m$^3$ |
| Surface treatments compacting filling (SMM7 only) | 0.2/m$^2$ |
| Surface treatments trimming slopping surfaces (SMM7 only) | 0.40/m$^2$ |
| Blinding bed, not exceeding 50mm thick, level to falls, cross falls or cambers, 40mm finished thickness | 0.50/m$^3$ |

### Example

Imported filling as bed over 50mm but not exceeding 500mm deep, average 350mm finished thickness – cost per m$^3$.

A medium-sized mechanical shovel will spread and level approximately 15m$^3$ of hardcore per hour; it is then consolidated in 150mm layers by a roller.

**Data**

Hardcore: £ 17.00 per m$^3$ delivered to site
Mechanical shovel including driver and fuel: £20 per hour
Roller including fuel and driver: £18 per hour

| | | |
|---|---|---|
| 1m$^3$ hardcore @ £17 per m$^3$ delivered to site and tipped | | 17.00 |
| Add 20% consolidation | | 3.40 |
| Hardcore per m$^3$ | | 20.40 |
| Laying and consolidating per hour | | |
| Mechanical shovel | 20.00 | |
| Roller | 18.00 | |
| Labour | 13.20 | |
| | £51.20 | |

| | |
|---|---|
| Per m$^3$ £51.20 / 15m$^3$ | 3.41 |
| | £23.81 |
| Add Profit & Overheads 15% | 3.45 |
| Cost per m$^3$ | **£27.26** |

## Underpinning

Underpinning is a technique used for existing structures that have suffered differential settlement, caused by proximity to trees, landfill or mine workings. It is carried out in small sections and rates for underpinning will vary considerably depending on the ground conditions, the condition of the existing building and the perceived risk. Underpinning existing structures is specialised work carried out in small quantities and the pricing of it is outside the realm of this pocket book.

## Section E: concrete work

Factors to be taken into account when pricing concrete work items are:

- Whether ready mixed concrete is to be used or if concrete is to be mixed on site. It is usually found that on sites where access is not a problem and where reasonably large quantities of concrete are required over a regular period that it is better and more economical to set up a batching plant to mix on site. On a restricted site or where small quantities of concrete are required then it is probably better to use ready mixed concrete, as it will be the cheaper alternative.
- Assuming that concrete is to be mixed on site, whether bagged or bulk cement is used. Bagged cement is delivered in 50kg bags and has to be unloaded and stored in a dry location, whereas bulk cement is stored in a silo next to the mixer and is cheaper. Bagged cement also tends to be more wasteful.
- If concrete is to be mixed on site the best position for the mixing plant from the point of view of transporting the mixed concrete around the site.
- The type and size of mixer to be used based on the output required.
- The method of hoisting, placing and compacting the mixed concrete and whether a tower crane is to be used. Concrete may be transported by dumper, barrow or pumped.
- The cost of any measures necessary to protect the poured concrete either

due to excessive drying out in hot weather or damage from frost and low temperatures.

- Concrete mixes are described by the volume of the ingredients. For example (1:2:4) refers to a mix that is, by volume:
  - 1 part Portland cement
  - 2 parts sand
  - 4 part aggregate.
- When water is added during the mixing process the materials combine and reduce in volume by approximately 40%, although the percentage will depend on the mix of the concrete.

Quotations for sand and gravel can be either in tonnes of per cubic metre they can be adjusted as follows:

| **Material** | **Tonnes/m$^3$** |
|---|---|
| Cement in bags | 1.28 |
| Cement in bulk | 1.28–1.44 |
| Sharp sand (fine aggregate) | 1.60 |
| Gravel (coarse aggregate) | 1.40 |

On rare occasions concrete may have to be mixed by hand (a back-breaking and labour-intensive process); 4 hours/m$^3$ labourer should be allowed for hand mixing concrete on a wooden board.

### *Labour constants*

The following constants are average labour constants for mechanical mixing, transporting up to 25 metres, placing and compacting:

| | **Hours/m$^3$ labourer** |
|---|---|
| **Plain in situ concrete:** | |
| Concrete in trench filling | 2.0 |
| Horizontal work exceeding 300mm thick in structures | 1.5 |
| Sloping work not exceeding 15$^0$ exceeding 300mm thick in structures | 3.0 |

The following constants are average labour constants for mechanical mixing, transporting up to 25 metres, placing and compacting, packing around reinforcement and into formwork if necessary:

| | Hours/$m^3$ labourer |
|---|---|
| **Reinforced in situ concrete:** | |
| Concrete in trench filling | 4.0 |
| Sloping work not exceeding 15$^0$ exceeding 300mm thick in structures. | 4.5 |
| Vertical work exceeding 300mm thick in structures (Walls) | 6.0 |
| Vertical work exceeding 300mm thick in structures (Columns) | 8.0 |

## EXAMPLE

Reinforced in situ concrete (1:2:4), horizontal work not exceeding 300mm thick in structures – cost per $m^3$.

**Data**
Portland cement: £85.00 per tonne delivered to site
Sand: £10.00 per tonne delivered to site
Coarse aggregate: £9.00 per tonne delivered to site

**Materials**

| | £ | £ |
|---|---|---|
| 1$m^3$ cement = 1400 kg cement @ £85.00 per tonne | 119.00 | |
| Unloading 1 hour/tonne @ £13.54 per hour | 18.96 | |
| 2$m^3$ sand = 3200 kg @ £10.00 per tonne | 32.00 | |
| 4$m^3$ aggregate = 5600kg @ £9.00 per tonne | 50.40 | |
| | 220.36 | |
| Add shrinkage 40% | 88.14 | |
| | 308.50 | |
| Add waste 2.5% | 7.71 | |
| Cost per 7$m^3$ | £316.21 | |
| Cost per $m^3$ ÷ 7 | £45.17 | 45.17 |
| **Mixing** | | |
| Assume 200 litre mechanical fed mixer @ £20.00 per hour | | |
| Output 4$m^3$ per hour – cost per $m^3$ | | 5.00 |
| **Placing** | | |
| 4 hours labourer @ £13.54 | | 54.16 |
| | | 104.33 |

| Add Profit & Overheads 15% | 15.65 |
|---|---|
| Cost/$m^3$ | **£119.98** |

**Mixer outputs in $m^3$ of concrete per hour**

Hand loaded

| 100 litre | 150 litre | 175 litre | 200 litre |
|---|---|---|---|
| 1.2 | 1.8 | 2.1 | 2.4 |

Mechanical feed

| 200 litre | 300 litre | 400 litre | 500 litre |
|---|---|---|---|
| 4.0 | 7.2 | 9.6 | 12.0 |

### *Reinforcement*

It is usual for bar reinforcement to be delivered to site, cut to length and bent in accordance with the bending schedules. Steel fixers are entitled to additional payments in accordance with the Working Rule Agreement. Reinforcement is usually fixed by black tying wire; 5% should be added for binding wire and rolling margin. Allow four hours per tonne labour for unloading.

### *Labour constants*

Labour cutting, bending and fixing reinforcement, per 50kg:

- 10mm diameter bars: 4.00 hours
- 10mm–16mm bars: 3.00 hours
- over 16mm bars: 2.75 hours.

### *Fabric reinforcement*

As an average 30m$^2$ per hour should be allowed for cutting and fixing fabric reinforcement. An allowance of 12.5–15% for waste and laps should be made and the areas in the bills of quantities will be net areas and exclude laps between sheets.

#### EXAMPLE

Mesh reinforcement Ref A252 weighing 3.95 kg/m$^2$ with 150mm minimum side and end laps – cost per m$^2$.

**Data**

A252 Mesh reinforcement, weighing 45.5 kg/m² size 4800 × 2400mm – £55.00 (11.52m²)

Assume 23m² (approximately 2 sheets)

| **Material** | £ |
|---|---|
| 23m² of fabric A252 reinforcement delivered to site | 110.00 |
| Add 15% laps and waste | 16.50 |
| | 126.50 |
| Labour $\frac{23}{30}$ hours steelfixer @ £17.76 | 13.62 |
| Cost per 23m² | 140.12 |
| Add 15% Profit & Overheads | 21.02 |
| | 161.14 |
| ÷ 23 cost per m² | £7.01 |

***Formwork***

Formwork supports and forms the poured concrete until such time as the material has reached its design strength and then the formwork can be struck, or taken down. Traditionally formwork, which has to bear considerable weight, was made from timber and erected by a carpenter, but latterly metal forms and props have become more commonplace as they can be used more often than the timber equivalent. Metal formwork is more expensive than using timber, but the time taken to erect is considerably less and uses less labour both skilled and unskilled. If timber formwork is used the amount to which the formwork can be reused has a considerable effect on the cost. Timber formwork to large areas such as floor slabs can be re-used more often than formwork to smaller items such as columns or beams. Formwork can also be used to produce fair face or textured finishes to the face of the concrete, or if the finished structure is to be hidden from view it can be left 'as struck'.

A typical build-up for metal formwork would be as follows:

- hire cost of metal formwork and props delivered to site for the required period;
- labour fixing and stripping;

- any treatment to the face of the formwork;
- a small allowance to cover end pieces, etc. which may have to be made from timber.

The following are average labour constants for fixing and stripping metal formwork and are in skilled hours per m²:

| | **Fix** | **Strip** |
|---|---|---|
| Soffits of horizontal work not exceeding 300mm thick, propping not exceeding 3m high | 1.80 | 1.00 |
| Soffits of sloping work, sloping one way | 2.50 | 1.20 |
| Vertical faces of walls | 2.00 | 1.00 |

The following are average labour constants for fixing and stripping metal formwork and are in skilled hours per metre:

| | **Fix** | **Strip** |
|---|---|---|
| Edges of horizontal work not exceeding 500mm | 0.20 | 0.10 |

### *Precast concrete units*

Items such as lintels, sills, copings and plank floors come to site as precast units ready to be placed into position. To the prime cost of the precast units the additional costs of the following items should be considered:

- unloading and stacking
- hoisting and bedding in place
- material for bedding
- waste (damaged or broken units) – 2.5%
- profit and overheads.

Allow 0.30 hour bricklayer and labourer per linear metre for hoisting and fixing a 225 × 150mm lintel up to 3 metres above ground.

## Section F: masonry

### *Mortar*

In addition to the bricks or blocks the cost of the mortar also has to be calculated. Mortar comes in a variety of mixes depending on location and type of bricks or blocks being used; mortar is usually made from cement and sand. Generally, the mortar should be weaker than the bricks or blocks, so if pressure is placed on a brick, it is the mortar that should fail without causing the brickwork to crack. NRM2 requires that the type of pointing should be stated and this is particularly relevant to facing brickwork. Pointed brickwork requires additional labour and therefore cost. Common pointing types are weather struck (or struck), flush and bucket handle joints.

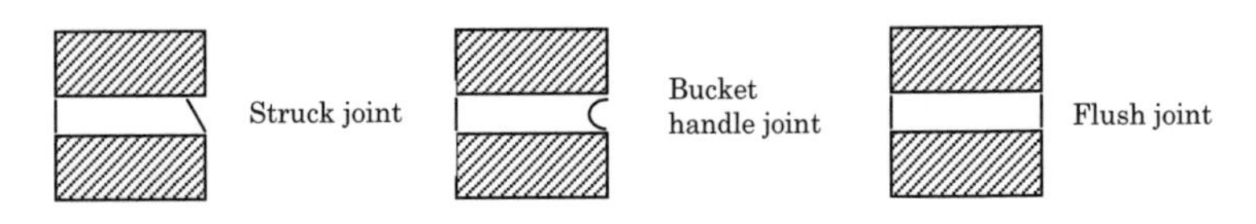


**Figure 5.1**

EXAMPLE

Cement mortar (1:3)

**Materials**

| | £ | £ |
|---|---|---|
| $1m^3$ cement = 1400 kg cement @ £85.00 per tonne | 119.00 | |
| Unloading 1 hour/tonne @ £13.54 per hour | 18.96 | |
| $3m^3$ sand = 4800 kg @ £10.00 per tonne | 48.00 | |
| | 185.96 | |
| Add shrinkage 25% | 46.49 | |
| | 232.45 | |
| Add waste 5% | 11.62 | |
| Cost per $4m^3$ | £244.07 | |
| Cost per $m^3$ ÷ 4 | £61.01 | 61.01 |
| Mixing | | |
| Assume 100 litre mixer @ £16.00 per hour | | |
| Output $4m^3$ per hour – cost per $m^3$ | | 4.00 |
| Cost per $m^3$ | | **£65.01** |

For small quantities of mortar mixed by hand on a board, allow 4 hours/$m^3$ labourer. The amount of mortar required per $m^2$ of half-brick wall is $0.03m^3$.

***Bricks***

The usual size for a clay brick is 215 × 102.5 × 65mm with a 10mm mortar joint making the nominal size 225 × 112.5 × 75mm. A 112.5mm-thick wall is referred to as a half-brick wall and there are 60 bricks per $m^2$ in a half-brick wall. Once upon a time bricks were delivered to the site loose in a lorry and tipped, causing a great deal of waste, now bricks are packed on pallets in polythene and unloaded by mechanical hoist thereby cutting down the amount of wastage and reducing the need for stacking. The amount of wastage, however, will vary considerably and depending on the nature and complexity of the work allow 7.5–12.5% for cutting and waste.

For facework, the number of bricks per m$^2$ will vary according to the specified brick bond as follows:

| | |
|---|---|
| Stretcher bond | 60 |
| Header bond | 120 |
| English bond | 90 |
| Flemish bond | 80 |

### *Labour constants*

The productivity of labour will depend again upon the nature of the work and in addition the organisation of the gang. A bricklaying gang is made up from bricklayers and labourers. It is the responsibility of the labourers to keep the bricklayers supplied with bricks and mortar so that they are able to maximise their output without having to keep breaking off to replenish their materials. It is usual to have two bricklayers to one labourer.

### Example

Walls one brick thick in common bricks in cement mortar (1:3) – cost per m$^2$.

**Data**

Common bricks £220.00 per thousand delivered to site

Assume 10m$^2$

| **Materials** | £ | |
|---|---|---|
| 1200 Common bricks @ £220.00 per thousand | 264.00 | |
| Add waste 7.5% | 19.80 | |
| Mortar | | |
| 10m$^2$ × 0.06 = 0.6m$^3$ @ £64.86 | 38.92 | |
| Add waste 5% | 1.95 | |
| | £324.67 | 324.67 |

**Labour**
Based on gang rate:

2 bricklayers and 1 labourer

| | |
|---|---|
| | 2 × £17.76 = £35.52 |
| | £13.54 |
| Hourly rate per gang | £49.06 |

1 bricklayer can lay 60 bricks per hour, therefore
output = 120 bricks per gang hour = 2 $m^2$ = £24.53/$m^2$

| | |
|---|---|
| 10$m^2$ @ £24.53/$m^2$ | 245.30 |
| Cost per 10$m^2$ | 569.97 |
| Add 15% Profit & Overheads | 85.50 |
| | 655.47 |
| ÷ 10 cost per $m^2$ | **£65.55** |

EXAMPLE

Walls half-brick thick in Oast Russett facings in stretcher bond in cement mortar (1:3) with weather struck joint one side – cost per $m^2$.

**Data**
Oast Russett facings: £820.00 per 1,000 delivered to site

Assume 10$m^2$

**Materials**

| | £ | £ |
|---|---|---|
| Facings | | |
| 600 facings @ £820.00 per 1,000 | 492.00 | |
| Add waste 7.5% | 36.90 | |
| Mortar | | |
| 10$m^2$ × 0.03 = 0.3$m^3$ @ £64.86/$m^3$ | 19.46 | |
| Add waste 5% | 0.97 | |
| | 549.33 | 549.33 |

**Labour**
2 bricklayers and 1 labourer

| | |
|---|---|
| | 2 × £17.76 = £35.52 |
| | £13.54 |
| Hourly rate per gang | £49.06 |

1 bricklayer can lay 60 bricks per hour, therefore
output = 120 bricks per gang hour = 2 $m^2$ = £24.54/$m^2$

| | |
|---|---|
| 10$m^2$ @ £24.54/$m^2$ | 245.40 |
| Cost per 10$m^2$ | 794.73 |
| Cost per $m^2$ | 79.47 |
| Add 15% profit & overheads | 11.92 |
| Cost per $m^2$ | **£91.39** |

**Sundry labour constants hours per $m^2$**

| **Item** | **Bricklayer** | **Labourer** |
|---|---|---|
| Forming cavity 50mm wide including building in wall ties at 5/$m^2$ | 0.10 | 0.05 |
| Damp-proof course not exceeding 300mm wide | 0.20 | 0.10 |

**Sundry labour constants hours per m**

| **Item** | | |
|---|---|---|
| Extra over walls for closing cavities with brickwork at opening perimeters | 0.35 | 0.18 |
| Deemed to be included items | | |
| Bonding half-brick wall to existing | 1.10 | 0.55 |
| Bedding and pointing frames | 0.05 | 0.025 |

**Sundry labour constants hours per no. in *existing buildings***

| **Item** | | |
|---|---|---|
| Holes for pipes 50mm diameter in half-brick wall | 0.25 | 0.15 |
| Ditto over 110mm diameter in half-brick wall | 0.30 | 0.25 |
| Ditto over 150mm diameter in half-brick wall | 0.55 | 0.35 |

***Extra over***

Some masonry items are required to be described as extra over and the estimator must calculate the extra cost of the labour and material involved with these items.

EXAMPLE

I layer pitch polymer damp-proof course not exceeding 300mm wide horizontal bedded in gauged mortar (1:1:6) – cost per m$^2$.

**Data**
Pitch polymer dpc 112.5mm wide, £11 per 20m delivered to site

Assume 10m

| | | £ |
|---|---|---|
| Material | | |
| 20m pitch polymer dpc @ £11.00/m | | 11.00 |
| Add 5% Waste and laps | | 0.55 |
| Labour | | |
| 0.10 hours bricklayer @ £17.76 | £1.78 | |
| 0.05 hours labourer @ £13.54 | £0.68 | |
| | **£2.46 per m** | |
| 10m @ £2.46 per m | | 24.60 |
| | | 36.15 |
| Add 15% Profit & Overheads | | 5.43 |
| Cost per 10m | | £ 41.58 |
| **÷ 10 cost per m** | | **£4.16** |

***Blockwork***

Blocks are heavier and larger than bricks; the usual size is 440mm × 215mm × 100mm thick, excluding the mortar joint. There are three main types of block; aerated, dense and hollow clay. Like bricks, blocks are delivered to site stacked on pallets for easy unloading and use. Larger than bricks, blocks need more time for transporting around the site and hoisting in place. Allow 5% waste.

EXAMPLE

Walls in lightweight concrete blocks (3.5KN) in gauge mortar (1:1:6) – cost per m$^2$.
Assume 10m$^2$

**Data**

440 × 215 × 100mm aerated concrete blocks: £9.50 per $m^2$ delivered to site

Gauge mortar: hydrated lime 0.6 tonnes/$m^3$, £25.00 per tonne

Gauge mortar (1:1:6)

**Materials**

| | £ | £ |
|---|---|---|
| 1$m^3$ cement = 1400kg cement @ £85.00 per tonne | 119.00 | |
| Unloading 1 hour/tonne @ £13.54 per hour | 18.96 | |
| 1$m^3$ hydrated lime = 600kg @ £25.00 per tonne | 15.00 | |
| Unloading 1 hour/tonne @ £13.54 per hour | 8.12 | |
| 6$m^3$ sand = 9600kg @ £10.00 per tonne | 96.00 | |
| | 257.08 | |
| Add shrinkage 25% | 64.27 | |
| | 321.35 | |
| Add waste 5% | 16.07 | |
| Cost per 8$m^3$ | £337.42 | |
| Cost per $m^3$ ÷ 8 | £42.18 | 42.18 |
| **Mixing** | | |
| Assume 100 litre mixer @ £16.00 per hour | | |
| Output 4$m^3$ per hour – cost per $m^3$ | | 4.00 |
| Cost per $m^3$ | | **£46.18** |
| Blocks | | |
| 10$m^2$ blockwork @ £9.50/$m^2$ | 95.00 | |
| Waste 10% | 9.50 | |
| Mortar 0.01$m^3$ per $m^2$ = 0.1$m^3$ @ £46.18 | 4.62 | |
| Add waste 5% | 0.23 | |
| | 109.35 | |
| Labour | | |
| 0.50/$m^2$ per gang per $m^2$ = 5 gang hours @ £49.06 | 245.30 | |
| | 354.65 | |
| Add Profit & Overheads 15% | 53.20 | |

| | |
|---|---|
| Cost per 10m² | 407.85 |
| Cost per m² | **£40.78** |

## Roofing

### *Asphalt work*

Asphalt is manufactured in 25kg blocks and then heated to melting point and applied on site. The following is the approximate covering capacity of 1,000 kg of asphalt:

| | |
|---|---|
| first 12mm thickness | 35 square metres |
| additional 3mm thickness | 150 square metres |

For example, the following is the weight of asphalt required per m² 19mm thick:

$$\text{first 12mm} = \frac{1{,}000\text{ kg}}{35\text{ square metres}} = 28.57\text{kg}$$

$$\text{next 6mm} = \frac{1{,}000\text{ kg}}{150\text{ square metres}} \times \frac{7\text{mm}}{3\text{mm}} = 15.56\text{kg}$$

**44.13kg**

Allow 2.5% waste

Asphalt is always applied on sheathing felt and this has to be included in the calculation. The finished surface must be protected against deterioration due to exposure to solar radiation; this is usually in the form of solar reflecting paint in the case of asphalt.

### *Labour constants: craftsman and labourer*

| | |
|---|---|
| 19mm two-coat asphalt roofing | 0.15 per m² |
| Skirtings, fascias and aprons 150–225mm girth (including internal angle fillet and labours) | 0.60 per m |

### *Felt roofing*

Felt roofing is the most common form of material for covering flat roofs. It is traditionally made from bitumen, but there are now more high-tech high-performance membranes available, based on polyester or similar material, that do not suffer from the main disadvantage of traditional materials, namely deterioration in 15–20 years. In the case of traditional bituminous felts, the material is supplied in rolls in a variety of weights and is laid usually in two or three layers. Felt is measured net but laid with side and end laps of 150mm and these must be taken into account in the calculation. The finished roofing must be protected against solar radiation, the most common approach being with the application of chippings. Allow 15% for laps and waste.

### *Pitched roof coverings*

The most common form of sloping roof coverings are slates and tiles; they are both fixed on softwood battens with nails which are in turn laid on underfelt. In Scotland in addition, sarking, treated sawn softwood boarding, is fixed directly to the rafters prior to the application of underfelt and battens. This practice is probably due to the harsher climate experience north of the border. Roof coverings are now generally delivered to site on pallets, pre-packed (see Figure 5.2) and usually ordered by the thousand.

Because the size of the slate or tile and the method of fixing affects the number of units required it is necessary to calculate the number of slates or tiles per m$^2$, although this type of information is readily available in manufacturer's literature that can be accessed online.

**Figure 5.2** Unloading roofing tiles

### *Slate roofing*

Slates come in a variety of sizes with two nail holes and can be either head nailed or centre nailed. Occasionally, roofers have to form the fixing holes in slates, on site using a special tool known as a zax. Head nailing tends to give better protection from the weather for the nails. A special shorter tile is used for the under eaves course; an undercloak course is required at verges in order to divert water back onto the roof.

Therefore, to calculate the number of centre nail slates size, 450mm × 300mm, laid with 100mm lap, apply the following:

$$\frac{\text{length of slate} - \text{lap}}{2} \times \text{width of slate}$$

$$\frac{0.450 - 0.100}{2} \times 0.300 = 0.0525$$

$$\frac{1\text{m}^2}{0.0525} = 19.05 \text{ say 19 slates per m}^2$$

The number of head-nailed slates per square metre is calculated using the following:

$$\frac{\text{length of slate} - \text{lap} - 25\text{mm}}{2} \times \text{width of slate}$$

Note: non-ferrous nails such as aluminum or copper should be used for fixing as this reduces nail fatigue.

Slates for roofing come in a variety of sizes, varying from 650 × 400mm to 250 × 150mm and very often are laid with diminishing courses towards the eaves and for this reason the number of slates required has to be carefully calculated once the details are known.

### *Synthetic slate*

Synthetic slates closely resemble their natural counterparts, but are much less expensive and lighter; the most common size is 600mm × 300mm. They are fixed as for centre-nailed natural slates, with a clip being used on the bottom edge to prevent lifting, with copper nails. Also available is a synthetic slate made from a mixture of resin and crushed natural slate. These are single-lap interlocking slates and fixed accordingly – see single-lap tiling below.

### *Tile roofing*

#### PLAIN TILES

Plain tiles are known, like slates, as double-lap coverings to prevent water from penetrating the coverings from run-off or capillary action. They are referred to as double-lap because for part of their area the slate or tile laps two others in the course below. The most common size for a plain tile is 265mm × 165mm.

#### INTERLOCKING TILES

All interlocking tiles are single lap which means that there is only one layer of tiles on the roof, apart from the overlaps. Interlocking tiles are not, with the exception of the perimeter tiles, nailed.

#### BATTENS

To calculate the linear quantity of batten per square metre, divide the gauge of the tiles into 1m:
Therefore for 100mm gauge:

$$\frac{1{,}000}{100} \times 1\text{m} = 10\text{m per m}^2 \text{ of roof}$$

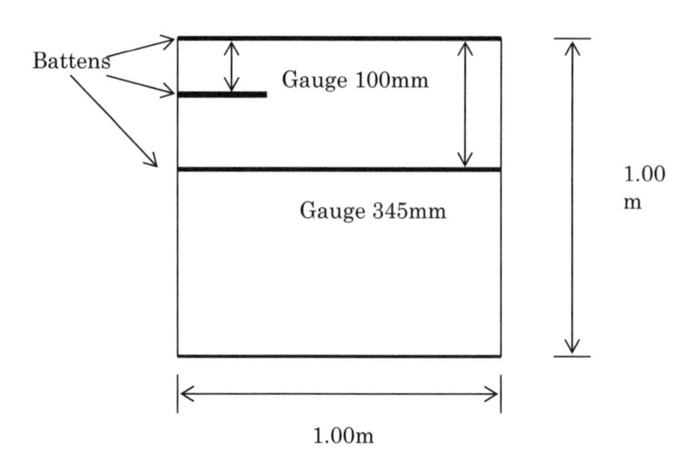


**Figure 5.3**

and for 345mm gauge:

$$\frac{1{,}000}{345} \times 1\text{m} = 2.9\text{m per m}^2 \text{ of roof}$$

There are a number of free calculators available on manufacturers' websites that calculate the quantity of tiles needed.

### Nails for tiling

Approximate number of 44mm nails per kilogram:

Copper: 290
Composition: 280

Note: tiles are usually nailed every fourth course.

### Nails for battens

Approximate number of nails per kilogram:

50mm × 10 gauge: 260
63mm × 10 gauge: 220

Allow 5% waste on nails

**Labour constants and materials for plain tiling per m²**

| Lap of tile | Number of tiles | Tile nails (Kg) | Battens (Lin m) | Batten nails (Kg) | Labour: tiler & labourer |
|---|---|---|---|---|---|
| 60 | 63 | 0.09 | 10.08 | 0.11 | 0.67 |
| 75 | 68 | 0.10 | 10.88 | 0.12 | 0.73 |
| 90 | 74 | 0.11 | 11.84 | 0.13 | 0.79 |
| 100 | 78 | 0.12 | 12.48 | 0.14 | 0.83 |

Fixing tile battens per 100m 2.50 hours roofer/1.25 hours labourer, including unloading.

### Example

Roof coverings 265 × 165mm Marley Eternit Hawkins Staffordshire Blue plain clay roofing tiles with 75mm head lap 40° pitch each tile nailed with 2

No. 44mm composition nails to 38 × 25mm treated sawn softwood battens at 100mm centres to one layer of felt to BS Type 1F with 100mm minimum laps fixed with galvanised clout nails – cost per $m^2$.

#### MATERIALS DELIVERED TO SITE

Hawkins Staffs Blue tiles £640.00 per 1,000 delivered to site and unloaded

| | |
|---|---|
| Underfelt | £12.00 per 15m × 1m roll |
| Nails for tiles | £2.00 per kg |
| Battens for tiles | £3.00 per 10m |
| Nails for battens | £1.00 per kg |

Assume $10m^2$

| **Materials** | £ | £ |
|---|---|---|
| Tiles (from Marley website) 60 per $m^2$ | | |
| 680 Hawkins Staffs Blue tiles @ £640 per 1,000 | 435.20 | |
| Add waste 10% | 43.52 | |
| Battens | | |
| 100m batten (see calculation above) | | |
| @ £3.00 per 10m | 30.00 | |
| Add waste 10% | 3.00 | |
| Nails for battens | | |
| 1.2kg @ £1.00 per kg | 1.20 | |
| Waste 5% | 0.06 | |
| Nails for tiles | | |
| 680 tiles × 2 = 1360 nails @ 280/kg = kg | | |
| 4.86 kg nails @ £2.00 | 9.72 | |
| Add waste 5% | 0.49 | |
| Underfelt | | |
| 15 × 1m roll = $15m^2$ | | |
| Cost per $10m^2 = \frac{10.00}{15.00} \times £12.00 =$ | 8.00 | |

| | | | |
|---|---|---|---|
| Add laps 10% | | 0.80 | |
| Add waste 5% | | 0.40 | |
| | | | 532.39 |
| Labour | | | |
| Tiles | | | |
| 7.30 hours roofer and labourer | £17.76 | | |
| | £13.20 @ £30.96 | | 226.01 |
| Tile battens | | | |
| 2.5 hours tiler | 17.76 | | 44.40 |
| 1.25 hours labourer | 13.59 | | 16.99 |
| | | | £819.83 |
| Add Profit & Overheads 15% | | | £122.98 |
| | | | £942.81 |
| ÷ 10 – cost/m² | | | £94.28 |

## Example

Roof coverings 420 × 330mm Marley Eternit Malvern concrete interlocking tiles with 75mm head lap 40° pitch with 38 × 25mm treated sawn softwood battens at 345mm centres fixed with clips including one layer of felt to BS Type 1F with 100mm minimum laps fixed with galvanised clout nails – cost per m².

Single-lap tiles laid to 345mm gauge as manufacturer's recommendations, therefore battens required is 2.90m per m².

Interlocking tiles are single-lap tiles fixed with nails or clips. They are generally made from concrete, with an applied finish and are heavier than slate or clay tiles and therefore should be used with caution when used as a replacement to lighter more traditional coverings on an existing roof. Their extra weight is a disadvantage from the site handling point of view, however they have greater coverage than slate or tiles and can be fixed more quickly, there being 9.8 interlocking tiles per m² compared with 60 for double-lap clay tiles; these factors combine to provide a cheaper alternative which is popular with speculative house builders.

**Materials delivered to site**

| | |
|---|---|
| Malvern concrete interlocking tiles | £450.00 per 1,000 |
| Underfelt | £12.00 per 15m × 1m roll |
| Nails for tiles | £2.00 per kg |
| Battens | £3.00 per 10m |

Nails for battens £1.00 per kg

| | £ | £ | |
|---|---|---|---|
| Assume $10m^2$ | | | |
| Tiles – from Marley website – 9.8 per $m^2$ | | | |
| 98 Malvern tiles @ £450 per 1,000 | 44.10 | | |
| Add waste 10% | 4.41 | | |
| Battens | | | |
| 29m batten @ £3.00/10m | 8.70 | | |
| Add waste 10% | 0.87 | | |
| Nails for battens | | | |
| 0.40 kg @ £1.00 per kg | 0.40 | | |
| Add waste 5% | 0.02 | | |
| Nails for tiles – nailed every 4th course | | | |
| $\frac{98}{4}$ = 25 nails say 30 to include waste | <u>0.20</u> | 58.70 | |
| Underfelt | | | |
| $15 \times 1m$ roll = $15m^2$ | | | |
| Cost per $10m^2 = \frac{10.00}{15.00} \times £12.00 =$ | | 8.00 | |
| Add laps 10% | | 0.80 | |
| Add waste 5% | | <u>0.40</u> | 67.94 |

Labour

Tiler and labourer can lay $3.5m^2$ of interlocking tiles per hour

| | | | |
|---|---|---|---|
| Tiles | | | |
| 2.86 hours tiler and labourer £17.76 <br> <u>£13.54</u> 31.30 | | | 89.52 |
| Tile battens | | | |
| 0.30 hours tiler | 17.76 | | 5.33 |

| | | |
|---|---|---|
| 0.15 hours labourer | 13.54 | 2.03 |
| | | £164.82 |
| Add Profit & Overheads 15% | | £24.72 |
| | | £189.54 |
| ÷ 10 – cost/m$^2$ | | £18.95 |

**Woodwork**

There are two general groups of wood: hardwood and softwood. These are defined by a number of characteristics. Hardwood is timber produced from broad-leaved trees and soft wood from coniferous trees. Structural timbers, such as floor and roof joists, are stress graded according to the proportion of defects. For softwood the gradings are divided into GS (general structural) and SS (special structural). Timber is further defined by species, e.g. Scots pine. Hardwoods have only one grading. Sizes for timber sections are stated in two ways; ex and finished. Unplanned or sawn timber is referred to as 'ex' once planed it becomes wrot timber and referred to as finished. Therefore timber sections may be referred to in two ways depending on whether the sizes quoted are 'ex' or finished:

50 × 25mm ex softwood batten
46 × 21mm finished softwood batten

The finishing process planes 2mm from each face. Softwood is sold by the linear, square and cubic metre.

For measurement purposes NRM2 divides timber into the following:

- 16 Carpentry/carcassing: structural/timber framing such as roof and floor joists, roof trusses, etc. Timber in this category is generally sawn softwood, treated to protect it against attack by decay and rot. Most modern roofs are constructed from roof trusses or trussed rafters, which are prefabricated and delivered to site ready for lifting into position by crane. The timbers, which are typically 80 × 40mm, are held together with special metal plates. Roof trusses have several advantages over traditional in situ roof construction, namely:
  - speed;
  - skilled labour is not required;
  - spans of up to 12m can be accommodated;
  - cost savings compared to traditional construction.

  Once hoisted into position it is necessary to secure the roof trusses with softwood binders and galvanised metal straps to prevent distortion.

- 22 General joinery first fixings: gutter boards and fascia, generally softwood with wrot faces.
- Unframed isolated trims/sundries: (second fixings): skirtings, architraves, either softwood or hardwood with a wrot finish, some ironmongery.
- 23 Windows, screens and lights and associated ironmongery.
- 24 Doors, shutters and hatches and associated ironmongery.
- 25 Stairs, walkways and balustrades.

### *Carpentry/carcassing*

**Sundry labour constants**

| **Item** | **Hours/m – carpenter** |
|---|---|
| Plates, sleepers | 0.10 |
| Partitions | 0.15 |
| Floor joists | 0.18 |
| Upper floor joists | 0.20 |
| Strutting between 200mm joists | 0.35 |

#### Example

100 × 75mm softwood treated wall plate – cost per m.

**Data**

Assume 10 linear metres

100 × 75mm sawn softwood treated plate – £5.00/m delivered to site and unloaded

| | | |
|---|---|---|
| **Materials** | | |
| 10m 100 × 75mm sawn softwood @£5.00/m | £50.00 | |
| Add waste 7% | £3.50 | |
| | £53.50 | |
| **Labour** | | |
| 1 hour carpenter per 1m | | £53.50 |
| 1 hour carpenter @ £17.76 | | £17.76 |
| | | £71.26 |
| Add Profit & Overheads 15% | | £10.69 |
| Cost per 10m | | £81.95 |
| ÷ 10 cost per m | | £8.20 |

## *General joinery: first fixings*

Sundry labour constants

| Item | Hours carpenter/m |
|---|---|
| Gutter boarding | 1.00 |
| 19mm fascia | 1.00 |
| Barge board | 0.40 |

### EXAMPLE

18mm flooring grade horizontal chipboard flooring exceeding 600mm wide – cost per m².

**Data**
£13.30 per 2400 × 600mm sheet delivered to site and unloaded

Assume 10m².

**Materials**
10m² 18mm chipboard flooring @ £13.30 per 1.44m² sheet

$$\frac{10.00}{1.44} \times £13.30 = \quad £92.36$$

| | |
|---|---|
| Nails 0.1kg/m² 1kg @ £2.00/kg | £2.00 |
| | £94.36 |
| Add waste 10% | £9.44 |
| | £103.80 |

**Labour**
0.50 hours carpenter/m²

| | |
|---|---|
| 5 Hours @ £17.76 | £88.80 |
| | £192.60 |
| Add Profit & Overheads 15% | £28.89 |
| | |
| Cost per 10m² | £221.49 |
| ÷10 cost per m | £22.15 |

Unframed second fixings
Sundry labour constants

| **Item** | **Hours/m – carpenter/joiner** |
|---|---|
| Architraves (19 × 50mm) | 0.10 |
| Skirtings (19 × 75mm) | 0.15 |
| Window boards | 0.30 |
| Allow 0.05kg nails per m | |

EXAMPLE

25 × 50mm chamfered softwood architrave – cost per m.

**Data**
25 × 50mm chamfered architrave delivered to site – £3.00/m

63mm oval nails delivered to site – £2.00/kg

Assume 10m

**Materials**

| | |
|---|---|
| 10m 25 × 50mm architrave @ £3.00/m | £30.00 |
| Nails 0.5kg nails @ £2.00/kg | £1.00 |
| | £31.00 |
| Add waste 7% | £2.17 |
| | £33.17 |
| Labour | |
| 1 hour carpenter/joiner @£17.76 | £17.76 |
| | £50.93 |
| Add Profit & Overheads 15% | £7.64 |
| Cost per 10m | £58.57 |
| Divide by 10 cost per m | £5.86 |

**Doors, shutters and hatches and associated ironmongery**
Sundry labour constants
Doors – allowances include for fixing and hanging on standard hinges

| **Item** | **Hours joiner each** |
|---|---|
| Standard door | 1.40 |
| Solid core doors | 2.00 |
| Door lining set | 3.00 |
| 1050 × 1200mm | |
| Standard casement | 1.10 |
| Staircases – straight flight | 5.00 |

EXAMPLE

40 × 762 × 1981 John Carr flush external door (ref 362001) – £50.90 delivered to site

| Materials | |
|---|---|
| 40 × 762 × 1981 external door | £50.90 |
| Labour 1.40 hours @ £17.76 | £24.86 |
| | £75.76 |
| Add 15% Profit & Overheads | £11.36 |
| Cost per door – supplied and fixed | £87.12 |

## *Finishes*

### IN SITU FINISHES

Most modern in situ plasterwork is carried out in lightweight gypsum plaster, known by its trade names Carlite or Thistle, although Carlite has now been merged into the Thistle brand. It is delivered to site in 25kg bags on pallets for ease of unloading. These retarded hemihydrate plasters are premixed and contain lightweight aggregate and gypsum plaster and require only the addition of water to make them ready for use. Thistle plasters come in a variety of types depending on where they are to be used and the background to which the plaster is applied. The finish grades are used neat, while the undercoat grades are usually mixed with sand. According to NRM2, work over 600mm wide is measured in square metres while work under 600mm wide is measured in linear metres.

It is very important that the correct type of plaster is used. The five types (four undercoats and one topcoat) available are as follows.

### UNDERCOATS

**Browning plaster** is an undercoat plaster for moderate suction solid backgrounds that have a good mechanical key, such as brickwork or blockwork. A slow-setting variety is available that gives greater time for application.

**Bonding plaster** is an undercoat plaster for low suction backgrounds, for example concrete or plasterboard or surfaces sealed with pva (a universal water-based adhesive).

**Toughcoat** is an undercoat plaster for solid backgrounds of high suction with an adequate mechanical key.

**Hardwall** is an undercoat plaster that provides a much harder and more durable finish and is also quick drying.

### *Finishing coats*

**Finishing plaster** is an ideal choice over sand and cement bases and can be used on still damp backgrounds.

**Board finishing plaster** is a one-coat plaster for skim coats to plasterboard.

**Multi-finish** is used where both undercoat and skim coat are needed on one job. Suitable for all suction backgrounds and ideal for amateur plasterers.

In addition, **universal one coat** is a one-coat plaster for a variety of backgrounds, suitable for application by hand or machine.

*The White Book* published by British Gypsum is a good source of reference for all types of plaster finishes. It can be viewed at the following web page:

http://www.british-gypsum.bpb.co.uk/literature/white_book.aspx

### *What is used in practice?*

For plasterboard ceilings: board finish plaster.

For blockwork walls: an 11mm-thick undercoat of browning plaster followed by 2mm skim coat of finishing plaster. The undercoat is lightly scratched to form a key.

Metal lathing: two undercoats are often required followed by a finishing coat.

Galvanised or stainless steel angle beads are used to form external angles in in situ plaster.

In situations where damp walls are plastered following the installation of an injection dpc, most gypsum plasters are not suitable as they absorb water and fail.

#### *Constants*

| Approximate coverage per tonne | $m^2$ |
|---|---|
| 11mm browning plaster | 135–155 |
| 11mm tough coat plaster | 135–150 |
| 11mm hardwall plaster | 115–130 |
| 2mm board finishing plaster | 410–430 |
| 2mm finishing plaster | 410–430 |
| 13mm universal one coat | 85–95 |

***Labour constants***

| **Item** | **Plaster and labourer/10m²** |
|---|---|
| Base coat | 2 hours |
| Finishing coat | 2.5 hours |

For ceilings increase by 25%

### *Sand/cement screeds*

The function of a floor screed is to provide a smooth and even surface for finishes such as tiling and is usually a mixture of cement and coarse sand, typically in the ratio of 1:3. It is usually a dry mix with the minimum of water added and has a thickness of between 38 and 50mm and is laid on top of the structural floor slab.

### *Board finishes*

Plasterboard is available in a variety of thicknesses, the most common being 9.5mm and 12mm, and is fixed to either metal or timber studs with screws. Standard sheets are 900 × 1800mm and 1200 × 2400mm. It is also possible to fix plasterboard to block work with plaster dabs. Boards can either be pre-finished and require no further work or finished with a skim coat of plaster. Boards with tapered edges and foil backing are also available.

### *Floor and wall tiling*

Most wall tiles are 100 × 100mm with spacers to ensure ease of fixing. Tiles require adhesive for fixing and grout for the joints. Thickness can vary and floor tiles can be found in a wider variety of sizes.

### EXAMPLE

13mm-thick two-coat plaster comprising 11mm-thick Thistle bonding coat and 2mm-thick Thistle multi-finish plaster on block walls over 600mm wide internally with steel trowel finish – cost per m².

Assume 10m²

**Data**

Bonding plaster – £347.00 per tonne delivered to site on pallet and unloaded

Multi-finish plaster – £210.00 per tonne delivered to site on pallet and unloaded

**Materials**

Approximate coverage of bonding plaster 11mm thick – 100m²/tonne
Approximate coverage of multi-finish plaster 2mm thick – 420m²/tonne

Bonding plaster

| | |
|---|---|
| $\frac{10.00}{100.00} \times$ £347.00 = | £34.70 |

Finishing plaster

| | |
|---|---|
| $\frac{10.00}{420.00} \times$ £210.00 = | £5.00 |
| | £39.70 |
| Add waste 10% | £3.97 |
| | £43.67 |

**Labour**

Plasterer and labourer

| | | | |
|---|---|---|---|
| Base coat | 2.00 hours | | |
| Finishing coat | 2.50 hours | | |
| | 4.50 hours @ | £17.76 | |
| | | £13.54 | |
| | | £31.30 | £140.85 |
| | | | £184.52 |
| Add Profit & Overheads 15% | | | £27.68 |
| Cost per 10m² | | | £212.20 |
| ÷10 cost per m² | | | £21.22 |

**EXAMPLE**

50mm thick cement and sand (1:4) level screeded bed over 600mm wide level and to falls only not exceeding 15° from horizontal to concrete base – cost per m².

**Data**

Portland cement: £85.00 per tonne delivered to site
Sand: £10.00 per tonne delivered to site

| **Materials** | £ |
|---|---|
| 1m$^3$ cement = 1400kg cement @ £85.00 per tonne | 119.00 |
| Unloading 1 hour/tonne @ £13.54 | 18.96 |
| | |
| 4m$^3$ sand = 6400 kg @ £10.00 per tonne | 64.00 |
| | 201.96 |
| | |
| Add shrinkage 25% | 50.49 |
| Cost per 5m$^3$ | 252.45 |
| Cost per m$^3$ ÷ 5 | 50.49 |
| | |
| Mixing | |
| Assume 200 litre mechanical fed mixer @ £20.00 | |
| per hour: output 4m$^3$ per hour – cost per m$^3$ | 5.00 |
| Cost per m$^3$ | 55.49 |
| Cost per m$^2$ – 50mm thick | 2.52 |
| | |
| Placing | |
| 0.30 Craftsman & labourer per m$^2$ | |
| 0.30 hours @ £17.76 | |
| £13.54 | |
| £31.30 | 9.39 |
| | 11.91 |
| Add Profit & Overheads 15% | 1.79 |
| Cost per m$^2$ | £13.70 |

### Example

100 × 100mm × 7mm thick white rustic wall tiles exceeding 600mm wide fixed to plaster with tile adhesive including grouting joints as work proceeds – cost per m$^2$ .

**Data**
White rustic wall tiles – £14.97 per pack (68 tiles)
Unibond wall tile adhesive and grout – £15.00 per 10 litres

5m$^2$ per 10 litres 1mm-thick adhesive including grout
Assume 1m$^2$

| **Materials** | £ |
|---|---|
| Tiles | |
| Coverage 95 tiles per m$^2$ | |

95 tiles @ £14.97 per 68 tiles

| | |
|---|---|
| 95/68 @ £14.97 | 20.91 |
| Add waste 10% | 2.09 |
| Adhesive and grout | |
| 2 litres @ £15.00 per 10 litres | 3.00 |
| Add waste 5% | 0.15 |
| | 26.15 |
| Labour | |
| 0.6 hour per $m^2$ @ £17.76 | 10.66 |
| | 36.81 |
| Add Profit & Overheads 15% | 5.52 |
| Cost per $m^2$ | £42.33 |

## Glazing

There are a number of different types of glass in a range of tints and patterns as follows.

### *Ordinary sheet glass*

Relatively cheap with a manufacturing process that produces a glass with a certain amount of distortion and because of this it tends to be used for garden sheds and greenhouses.

### *Float glass (plate)*

A superior type of glazing available in a variety of thicknesses from 3mm. More expensive than sheet glass.

### *Patterned (obscure glass)*

Made from flat glass with a design rolled onto one side, used in bathrooms and similar locations.

Other commonly used types of glazing are:

- energy efficient glass
- self-cleaning glass
- toughened (safety glass)
- laminated glass
- wired glass.

Waste can be substantial and should be allowed for at the rate of 10% for cutting glass to size and an additional 10% for fixing. Glass is fixed into place with a combination of glazing compound and glazing beads on which wastage should be allowed for at 5%. Allow 0.20 kg of glazing compound per metre of bedding.

### EXAMPLE

3mm clear float glass and glazing to wood with glazing compound in panes size 900 × 900mm – cost per m$^2$.

**Data**

| | |
|---|---|
| Glass cut to size and delivered to site | £15.00/m$^2$ |
| Glazing compound | £0.50/kg |

Assume 10m$^2$

| | | |
|---|---|---|
| **Materials** | | |
| 10m$^2$ 3mm clear float glass @ £15.00 | 150.00 | |
| Add waste 10% | 15.00 | |
| 40m of compound = 8kg compound @ £0.50 | 4.00 | |
| Waste on compound 5% | 0.20 | |
| | £169.20 | 169.20 |
| **Labour** | | |
| 10m$^2$ glass @ 15m$^2$ per hour | | |
| 0.67 hours glazier @ £17.76 | | 11.90 |
| | | 181.10 |
| Add Profit & Overheads 15% | | 27.16 |
| Cost per 10m$^2$ | | 208.26 |
| ÷ 10 cost per m$^2$ | | £20.83 |
| Cost per pane 900 × 900mm | | £16.87 |

## Plumbing

A large section covering a wide range of items of a specialist nature; only a limited number of items have been included here.

| **Rainwater installations** | **Hours – craftsman** |
|---|---|
| 110mm plastic rainwater pipes | 0.50 per 3m length |
| 65mm plastic rainwater pipes | 0.50 per 3m length |
| Fittings | 0.10 each |
| | |
| **Waste installations** | |
| 32mm plastic waste | 0.30 per 3m length |
| 110mm plastic soil pipe | 0.35 per 3m length |
| Fittings | 0.15 each |
| | |
| **Hot and cold water installations** | |
| 15mm light gauge copper tube with capillary joints fixed with pipe clips @ 1m centres | 0.20 per metre |
| 28mm ditto | 0.25 per metre |
| Fittings | 0.40 each |
| Pipe ancillaries | 0.80 each |
| | |
| **Sanitary fittings** | |
| Lavatory basin including waste and pair taps | 2.50 per unit |
| Bath including waste and pair taps | 2.80 per unit |
| Close-coupled WC | 2.25 per unit |
| Shower tray with waste, trap, mixer and shower fitting | 6.00 per unit |

## Decoration

Decoration, or more particularly, painting is one of the building trades that has changed radically over the last 30–40 years, with the introduction of painting systems rather than oil-based paints. In its widest sense, decoration includes painting on a variety of surfaces and paper hanging.

The traditional approach to painting wood is KPS 3 oils or knot, prime, stop and apply two undercoats and one gloss coat to external surfaces and one undercoat and one gloss coat to internal surfaces. Knotting is a shellac/methylated spirit based sealer that is applied to knots in the wood, thereby preventing the sap in the knot from seeping out and bubbling through the paint finish blistering the surface. Prime is a coat of primer applied to the bare wood and stopping was traditionally linseed oil putty applied to all small holes and cracks in the wood. It is a popular belief that paint fills small holes and cracks, but that is not the case.

All these products are still available today but paints have moved away from being oil-based to solvent-based or water-based and wood filler is used in preference to putty as it does not dry out and crack so easily.

In practice, it is now usual to apply a coat of primer or combined primer/undercoat followed by only one gloss coat whether internal or external, except in very exposed or extreme conditions.

Increasingly in some locations, public staircases, etc., paint is applied by spraying. Portaflek and Multiflek finishes were very popular in the 1970s, but now they have been superseded by compliant water-based versions: Aquaflek and Aquatone. The advantage of spray application is that it is considerably quicker than using a brush or roller; the disadvantages are that a considerable amount of time is required for masking, protection and cleaning.

When pricing painting and decorating it is important to study the specification or preambles to the section as this will contain information relating to the preparation work that is expected. More than most trades this can be extensive and include washing down, burning off existing paint and stripping existing paper, etc.

Two peculiarities of estimating painting and decorating are brush money; an allowance given to cover the cost of new brushes, rollers, etc., and an allowance to cover cleaning overalls, etc. Both of these allowances are covered in the National Working Rule Agreement mentioned previously in this chapter.

### *Wallpaper*

Wallpaper is specified in 'pieces' which is more commonly called a roll and the standard UK roll or piece is 520mm wide × 10.5m long. Some European papers are a little narrower and many American rolls are roughly twice as wide. The amount of paper required will depend on the pattern, the size of the room and the number of openings, etc. Once again the surface to be papered may need to be prepared, for example new plasterboard will require at least two coats of dry wall primer, to adjust the suction whereas old plastered walls can become very dry and powdery, often referred to as 'blown' and need a coat of plaster sealer. Paper hanging to ceiling takes approximately 50% longer than walls! Newly plastered walls will required a coat of size, usually dilute paper adhesive, to adjust the sunction.

It is common to have wallpaper specified in the bills of quantities as a prime cost or provisional sum and the labour for hanging measured as a hanging/fix only item.

| **Sundry labour constants** | **Skilled hours per 100m²** |
|---|---|
| Knotting and stopping | 4.5 |
| Priming | 11.00 |
| Undercoat | 14.00 |
| Gloss coat | 18.00 |
| Emulsion paint | 8.00 |
| Preparation of surfaces (rubbing down between coats, etc.) | 4.00 |

| **Wallpaper** | **Skilled hours per 10m²** |
|---|---|
| Strip off old paper to walls | 1.00 |
| Applying size | 0.50 |
| Lightweight paper | 2.00 |
| Heavyweight paper | 2.50 |

EXAMPLE

Prepare and two coats of emulsion on general surfaces of plaster internal – cost per m².

**Data**
Emulsion paint – £12.00 per 2.5 litres

Assume 100m²

| **Materials** | | £ |
|---|---|---|
| Coverage | | |
| 1st coat 15m² per litre | 100 ÷ 15 = 6.67 | |
| 2nd coat 8m² per litre | 100 ÷ 8 = 12.50 | |
| | 19.17 say 19 litres | |
| 19 litres of emulsion @ £12 per 2.5 litres | | 91.20 |
| **Labour** | | |
| Preparation | 5 hours | |
| Emulsion (2 coats) | 16 hours = 21 hours @ £17.76 | 372.96 |
| Add brush allowance 2% | | 7.46 |
| | | 471.62 |
| Add Profit & Overheads 15% | | 70.74 |
| Cost per 100m² | | 542.36 |
| ÷ 100 cost per m² | | £5.42 |

## Drainage

### *Excavation*

Excavation to drain trenches can be based on the labour constants given for surface trenches except that allowance has to be made in grading the bottoms of the trench to the correct gradient. Drain trenches are measured in linear metres and are deemed to include all necessary earthwork support, consolidation of bottom of trench, trimming excavation, filling and compaction of general filling materials and removal of surplus excavated materials. The following is a guide to the width of a drain trench based on pipes up to 200mm diameter:

| Depth of trench | Width |
|---|---|
| Average depth up to 1m | 500mm |
| Average depth 1 to 3m | 750mm |
| Average depth over 3m | 1,000mm |

The actual width, though, is at the discretion of the contractor.

For pipes over 200mm diameter the width of the trench should be increased by the additional diameter of the pipe over 200mm.

#### EXAMPLE: DRAIN RUNS

The quantities of excavation, earthwork support, etc., are calculated for each category of pipe and depth. Once calculated they can be applied to any project.

The following items are deemed to be included with drain runs:

> earthwork support, consolidation of trench bottoms, trimming excavations, filling with and compaction of general filling materials, disposal of surplus excavated materials, disposal of water, building in the end of pipes and bedding and pointing and the length of pipe with manhole walls.
>
> (Source: NRM2)

Backfilling with material arising from the excavations in deemed to be included, however backfilling in selected material must be identified and described. The exact nature of the backfilling required will depend on the pipes being used.

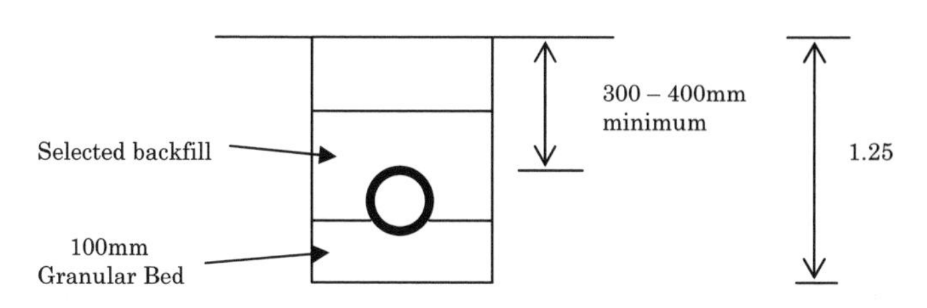


**Figure 5.4** Typical section through pipe trench

### *Pipe beds, haunchings and surrounds*

Once laid in the trench, protection to the drain pipes is provided by beds under the pipe, haunchings to the sides of the pipe or complete surrounds to the pipes depending on the location of the pipes and the specification. If concrete is used, any formwork is deemed to be included in the item.

### *Drain pipes*

The traditional material for below ground drainage pipes and fittings is clay; however, this material has now been largely replaced with UPVC (plastic) pipes. Clay pipes almost exclusively referred to as vitrified pipes, are still used in certain situations and are manufactured in 600mm lengths. The big advantage of UPVC pipes is that they weigh considerably less than vitrified clay pipes and come in 3-metre lengths, making handling, laying and jointing easier and quicker. Pipes of both materials are jointed with flexible push-fit joints that help to prevent failure.

#### EXAMPLE

110mm Floplast pipe laid in trenches average 1.5m deep with coupling sockets joints on 100mm granular bed and surrounded with 150mm thick selected excavated material – cost per m.

**Excavation**

Assume 10 linear metres

| | | | |
|---|---|---|---|
| Excavation | 10.00 | | |
| | 0.75 | | |
| | 1.25 | 9.38m³ @ £5.00 | 46.90 |
| Earthwork support | 2/10.00 | | |
| | 1.25 | 25.00m² @ £2.00 | 50.00 |
| Compaction to bottom of trench | 10.00 | | |
| | 0.75 | 7.50m² @ £1.20 | 9.00 |
| General backfill | 10.00 | | |
| | 0.75 | | |
| | 0.80 | 6.00m³ @ £2.00 | 12.00 |
| Remove surplus | 10.00 | | |
| | 0.75 | | |
| | 0.45 | 3.38m³ @ £4.53 | 15.31 |
| Cost per 10m | | | £133.21 |
| ÷ 10 cost per m – excavation | | | £13.31 |

**Drain pipes**
110mm Floplast pipes: £10.41 per 3m
Coupling sockets: £2.19 each

Assume 30m

**Materials**

| | | |
|---|---|---|
| 10 Floplast pipes @ £10.41 each | 104.10 | |
| 10 coupling sockets @ £2.19 | 21.90 | 126.00 |

**Labour**

| | |
|---|---|
| Laying and jointing pipes – 0.1 hours per m | |
| 3 hours @ £17.76 | 53.28 |
| | 179.28 |
| Add Profit & Overheads 15% | 26.89 |
| Cost per 30m | 206.17 |
| ÷ 30 Cost per m | £6.87 |

**Bed and surround to pipe**
Assume 10 linear metres
100mm granular bed

**Materials**

| | |
|---|---|
| 10.00 × 0.75 × 0.10 = 0.75m³ selected granular bed @ £6.00 per m³ | £4.50 |
| 10.00 × 0.75 × 0.30 = 2.25m³ selected excavated material surround @ £3.00 per m³ | £6.75 |

**Labour**

| | | |
|---|---|---|
| 10m bed and surround pipes @ 0.2 hour per m | | |
| | 2 hours @ £17.76 | £35.52 |
| | Cost per 10m | £46.77 |
| ÷ 10 cost per m | | £4.68 |

**Summary**

| | |
|---|---|
| Excavation | £13.31 |
| Pipework | £6.87 |
| Bed and surround | £4.68 |
| Total | £24.86 per linear metre |

### *Manholes and inspection chambers*

Traditionally the construction of manholes was a labour intensive operation involving many trades. Although still constructed from engineering bricks it is more common to use preformed manholes made from pre-cast concrete, plastic and clay. The labour constants for work to manholes can be taken from the respective trades with the addition of about 25% to take account of working in confined spaces. Inspection chambers are essentially the same as manholes but are usually shallower.

### External works

External works include paving, planting fencing, etc., and trades associated with this element are measured and should be priced in accordance with the constants used in the appropriate trade elsewhere in the section.

## PRO RATA PRICING

One of the advantages of having a bill of quantities is that the degree of detail contained in the document can be used as a basis for the valuing of variations during the post-contract stage of a project. As discussed in the previous pages bill rates are composed from the following:

- labour
- material
- plant
- profit and overheads.

The technique of pro rata pricing involves disassembling a bill rate and substituting new data in order calculate a new rate that can be used for pricing variations – see Chapter 7.

## MANAGING THE PRICING PROCESS

A successful contractor must have a clear policy regarding tendering policy as many contractors have perished due to the lack of one. The policy must spell out to the senior management the type and the value of the contracts that the contractor will tender for. But surely all potential projects are good news and should be considered? The answer is no, it is far better to 'stick to the knitting' rather than attempting to win every contract that is offered. A contractor should have a system to assess the risk that a new contract may pose to the organisation. The factors that should be considered before accepting a place on a tender list are:

- The definition of the brief.
- The type of project: house building, commercial, etc.
- The type of client: private, public sector, Public–Private Partnerships, etc.
- The contract: is the contract a mainstream contract such as JCT forms or NEC; has the contract been modified?
- The maximum and minimum value of potential contracts; large value contracts will require financial backing in order to maintain cash flow.
- Geographical location: the supply chains could struggle if stretched.
- The number and the significance of PC and provisional sums: work covered by PCs and provisional sums is generally outside the control of the main contractor and could be difficult to control.
- The time of year: will the contract be starting at the beginning of the winter months?

Once risk has been identified, the response of a contractor should have systems for:

- rejection
- amelioration
- transfer
- acceptance.

In addition a contractor will also have to consider its current and future commitments and any problems in obtaining the required labour and materials. It may be that the contractor will seek to build into the bid figure a provision or contingency for potential risks.

## TENDER ADJUDICATION

Tender adjudication takes place after the bills of quantities are priced and usually involves the senior management of the contracting organisation who will, amongst other matters, decide the level of profit required, sometimes referred to as the margin. This figure, together with a percentage to cover overheads, will be added to the rates calculated by the estimator. The level of profit is kept a closely guarded secret but historically is seldom above 4–5% in the UK.

Having taken the decision to submit a tender the contractor will take delivery of the tender documents and proceed to prepare the bid. This will involve:

- carefully examining the drawings and bills of quantities for accuracy and then prepare a timetable for the preparation of the bid;
- visiting the site, particularly appropriate for alteration work, or where the site has restricted access or working;
- preparing any method statements: method statements are an analysis and description of how the contractor proposes to carry out particularly complicated parts of the works.

## E-TENDERING

Electronic tendering enables the traditional process not only to be made more efficient but also to add significant value. It can provide a transparent and paperless process allowing offers to be more easily compared according to specific criteria. More importantly by using the internet tendering opportunities become available to a global market.

E-procurement is the use of electronic tools and systems to increase efficiency and reduce costs during each stage of the procurement process. Since Autumn 2002 there have been significant developments in e-procurement; legislative changes have encouraged greater use throughout the EU; new techniques such as electronic reverse auctions have been introduced, not, it has to be said, without controversy. In addition, the UK government has launched a drive for greater public sector efficiency following HM Treasury's publication of the Gershon Efficiency Review: Releasing Resources to the Frontline, in July 2004 and e-procurement is seen to be at the heart of this initiative.

The stated prime objective of electronic tendering systems is to provide central government, as well as the private sector, with a system and service that replaces the traditional paper tendering exercise with a web-enabled system that delivers additional functionality and increased benefits to all parties involved with the tendering exercise. The perceived benefits of electronic procurement are:

- efficient and effective electronic interfaces between suppliers and civil central governments, departments and agencies, leading to cost reductions and timesaving on both sides;
- quick and accurate pre-qualification and evaluation, which enables automatic rejection of tenders that fail to meet stipulated 'must have criteria';
- a reduced paper trail on tendering exercises, saving costs on both sides and improving audit;
- increased compliance with EU procurement Directives, and best practice procurement with the introduction of a less fragmented procurement process;
- a clear audit trail, demonstrating integrity;
- the provision of quality assurance information – e.g. the number of tenders issued, response rates and times;
- the opportunity to gain advantage from any future changes to the EU procurement Directives;
- quick and accurate evaluation of tenders;
- the opportunity to respond to any questions or points of clarification during the tendering period;
- reduction in the receipt, recording and distribution of tender submissions;
- twenty-four-hour access.

Figure 5.5 illustrates the possible applications of e-procurement to projects that are covered by the EU public procurement Directives.

Benefits of electronic tendering and procurement of goods and services are said to be wider choice of suppliers leading to:

- lower cost;
- better quality;
- improved delivery, reduced cost of procurement (e.g. tendering specifications are downloaded by suppliers rather than by post); and
- electronic negotiation and contracting and possibly collaborative work in specification can further enhance time, cost-saving and convenience.

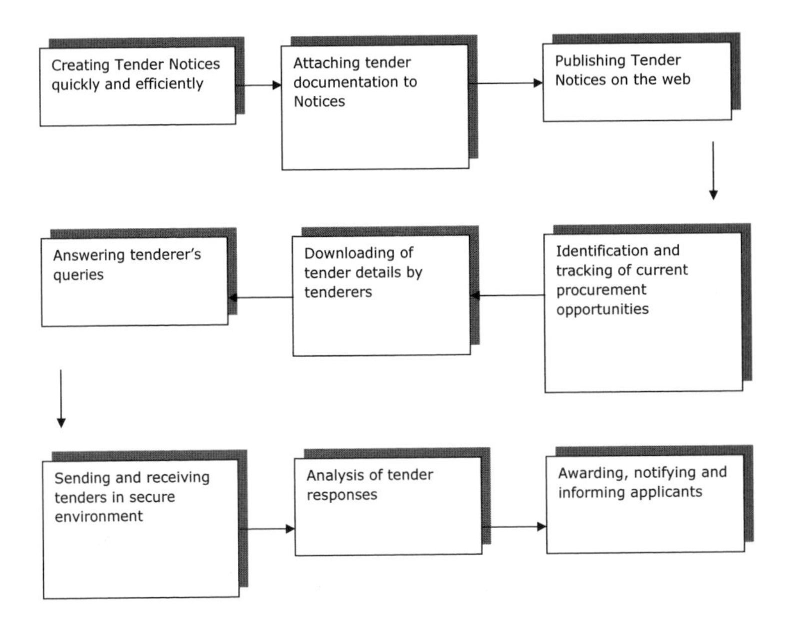


**Figure 5.5**

For suppliers the benefits are:

- more tendering opportunities, possibly on a global scale;
- lower cost of submitting a tender; and
- possibly tendering in parts, which may be better suited for smaller enterprises or collaborative tendering.

Lower costs can be achieved through increased efficiency; however, a survey carried out by e-Business Watch in 2002 of over 6,000 organisations found that nearly 60% of those surveyed perceived that face-to-face interaction was a barrier to e-procurement while online security still was a major concern.

In autumn 2005 the RICS produced a guidance note on e-tendering in response to the growth in the preparation of tender documents in electronic format. Figure 5.6 sets out their recommended approach to the e-procurement process while Figure 5.7 maps the way by which contract documentation can be organised for the e-tendering process.

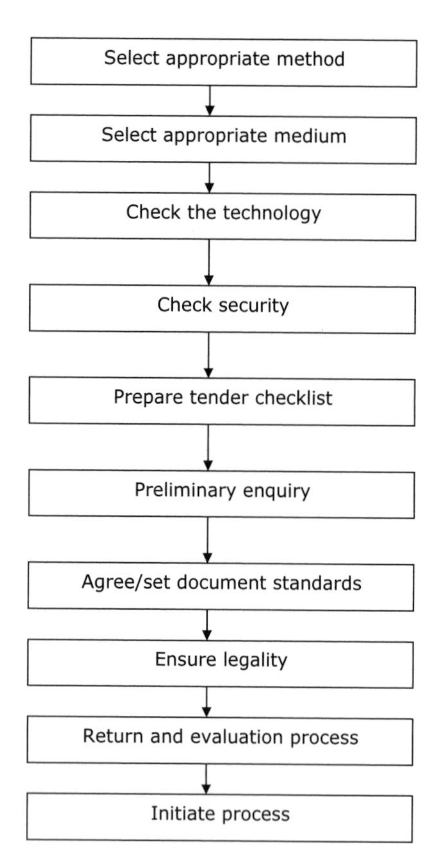


**Figure 5.6** E-procurement process
(Source: Adapted from RICS)

## E-AUCTIONS

An online auction is an internet-based activity, which is used to negotiate prices for buying or selling direct materials, capital or services.

Online auctions can be used to sell: these are called forward (or seller) auctions and closely resemble the activity on websites such as eBay; the highest bidder wins. Now some companies are starting to use reverse auctions where purchasers seek market pricing inviting suppliers to compete for business on an online event. Auctions can either be private/closed where there are typi-

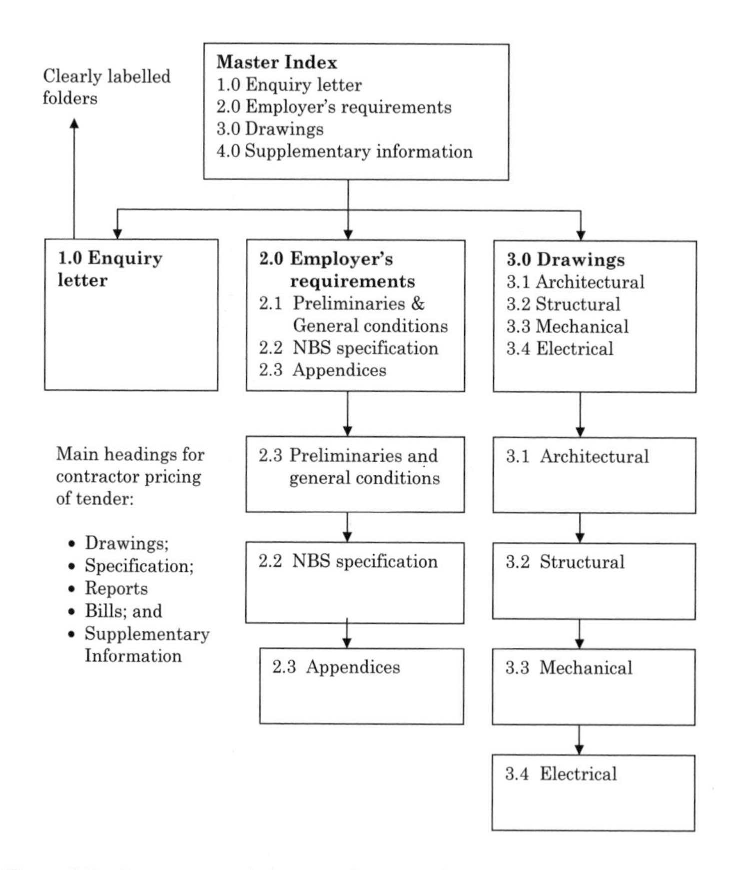


**Figure 5.7** Organisation of contract information for e-procurement (Source: RICS)

cally few bidders who have no visibility of each other's bids, or open, where a greater number of participants are invited. In this case participants have visibility of either their rank or the bidding itself. When used, the technique can replace the conventional methods of calling for sealed paper tenders or face-to-face negotiations.

Online auctions are said to offer an electronic implementation of the bidding mechanism used by traditional auctions and systems may incorporate integration of the bidding process together with contracting and payment. The sources of income for the auction provider are from selling the technology platform, transaction fees and advertising. Benefits for suppliers and buyers are increased efficiency and time savings, no need for physical transport until the deal has been established, as well as global sourcing.

There have been some strong objections to e-auctions and in particular reverse e-auctions from many sections of the construction industry. Amongst quantity surveyors the perception of reverse e-auctions is very negative with 90% feeling that they reduce quality and adversely affect partnering relationships.

## REVERSE AUCTIONS

The reverse e-auction event is conducted online with pre-qualified suppliers being invited to compete on predetermined and published award criteria. A reverse auction can be on any combination of criteria, normally converted to a 'price equivalent'. Bidders are able to introduce new or improved values to their bids in a visible and competitive environment. The procedure duration of the event will be defined before the start of the reverse auction commences. There will be a starting value that suppliers will bid against until the competition closes.

Three characteristics that need to be present to have a successful reverse auction are as follows:

- the purchase must be clearly defined;
- the market must be well contested; and
- the existing supply base must be well known.

These three factors are interdependent and together form the basis for an auction that delivers final prices as close as possible to the true current market price. For the buyer and the supplier, a clearly defined scope of work is essential, without this it becomes very difficult to accurately bid for the work.

# 6

# Contract procedure, administration and organisation

This chapter of the book has been sub-divided for easy reference into the following parts:

- **Part 1** Contract procedure: a review of the popular forms of contract in current use, insurances, bonds, guarantees, collateral warranties.
- **Part 2** Contract administration: a review of the application of contract conditions during the site operations stage and the final account.
- **Part 3** Contract organisation: planning, site layout and organisation.

## PART 1: CONTRACT PROCEDURE

In 2007 the RICS published the latest in a series of reports analysing the contracts that are being used. The report, the eleventh in the series, was compiled on behalf of the RICS by Davis Langdon and produced the following findings:

- 97.5% of all projects were let using a standard form of contract.
- 79% of all contracts employed were one of the JCT standard forms, up from 78% from the previous (2004). This level of usage reflects the competition that the JCT is experiencing from rival standard forms such as the NEC suite of contracts.

In more detail, the distribution of methods of procurement, by number of contracts, are as follows:

- Lump sum – firm BQ – 20%
- Lump sum – specification and drawings – 47%
- Lump sum – design and build – 22%
- Partnering agreements – 2%.

When considered by the value of contracts, the distribution of methods of procurement is as follows:

- Lump sum– firm BQ – 13.2%
- Lump sum – specification and drawings – 18.2%
- Lump sum – design and build – 32.6%
- Partnering agreements – 15.6%.

### Types of contract

The following are the main types of traditional contract:

- **Lump sum contracts** – where the contract sum is known before work starts on site and the contractor agrees to undertake a defined amount of work for a specific amount. This type of contract is often based on a firm bill(s) of quantities and drawings.
- **Measurement contracts** – where the contract is assessed and remeasured as on a previously agreed basis. This type of contract can be based on approximate bill(s) of quantities and drawings.
- **Cost reimbursement contracts** – where contractor is reimbursed on the basis of the prime cost of labour materials and plant plus an agreed percentage addition to cover overheads and profit.

In addition the following types of contract have become popular in recent times:

- **Design and build** – in a number of variant forms, where the contractor both designs and builds a project.
- **Management contracts** – can take a variety of forms but involve a management contractor managing the works although the contractor does not actually carry out any of the work.

The type of contract will to a large extent influence the choice of contract.

### Standard forms of contract

The contracts used in the UK construction industry are not bespoke documents; instead they are standard documents or forms drafted by organisations whose membership is drawn from industry and the professions. The forms should not be altered or amended, as this practice can lead to major problems if the amended portion becomes the subject of legal action. There is, however, provision within the standard forms to adapt them to suit particular projects by completion of articles of agreement and appendices. The contracts can be lengthy documents although in practice most quantity surveyors will only become fully familiar with key clauses. Several bodies publish standard forms of contract as follows.

### *The Joint Contracts Tribunal (JCT)*

The JCT was established in 1931 and has for seventy-five years produced standard forms of contracts, guidance notes and other standard documentation for the use of the construction industry. Members of the JCT include the RICS, the RIBA, the British Property Federation, Local Government Association, National Specialist Contractors Council, the Contractors Legal Group and the Scottish Building Contract Committee Ltd.

This body produces several standard forms including the flagship JCT Standard Building Contract 2011 Edition, referred to as JCT(11). Contracts published by the JCT include:

- JCT Standard Building Contract (SBC) 2011 Edition: With Quantities, Without Quantities and With Approximate Quantities together with Standard Subcontract Agreements. Note: the Local Authorities Edition is no longer available since the 2005 Edition;
- Intermediate Building Contract (IC 2011)/With contractor's design together with Standard Subcontract Agreements;
- Minor Works Contract 2011 (MN 2011)/With contractor's design;
- Design and Build Contract together with Subcontract Agreement;
- Major Project Construction Contract;
- Management Building Contract;
- Prime Cost Contract;
- Measured Term Contract;
- Construction Management Trade Contract;
- Collateral warranties.

### JCT Standard Building Contract 2011 Edition

To coincide with the publication of the 2005 Edition the JCT ended its long association with the RIBA's publishing house and the forms of contract are now published by Sweet & Maxwell; also a digital service is available alongside the hard copy format. In addition the format and appearance of the JCT forms were given a facelift in order to simplify the interpretation of clauses. Since the last edition of the JCT Standard Form in 2005 competition from other forms of contract has become more acute, in particular the New Engineering Contract and the Project Partnering Contract 2000 and this pressure has forced the JCT to update the structure and the clarity of the JCT form. Over the last two editions, one of the major changes is the elevation of mediation as a form of dispute resolution. The 2011 Edition states that mediation ought to be considered as the very first condition in the dispute resolution clause. In addition a number of familiar JCT phrases have been amended in the 2005 and 2011 Editions, for example 'Defects Liability Period' becomes 'Defects Rectification Period' and 'Certificate of Completion of Making Good Defects' has become 'Certificate of Making Good'. Since JCT(05) nominated subcontractors have been replaced by named subcontractors.

**With Quantities** – this edition of the JCT 2011 is for use where the design team is able to fully detail the project and the time is available for a detailed bill of quantities to be prepared. JCT 2011 is supplied with the facility for an element of design work to be carried out by the contractor – the Contractor's Designed Portion (CDP). The JCT 2011 form makes it obligatory for the contractor to obtain and maintain design liability insurance when design work is to be carried out by the contractor. Limits of responsibility need clear definition.

**Without Quantities** – this edition is for use when a bill of quantities is not to be part of the contract documentation, primarily due to lack of time or detailed information. Nevertheless a set of drawings and specification is provided. The contractor is required to submit a schedule of rates as a basis for valuing variations, etc.

**With Approximate Quantities** – is similar to the Standard Form Building Contract With Quantities but is for use on projects where it is necessary to make an early start and for which adequate contract documents cannot be prepared before the tender stage. No lump sum price can be given for the work, but a tender sum indicating the anticipated final price can be calculated based on approximate measurement of the likely work involved. As the contract progresses, the work is completely remeasured and the remeasurement priced on the basis of the rates set out in the bills of approximate quantities.

The only difference between bills of approximate quantities and firm bills of quantities is that they are prepared from less complete design information and so can be produced at an earlier date. Their preparation must still be in accordance with the *Standard Method of Measurement* or NRM2 as from January 2013.

Some of the principal clauses in the JCT(11) will be discussed later in this section.

## Intermediate Building Contract (IC 2011)

As a contract the intermediate form of contract has tried to bridge the gap between the full JCT (2011) and the Minor Works Contract. It is available in one version for local authorities and private clients, with or without quantities.

Appropriate:

- where the proposed works are to be carried out for an agreed lump sum;
- where an architect or contract administrator has been appointed to advise on and to administer its terms;
- where the proposed building works are of simple content involving the normal, recognised basic trades and skills of the industry, without building service installations of a complex nature or other complex specialist work; and
- where the works are adequately specified, or specified and bills of quantities prepared prior to the invitation to tender.

This form provides more detailed provisions and more extensive control procedures than the Agreement for Minor Building Works. A sectional completion supplement allows its use where the project is divided into sections.

## Minor Works Contract 2011 (MN 2011)/With Contractor's Design

The most widely used form within the JCT suite of contracts reflecting the large number of relatively small value contracts carried out.

Appropriate:

- where the work involved is simple in character;
- where the work is designed by or on behalf of the employer;
- where the employer is to provide drawings and/or a specification and/or work schedules to define adequately the quantity and quality of the work; and
- where a contract administrator is to administer the conditions.

Can be used:

- by both private and local authority employers.

Not suitable:

- where bills of quantities are required;
- where provisions are required to govern work carried out by named specialists;
- where detailed control procedures are needed;
- where the contractor is to design discrete part(s) of the works, even though all the other criteria are met – consider the Minor Works Building Contract With Contractor's Design (MWD).

**Design and Build Contract together with Subcontract Agreement**

Appropriate:

- where detailed contract provisions are necessary and the Employer's Requirements have been prepared and provided to the contractor;
- where the contractor is not only to carry out and complete the works, but also to complete the design; and
- where the employer employs an agent (who may be an external consultant or employee) to administer the conditions.

Can be used:

- where the works are to be carried out in sections;
- by both private and local authority employers.

Where the contractor is restricted to design small discrete parts of the works and not made responsible for completing the design for the whole works, consideration should be given to using one of the JCT contracts that provide for such limited design input by the contractor and the employment of an architect/contract administrator.

**Design and Build Subcontract (DBSub/A and DBSub/C)**

Appropriate:

- for use with the Design and Build Contract; and
- for subcontract works whether or not they include design by the subcontractor.

Can be used:

- where the subcontract works and/or main contract works are to be carried out in sections;
- for subcontract works that are to be carried out on the basis of an adjusted subcontract sum (adjustment for variations, etc.) or by complete remeasurement of the subcontract works.

## Major Contracts Construction Contract (MP) 2011

In June 2003, the Joint Contracts Tribunal launched another standard form of contract, the Major Project Form, rebadged the Major Contracts Construction Contract in 2005, in response to an apparent demand by clients who sought a simpler contract which would avoid the perceived necessity for either creating bespoke contracts or radical redrafting one of the current available standard forms. The Major Contracts Construction Contract is considerably shorter than any of its contemporary JCT contracts, for example it is 80 per cent smaller than the With contractor's Design Contract as users require a thorough understanding of both the building and general contractual procedures.

As is frequently the case, the consequence of producing a shorter and simpler contract is that there is a greater need for the interpretation of its terms. Perhaps in recognition of this the JCT has also published a 24-page set of guidance notes for use with the the Major Contracts Construction Contract.

The stated aim of the the Major Contracts Construction Contract is that, having defined the 'Employer's Requirements', the employer should then allow the contractor to undertake the project without the contractor being reliant upon the employer for anything else other than access to site, review of design documents, and payment. In particular, there is no requirement for the employer to issue any further information to the contractor, as design information beyond that contained in the Requirements will be produced by the contractor.

### *Key features*

#### Attestation provision

The contract is to be executed as a deed and therefore provides a limitation period of 12 years.

#### Clause 1

The contractor is responsible for completing the project in accordance with the contract, which is defined in clause 29 as: 'The Contract Conditions, the

Appendix, the Third Party Rights Schedule, the Requirements, the Proposals and the Pricing Document'.

## Design

The key document is the Employer's Requirements, which needs to be sufficiently detailed to set out what it expects the contractor to design. Clause 5.1 expressly excludes the contractor's responsibility for the contents of the Requirements or the adequacy of the design contained therein. Clearly, failure to adequately define the design parameters could result in dispute.

## Design standards and procedure

The contractor is liable for a standard duty of using reasonable skill and care with clause 5.3 expressly excluding any 'fitness for purpose' provision.

The JCT in recognition of those Employers requiring a 'fit for purpose' obligation set out in the guidance notes page 7 a slightly watered down alternative, which requires the contractor to warrant that the project will be 'suitable for the purpose stated in the Requirements'.

Clause 6 expects any design documents, which are widely defined, to be prepared and submitted to the employer. The employer has 14 days for approval and comment.

## Discrepancies

Only discrepancies within the Requirements and changes in the Statutory Requirements after the Base Date will give rise to a Change. Other discrepancies between or contained within the documents will not give rise to a Change.

## Possession and completion

The contractor does not have exclusive possession of the site, which suggests that the employer may undertake works at the same time with the proviso that the contractor has sufficient access to areas of the site to complete his works. Limitations in this regard must be set out in the Requirements.

The employer, with consent, may also take over any part or parts of the project prior to Practical Completion. Unlike other JCT Contracts, clause 39 provides a definition for Practical Completion, although whether the definition provided will reduce disputes regarding this issue remains to be seen.

## Time

To preserve the liquidated damages provision the Major Contracts Construction Contract provides an extensive list of relevant delay events, but this list does not include exceptional weather conditions, industrial disputes, and the inability to obtain labour and/or materials, or delays in statutory approvals. These excluded items are therefore at the risk of the contractor.

The Major Contracts Construction Contract adopts the following principles for making an extension of time:

- the employer should implement any agreement reached regarding changes, acceleration or costs savings;
- regard must be given to any failure by the contractor of clause 9.3 i.e. using reasonable endeavours to prevent or reduce delay to the works; and
- a fair and reasonable adjustment should be given regardless of any concurrent culpable delay.

## Acceleration and bonus

The contract expressly provides for acceleration by agreement.

The employer is also liable to paying the contractor an optional agreed bonus payment if the project is practically complete before the completion date.

## Pre-appointed consultants and specialists

If adopted by the employer he can name consultants in the Appendix which will be novated to the contractor on a Model Form which must be contained in the Requirements.

Contract clause 18.4 provides that the contractor will accept responsibility to the employer for services previously performed by the consultant and clause 18.6 prevents the contractor (without permission) altering the terms of the consultant's engagement.

## Cost savings

The contractor is 'encouraged' to suggest amendments to the Requirements and/or Proposals and the contractor will directly benefit on an agreed proportion basis as identified in the appendix to the contract.

### Variations: changes

The valuation of any change shall preferably be agreed before any instruction is issued or they will be determined on the basis of a fair valuation that takes into account all consequences of the change including any loss and/or expense, which may be incurred.

### Valuation

Valuations will be made monthly, however the Pricing Document provides for a range of payment options based upon, interim payments, stage payments, scheduled payments or indeed any other method of payment the parties wish to make.

### Third party rights

This is a major innovation albeit it remains optional. It provides for the contractor to effectively warrant their obligations under the contract to other defined third parties, which relies on the relatively untried 'Contracts (Rights of Third Parties) Act 1999'.

### Disputes

Clause 35 deals with the resolution of disputes and offers mediation by agreement, statutory adjudication under the scheme and finally any difference or dispute being decided through litigation. The contract does not provide an option for arbitration.

### What's missing?

Some of the common conditions which are missing from this form include:

1. nominated subcontractors
2. payment for materials either on or off site
3. no provision for any fluctuations
4. no retention
5. no oral mechanism for variations
6. no priority of documents provision
7. no specific insurance requirements
8. no lengthy provisions for VAT, HM Revenue and Customs, etc.
9. no insurance provisions.

### The contract documentation

The MPF consists of six distinct documents including a set of guidance notes; with the latter confirming they are not a definitive guide to the terms of contract or its interpretation.

**1. The Conditions**

The contract is executed as a deed and consists of 8 main sections and 39 individual clauses. Clause 39 provides a useful definitions clause for terms used throughout all the contract documents.

**2. The Appendix**

The appendix is similar to other JCT formats and contains all the relevant and necessary details required by the contract terms.

**3. The Third Party Rights Schedule**

This is a considerable (optional) step in adopting the provisions under the contracts (Rights for Third Parties) Act 1999, but remains limited to the main contractor providing a warranty to funders, purchasers and/or tenants.

**4. The Pricing Document**

Provides information for determining the manner in which the contractor is to receive payments in respect of the contract sum and includes the contract sum analysis and pricing information.

This information will need to be prepared in the knowledge that it will also govern the valuation of any changes.

**5. The Requirements**

These are the Employer's Requirements and this is what the contractor will have to meet to discharge its obligations under the contract.

**6. The Proposals**

These are the Contractor's Proposals which confirm how the contractor proposes to meet the Requirements.

## JCT Management Building Contract 2011

Appropriate:

- for large-scale projects requiring an early start on site, where the works are designed by or on behalf of the employer but where it is not possible to prepare full design information before the works commence and where much of the detail design may be of a sophisticated or innovative nature requiring proprietary systems or components designed by specialists;

- where the employer is to provide the management contractor with drawings and a specification; and
- where a management contractor is to administer the conditions.

The management contractor does not carry out any construction work but manages the Project for a fee. The management contractor employs works contractors to carry out the construction works.

Can be used:

- where the works are to be carried out in sections,
- by both private and local authority employers.

### JCT Prime Cost Contract 2011

Available in one version for local authorities and private clients.

Appropriate for use where the employer wants the earliest possible start. There may be insufficient time to prepare detailed tender documents, or circumstances such as an inability accurately to define the work may make their use inappropriate, necessitating the appointment of a contractor simply on the basis of an estimate of the total cost.

### Measured Term Contracts 2011

Available in one version for local authorities and private clients. Appropriate for use by employers who have a regular flow of maintenance and minor works, including improvements, to be carried out by a single contractor over a specified period of time and all under a single contract.

### Construction Management Trade Contract 2011

Appropriate:

- where the client is to enter into direct separate trade contracts; and
- where a construction manager is appointed under the client; and
- where a Construction Management Agreement is to administer the conditions on behalf of the client.

Can be used:

- where the works are to be carried out in sections.

## Collateral warranties

The principals of collateral warranties are discussed at the end of this section. The JCT(11) now has a series of collateral warranties for use with funders, purchasers, tenants and employers for both main contractors and subcontractors.

## Other forms of contract

### *Association of Consultant Architects (ACA)*

ACA is the national professional body representing architects in private practice – consultant architects – throughout the UK. Founded in 1973, it now represents some of the country's leading practices, ranging in size from one-person firms to very large international organisations. The ACA's most widely used contracts are:

- ACA Form of Building Agreement 1982, Third Edition 1998 (2003 Revision), and
- PPC2000 The ACA Standard Form of Contract for Project Partnering (Amended 2003).

The ACA Form of Building Agreement is a lump sum contract that has both fixed and fluctuating price versions. Unlike the JCT 2005 the ACA contract is a one-fit-suits-all form and is suitable for projects of all sizes based on a bill of quantities or specification. The ACA's PPC2000 is the first multi-party partnering contract to be produced following the government's 'Rethinking Construction' report and initiative and it is claimed that it:

- is a non-adversarial construction contract that provides the foundation for the partnering process; and
- can be applied to any type of partnered project in any jurisdiction.

### *PPC2000*

**Team-Based Multi-Party Approach** – PPC 2000 allows the client, the constructor and all consultants and key specialists (i.e. subconsultants, subcontractors and suppliers) to sign a single Partnering Contract. This avoids the need for several two-party professional appointments and a separate building contract and/or partnering agreement, and substantially reduces project paperwork. This single, integrated contract encourages a team-based commitment to the project, and should reduce the temptation to hide behind unconnected two-party agreements. Additional members can join the partnering team by signature of joining agreements.

**Integrated Design/Supply/Construction Process** – PPC 2000 provides for the early selection of a project partnering team and the collaborative finalisation of designs, prices and members of the supply chain (clauses 8, 10 and 12). It covers the full duration of the partnering relationships, and thereby encourages the contributions of the constructor and specialists during the key period prior to start on site, as well as during supply and construction.

**Egan Objectives** – PPC 2000 expressly recognises the recommendations of 'Rethinking Construction' and links these to the objectives of the partnering team on each project. Achievement of these objectives is measured against agreed key performance indicators (clauses 4 and 23).

**Supply Chain Partnering** – PPC 2000 provides for finalisation of the supply chain on an open-book basis, encouraging partnering relationships with all specialists, and includes provision for key specialists to become full members of the partnering team (clauses 10 and 12). It is compatible with the wide variety of subcontracts used by constructors, but those subcontracts must not conflict with the PPC2000 terms.

**Core Group** – PPC 2000 provides for a core group of key individuals representing partnering team members, who operate an early warning system for problems and who undertake regular reviews of progress and performance (clause 3). 4 PPC2000 [Amended 2003].

**Controls** – PPC 2000 provides for a partnering timetable to govern the contributions of all partnering team members to partnered activities, including development of designs, prices and the supply chain, and for a project timetable to govern their activities after commencement on site (clause 6).

**Incentives** – PPC 2000 provides for agreement of profit, central office overheads and site overheads (clause 12), with encouragement for partnering team members to agree shared savings and shared added value incentives (clause 13). Payments can also be linked to performance against KPIs (clause 13.5); value engineering and value management exercises are expressly recognised (clause 5.1).

**Risk Management** – PPC 2000 provides a clear system for reducing, managing and sharing risks and for agreeing changes openly and equitably in advance (clauses and 17 and 18). Risk management is a duty of partnering team members, and there is a facility to agree the balance and sharing of risk appropriate to each project (clause 18.1).

**Non-Adversarial Problem Resolution** – PPC 2000 provides for a problem-solving hierarchy of increasingly senior individuals within each partnering team member's organisation, working to strict time limits (clause 27.2), with further reference of a problem to the core group (clause 27.3). It also includes a facility for conciliation or other forms of alternative dispute resolution (clause 27.4). These options are without prejudice to partnering team members' legal right to refer a dispute to adjudication (clause 27.5).

**Partnering Adviser** – PPC 2000 recognises the role of partnering adviser recommended in the Construction Industry Council guide, an individual with relevant experience who can guide the partnering process, who can document the relationships, commitments and expectations of partnering team members and who can provide an additional facility for problem resolution (clauses 5.6 and 27.4).

## NEC form of contract

The NEC describes itself as:

> a modern day family of contracts that facilitates the implementation of sound project management principles and practices as well as defining legal relationships. Key to the successful use of NEC is users adopting the desired cultural transition. The main aspect of this transition is moving away from a reactive and hindsight-based decision-making and management approach to one that is foresight-based, encouraging a creative environment with proactive and collaborative relationships.

The NEC was launched by the ICE in 1993 with the 2nd Edition in 1995 and the 3rd Edition in 2005. The boxed set contains a total of 23 documents that together make up the new and extended family. It is now the most widely used contract in UK civil engineering and is often used by government departments such as the Highways Agency and by local authorities. It has been used on major projects and procurement initiatives such as the Channel Tunnel Rail Link, Heathrow T5, NHS Procure 21 and the Eden Project. The overall structure is quite unlike that of the JCT and other standard forms, as follows.

### *Flexibility*

- It is intended to be suitable for all the needs of the construction industry.
- It provides for a variety of approaches to risk allocation.

- It is adaptable for some design, full design or no design responsibility, and for all current contract options including target, management and cost reimbursable contracts.
- The simple wording of the documents is deliberately chosen, and lends itself to ready translation into other languages.

### *Clarity and simplicity*

- The ECC is written in ordinary language, using short sentences with bullet points.
- Imprecise terms such as 'fair' and 'reasonable' have been avoided.
- Legal jargon is minimised.
- The actions required from the parties are said to be 'defined precisely', with the aim of avoiding disputes.
- Flow charts are provided to assist usage.

### *Stimulus to good management*

This is stated to be the most important characteristic of the ECC.

- The ECC is a manual of management procedures, not just a contractual document.
- The aim is to present the purchaser's project manager with options for overcoming problems as they become apparent.
- An 'early warning procedure' places obligations on all parties to flag up problems which could affect time, money or performance of the works.
- 'Compensation events' are the method of dealing with problems of both time and money.
- A schedule of 'actual cost' is used.
- Changes are based on quotations prior to commitment wherever possible.
- The programme must be kept up to date at all times, to reflect changes.
- The aim is to highlight and resolve problems in a proactive way as the job proceeds.
- 'End of job' disputes should be reduced in consequence.

The documents making up the contract are:

- Form of tender
- Schedule of contract data
- Core clauses – see below

- Optional clauses – see below
- Schedule of actual cost.

The seven procurement routes now available are:

1. Priced contract – with either activity schedule or bill of quantities.
2. Target contract – with either activity schedule or bill of quantities.
3. Cost reimbursable contract.
4. Management contract.
5. Short contract – suitable for minor works and simple projects.
6. Term service contract.
7. Framework contract.

In addition to the NEC3 Engineering and Construction Contract (the black book) the six main core clause Options A–F are:

**Option A** Priced contract with activity schedule.
**Option B** Priced contract with bill of quantities.
(Note – the bill is used only for interim payments, not for valuation of compensation events)
**Option C** Target contract with activity schedule.
(Lump sum quoted by tenderer as target)
**Option D** Target contract with bill of quantities.
(Subject to remeasurement)
**Option E** Cost reimbursable contract.
**Option F** Management contract.

Other documents include:

- Contract guidance notes.
- Contract flow charts.
- Professional service contract.
- Short contract.
- Subcontracts and flow charts.
- Adjudicator's contract.
- Term Service contract, guidance notes and flow chart.
- Framework contract, guidance notes and flow charts.
- Procurement and contract strategies.

One of these six major options must be chosen with any number of the following secondary options:

| | |
|---|---|
| X1 | Price adjustment for inflation |
| X2 | Changes in the law |
| X3 | Multiple warranties |
| X4 | Parent company guarantees |
| X5 | Sectional completion |
| X6 | Bonus for early completion |
| X7 | Delay damages |
| X12 | Partnering |
| X13 | Performance bond |
| X14 | Advanced payment to the contractor |
| X15 | Limitation of the contractor's liability for his design to reasonable skill and care |
| X16 | Retention |
| X17 | Low performance damages |
| X18 | Limitation of liability |
| X20 | Key performance indicators |
| Y(UK)2 | The Housing Grants, Construction and Regeneration Act 1996 |
| Y(UK)3 | The Contracts (Rights of Third Parties) Act 1999 |
| Z | Additional conditions of contract. |

The 'ordinary language' sometimes appears curiously naive in its assumption of compliance by the parties. It must not be forgotten that the structure and language of traditional contracts has evolved organically, over a number of years, and has been refined and clarified by ongoing case law. The wording of the contract can arouse feelings of suspicion among design teams using the contract for the first time.

The language used and the new terms introduced have proved to be a double-edged sword; while intending to promote clarity and simplicity, in the opinion of some commentators, these have led to increased uncertainty and have been cited as being amongst the main reasons for a general reluctance to use the contract. It is often said that one of the greatest difficulties is with the use of the present tense in the form and the resultant difficulty in deciding whether a provision is descriptive or prescriptive. Clause 10.1 addresses this problem to an extent in that it requires the parties to 'act as stated in this Contract' thereby providing that descriptive provisions are in fact obligations. A greater issue is perhaps the brevity of the form, when compared to other standard contracts. Whilst in principle any attempt to reduce the length of contracts should be applauded, where this is achieved at the expense of clarity then the drafting cannot be considered to have been successful.

### *Subcontract works*

The standard form of subcontract broadly mirrors the rules set down in the main contract. Most key features summarised above will therefore apply. The time for giving notice has been reduced even further, in order that the contractor can comply with his own time limits. The use of the standard conditions is not mandatory. However, if the contractor does not intend to use the standard form, he is obliged to submit the proposed conditions to the project manager for approval. Presumably, the intention is to dissuade the contractor from imposing his own onerous, in-house terms.

The ECC therefore differs significantly from forms such as the JCT, ICE and FIDIC as it provides the parties with flexibility to tailor the contract to reflect the desired risk allocation and other parameters of that particular project. More than this though, it also contains a number of concepts which differ significantly from those found in other standard form contracts. These include:

- the roles of the project manager and supervisor;
- the fact that the programme is a contractual document prepared to specific parameters with an obligation to update regularly and on the occurrence of specific events;
- the provision of a number of pricing method options allowing contemporaneous costing of changes;
- the provision for Compensation Events;
- the introduction of the new concept of Work Information and Site Information;
- the allocation of design responsibility which is fully flexible in terms of extent, and also the contractor's co-ordination role;
- the provision for a joint obligation to notify defects and control of the correction period; and
- the introduction of the Scope of Contract Works and Weather Measurement (in the context of changes).

## GC/Works range of contracts

The GC/Works family of contracts are standard government forms of contract intended for use in connection with government construction works. These contracts are published by the Stationery Office for the Property Advisors for the Civil Estate (PACE) the latter having responsibility for the management of the GC/Works contracts since April 1996. It is claimed that all of contracts in the GC/Works suite are fully in line with the principles of the Latham

Report and meet the requirements of the Housing Grants, Construction and Regeneration Act 1996. The contracts are written in plain English and are accompanied by a comprehensive commentary in an attempt to make interpretation easy. The GC/Works suite comprises the following contracts:

GC/Works/1(1998 and 1999) – for use on major building and civil engineering projects. Six versions are available together with Model Forms and a Commentary volume:

- **GC/Works/1 With Quantities (1998)** is used with bills of quantities where all or most of the quantities are firm and are not subject to remeasurement, giving a lump sum contract subject to adjustment for variation orders.
- **GC/Works/1 Without Quantities (1998)** is for use when lump sum tenders are to be invited on the basic of specification and drawings only, without bills of quantities, but supported by a schedule of rates prepared by the contractor in order to value variations.
- **GC/Works/1 Single Stage Design and Build (1998)** This form of contract is sufficiently flexible to allow varying amounts of design input from the contractor.
- **GC/Works/1 Two Stage Design and Build Version (1999)** is intended to support a two-stage procedure with a separate design stage. It is a lump sum contract but at the time the contract is entered into the contract sum is not known owing to the lack of design information on which to base the contract sum. A design fee is agreed and included in the contract sum and the contractor submits a pricing document to help arrive at the contract sum. The employer has the right not to proceed with the contract should the outcome be unfavourable.
- **GC/Works/1 With Quantities Construction Management Trade Contract (1999). GC/Works/1 Without Quantities Construction Management Trade Contract (1999).** There is no single main contractor therefore the employer enters into a number of direct contracts with several contractors. The employer instructs a construction manager to run the contract.
- **GC/Works/1 Model Forms and Commentary (1998)** provides model forms (on disk) for use with the above contracts together with a commentary on each of the conditions.
- **GC/Works/2 Contract for Building and Civil Engineering Minor Works (1998)** is for use when lump sum tenders are to be invited on the basis of specifications and drawings only, without a bill of quantities. The typical values appropriate for this contract would be between £25,000 and £200,000. The contract is also suitable for demolition works of any value.

- **GC/Works/3 Contract for Mechanical and Electrical Engineering Works (1998)** is for use when lump sum tenders are invited on the basis of specification and drawings for mechanical and electrical engineering works.
- **GC/Works/4 for Building, Civil Engineering, Mechanical and Electrical Small Works (1998)** is for use when lump sum tenders are invited on the basis of specification and drawings. It should be used on work up to a value of £75,000.
- **GC/Works/5 General Conditions for the Appointment of Consultants (1998)** is to be used to procure consultancy services in connection with construction works on a single project basis.
- **GC/Works/5 General Conditions for the Appointment of Consultants: Framework Agreement (1999)** is to be used to procure consultancy services for a period of between 3 and 5 years on a 'call off' basis.
- **GC/Works/6 General Conditions of Contract for a Daywork Term Contract (1999)** is intended for work of a jobbing nature based on a schedule of rates for a period of between 3 and 5 years.
- **GC/Works/7 General Conditions of Contract for Measured Term Contracts (1999)** is for term contracts based on a schedule of rates.
- **GC/Works/8 General Conditions of Contract for a Specialist Term Contract for Maintenance and Equipment (1999)** is for use where specified maintenance of equipment is required and can be costed per task.
- **GC/Works/9 General Conditions of Contract for Operation, Repair and Maintenance and Electrical Plant, Equipment and Installations (1999)** is a lump sum maintenance term contract for fixed mechanical and electrical plant. The term of the contract is between 1 and 5 years.
- **GC/Works/10 Facilities Management Contract (2000)** is a term contract for a period of between 3 and 5 years.
- **GC/Works Subcontract** is for use with the major CG/Works forms.

### ICE Conditions of Contract

This is a family of standard conditions of contract for civil engineering works and is produced by the Conditions of Contract Standing Joint Committee (CCSJC); the ICE Conditions of Contract are jointly sponsored by ICE, the Civil Engineering Contractors Association (CECA) and the Association of Consulting Engineers (ACE). The ICE Conditions of Contract, which have been in use for over fifty years, were designed to standardise the duties of contractors, employers and engineers and to distribute the risks inherent in civil engineering to those best able to manage them.

The information below relates to the latest versions of each contract. Earlier editions are available and are used but it is recommended that the

Conditions of Contract set out below, with their respective guidance notes, should be used:

- Measurement Version, 7th Edition.
- Design & Construct, 2nd Edition.
- Term Version, 1st Edition.
- Minor Works, 3rd Edition.
- Partnering Addendum.
- Tendering for Civil Engineering Contracts.
- Agreement for Consultancy Work in Respect of Domestic or Small Works.
- Archaeological Investigation, 1st Edition.
- Target Cost, 1st Edition.
- Ground Investigation, 2nd Edition.

It is recommended that all clauses are incorporated unaltered because they are closely interrelated and any changes made in some may have unforeseen effects on others. Guidance notes have been prepared specifically to assist the users of the ICE Conditions of Contract in the preparation of contract documents and the carrying out of the contract works. They do not purport to provide a legal interpretation but they do represent the unanimous view of the CCSJC on what constitutes good practice in the conduct of civil engineering products. Guidance notes are published as separate documents for:

- Measurement Version.
- Design and Construct.
- Minor Works.
- Term Version.
- Ground Investigation.
- Archaeological Investigation.
- Target Cost.

In August 2009 the ICE council decided to formally recommend the NEC3 in preference to the ICE Conditions of Contract, the effect of this decision is that from this date ICE7 will not be updated or amended and gradually cease to be used.

**Fédération Internationale des Ingénieurs-Conseils (FIDIC)**

Prior to 1998, Fédération Internationale des Ingénieurs-Conseils (The International Federation of Consulting Engineers, FIDIC) published three forms of building and engineering contracts: for civil engineering works (known as the Red Book), for electrical and mechanical work (known as the Yellow Book), and for design-build (known as the Orange Book). In September 1999, FIDIC published four new first editions:

- **Short Form of Contract (The Green Book)** which might be suitable for a small contract (say, under US$500,000) if the construction time is short (say, less than 6 months), or for relatively simple or repetitive work (say, dredging), irrespective of whether the design is provided by the employer or the contractor, and of whether the project involves civil, electrical, mechanical and/or construction works.
- **Conditions of Contract for Construction (The Red Book)** ('the Construction Book' or 'CONS'), which is recommended for building or engineering works where most of the design is provided by the Employer. However, the works may include some contractor-designed civil, mechanical, electrical and/or construction works.
- **Conditions of Contract for Plant and Design-Build (The Yellow Book)** ('the Plant & D-B Book' or 'P&DB'), which is recommended for the provision of electrical and/or mechanical plant, and for the design and execution of building or engineering works. The scope of this book thus embraces both old Yellow and Orange Books, for all types of contractor-designed works.
- **Conditions of Contract for EPC/Turnkey Projects (The Silver Book)** ('the EPC Book' or EPCT'), which might be suitable for the provision on a turnkey basis of a process or power plant, of a factory or similar facility, or of an infrastructure project or other type of development, where:
  - a higher degree of certainty of final price and time is required, and
  - the contractor takes total responsibility for the design and execution of the project.

## Institution of Chemical Engineers (IChemE)

For over 40 years IChemE has published *Forms of Contract* written specifically for the process industries. Process plants are judged by their performance in operation. Process industry contracts must, therefore, also be performance-based. This is the underlying philosophy of IChemE's *Forms of Contract* suite. It is a fundamental difference from much of the wider construction industry. The IChemE contracts are used extensively not only in the process industries, which themselves range from oil and gas to pharmaceuticals, food and fibres and many more, but also increasingly in a wide range of other industries both in the UK and abroad where their philosophy of fairness between the parties and teamwork in project execution are recognised as beneficial. All the contracts are internally consistent and have extensive guide notes to help in their completion and operation. They are supported by both User Guides and training courses, as well as an active User Group. Originally the IChemE contracts were drafted to operate under English law. Their growing international use has led to the preparation of a further suite of Forms of Contract written for international use under a wide range of legal systems.

### *IChemE UK Contracts: current editions*

Lump Sum: The Red Book, 4th Edition, 2001.
Reimbursable: The Green Book, 3rd Edition, 2002.
Target Cost: The Burgundy Book, 1st Edition, 2003.
Minor Works: The Orange Book, 2nd Edition, 2003.
Subcontracts: The Yellow Book, 3rd Edition, 2003.
Subcontract for Civil Engineering Works: The Brown Book, 2nd Edition, 2004.

### *International contracts*

The International Red, International Green, International Burgundy and International Yellow Books, published November 2007 have, as in all IChemE contracts, extensive guide notes with draft forms and notice included.

The International Contracts, while immediately recognisable to users of the UK forms in their arrangement, general drafting and philosophy of co-operation and teamwork, have been fully revised to reflect the particular needs of international contracting in many and various jurisdictions. IChemE has also been rigorous in using common language across all forms wherever possible and appropriate.

Lump Sum: The International Red Book, 2007.
Reimbursable: The International Green Book, 2007.
Target Cost: The International Burgundy Book, 2007.
Subcontracts: The International Yellow Book, 2000.

## The Scottish Building Contract Committee (SBCC)

In Scotland the Scottish Building Contract Committee (SBCC) was established in April 1964 as a result of the decisions of the McEwan Younger Committee which in turn followed the recommendations of the Emmerson report, that closer contacts be established with central bodies in London and that a working party be appointed in Scotland. The McEwan Younger Report 'The Organisation and Practices for Building and Civil Engineering' was issued in December 1964 and SBCC adopted those recommendations and produced its first contract incorporating the 1963 Edition of the RIBA Form of Contract. The purpose of the SBCC was defined as providing adjustment to English Conditions of Contract, reflecting Scottish practice and providing for differences in Scots and English law. The Scottish Building Contract Committee (SBCC), a member of JCT, has in the past prepared and published Scottish Supplements to many JCT forms of contract, for use where Scots law is to apply. SBCC now publishes an integrated document, that is a JCT contract which has been amended formally for use where Scots law is to apply. Its aims are:

- amending, drafting and publishing forms of building contracts for use in Scotland;
- promotion of best practice in building contracts in Scotland;
- drafting and publication of guidance and practice notes, lecture notes and commentaries on relevant contractual matters;
- nomination of arbiters, mediators or third party tribunals to facilitate dispute resolution; and
- attendance at national and other committees to promote best practice in building contracts, including the Joint Contracts Tribunal.

The SBCC does not sell its documents but acts as wholesaler to its three selling bodies, the Royal Incorporation of Architects in Scotland, the Royal Institution of Chartered Surveyors and Scottish Building. Contracts drafted by the SBCC include:

- Standard Building Contract with Quantities for use in Scotland.
- Standard Building Contract without Quantities for use in Scotland.
- Minor Works Building Contract for use in Scotland.
- Design and Build Contract for use in Scotland.

**Note:** the Intermediate form of contract is not available in Scotland.

## Insurances, bonds, guarantees, collateral warranties

### *Inherent defects insurance*

An inherent defect is a defect that exists, but remains undiscovered, prior to the date of practical completion but later manifests itself by virtue of actual physical damage which may not have been reasonably discovered previously.

One of the recommendations of the Latham Report was that a system of compulsory inherent or latent defect insurance should be introduced into the UK. In the system currently used in the UK the client, if they consider that they have a claim for a defect, has to resort to the courts in order to try to recover damages. However, the system is not easy for clients to engage with for a number of reasons, namely:

- Is the quantum of the claim worth it? Assuming that the client wins, will the expense involved be more than the eventual award?
- There are many legal traps along the way for example:
  - Is the claim in time? The Statute of Limitations applies to construction contracts.
  - What is the exact cause? Many experts may have to be employed by the client in order to determine the cause of a defect.
  - What participants were involved? The construction industry comprises a number of itinerant subcontractors and supplies and it may be impossible to locate the party who is to blame. The system currently in the UK is one of joint and several liability, which means that lawyers usually pursue the parties with the ability to pay, namely; local authorities and professionals with professional indemnity insurance.

Clearly, the burden of proof is very much on the client and it can be a very long, costly road trying to assemble the facts of the case and the parties involved. The current system can also create 'black holes' of liability in the case of leases with full repair clauses, making letting and disposal of buildings difficult as clients are unwilling to sign a lease knowing that the property could contain latent defects that have to be rectified.

Under the present insurance system of separate Professional Indemnity and Contractors All Risks, for the consultants and contractors respectively, the issue of defective workmanship is not covered. However, the contractor is the party essentially responsible for workmanship and so liability rests with the builder. The tension rises out of the fact that the contractor cannot contribute to the design and could feel that buildability and quality could be a problem within the project but there are no means whereby they can effectively be involved in such issues. The current system can therefore be seen to not deliver value or security to the client. The origin of the defect could be: poor workmanship, unsuitability of materials or poor design, structural failure, etc. A number of insurance companies in the UK are now writing policies to cover the cost of remedial work to correct latent defects in finished construction projects for a period of 10–12 years after practical completion.

The difficulties facing insurers and preventing the wider use of latent defects insurance are as follows:

- lack of statistics;
- lack of experienced underwriters;
- the potential cost of failing to underwrite properly;
- difficulties in defining the cover;
- lack of demand;
- the poor image of the UK construction industry with the lowest productivity and skills levels of all European states. Low labour costs tend to equate with low productivity.

The proponents of latent defects insurance argue that the presence of latent defect insurance helps to promote an environment where the design team operate in a non-adversarial relationship and in addition helps to promote buildability and predictability. In addition the present system of professional liability reserves the function of the design to the employers' professional advisors (architects, engineers, etc.) and reduces the function of the contract to that of the simple execution of drawings, specifications and instructions.

Latent defects insurance has been available in the UK for about 20 years but only became popular from 1989 onwards when a number of UK owned insurers brought out their own policies. Until recently the cover was limited to the repair of damage to a building caused by an inherent defect in the main structure of the building. More recently the extent of cover offered has been widened although many believe that cover remains quite limited. It is common practice in the UK for policies to have an excess that is a proportion of any claim to be paid by the insured, of £25,000 for structural elements. The system is monitored by a series of technical audits. The technical auditors:

- monitor the design and construction method;
- check that there is a clear demarcation of responsibilities within the design team;
- visit the site at regular intervals;
- issue a certificate of approval to the insurance company at practical completion;
- carry out triennial audit health checks on mechanical and electrical installations.

In practice the range of situations encountered by the technical auditors range from the discovery of errors in the calculations for components to works on site by the contractor that contravened good practice.

The basic inherent defects policy covers actual physical damage to the whole of the building, but importantly, only that caused by inherent defects that originated in the structural elements, such as foundations, external walls, roof, etc. In addition, optional cover can be added for non-structural parts such as, weather proofing and mechanical and electrical services.

Typically a policy covers the following items:

- the cost of repairing damage in the main structure;
- the cost of remedial action to prevent imminent damage;
- professional fees;
- cost of debris removal and site clearance;
- extra cost of reinstatement to comply with public authority requirements.

The policy is a first party material damage policy which essentially means that there is no need to prove negligence by a third party and cover may be freely assigned to new owners, lessees or financiers.

The benefits for construction projects and clients if latent defects cover is in place are:

- there is no need to rely on the project team's professional indemnity insurance;
- the potential for confrontation is reduced;
- there is more peace of mind for all the project team;
- less time and money is spent on arguing about contract conditions and warranties;
- innovation is encouraged;
- everyone can concentrate on getting the actual design and construction right;
- it can be a major advantage when negotiating a sale or letting.

The basic structural cover plus weatherproofing for 12 years can be obtained at rates ranging from 0.65% to 1% of the total contract value, rising to 2% for completed buildings. Total cover including all the options can cost from between 1% to 2%. The premiums are usually paid by instalments and include the fees of the technical auditors described below. For example, a project with contract price £2m cover could cost up to £20,000.

### *Subrogation*

A definition of subrogation is 'the right of an insurer (the insurance company) to stand in the place of an insured (the client) in order to exercise the rights and remedies which the insured has against third parties for the partial or full recovery of the amount of a claim'. In other words the insurance companies may pursue what they perceive as negligent members of the project team in order to mitigate their losses. The client will avoid this route, but not so the project team and therefore may be subject to a second wave of litigation. What is the incentive therefore for the team to integrate and more specifically for the contractor to innovate and contribute to the design?

A way of averting the above situation is through the use of subrogation waivers. In this event the insurance company, for the payment of an additional premium, will not enforce its claim against, say, an architect, surveyor or engineer, and/or contractor in the event that the latent defect is caused by their negligence. To some it is thought that if waivers are not used then there is little or no likelihood of eradicating the back-watching, trail covering adversarial culture within which the design consultants and the contractor will often find themselves operating.

### Insurances: injury, damages and insurance

As previously stated, the construction process in general and construction sites in particular are the scene of many deaths and serious injuries. Section 6 of the JCT 2011 deals with indemnities and insurances to persons, property, the works and in addition professional indemnity insurance in cases where contractor design in involved.

### Injury to persons and property

The clauses referring to injury to persons and property have been significantly reworded in the JCT(11), contractors should confirm that apprenticed staff are covered and consider whether self-employed individuals are covered. Clauses 6.1 and 6.2 require the contractor to indemnify the employer for any injury to persons or property that occurs during the carrying out of the works. This clause has the effect of protecting the employer from any claim that may be made for any injury to persons or property. It should be noted that some items are specifically excluded from the indemnity provisions and these are set out in clause 6.3. For example sections of the works for which a practical completion certificate has been issued, which becomes the property of the employer.

Section 6 of JCT(11) continues with a requirement for the contractor to take out insurance to cover the items covered in clauses 6.1 and 6.2. The contractor must allow the employer to inspect the insurance policy and if it is considered to be inadequate, then the employer can take out his/her own policy and deduct the cost from any money due to the client.

### *Insurance of the works*

Insurance of the works is dealt with by clause 6.7 of JCT(11) and gives the parties to the contract three options. Only one option should be used; Option A and B are for new buildings, whereas Option C is for alterations and works to existing buildings.

Options A and B are similar, except that in Option A it is the contractor who takes out the policy for all risks insurance, whereas in Option B it is the employer who takes out the policy, again in joint names. Therefore the main difference between the two options is that in the case of a claim it is either the contractor or employer who receives the insurance monies and arranges to make good any damage, etc.

It is a sign of the times that a major change in JCT(11) is a definition of terrorism which greatly widens the range of events that ought to be covered by insurance.

### *CDP Professional indemnity insurance*

Clauses 6.11/12 of JCT(11) are comparative clauses to the JCT contract and require the contractor to maintain professional indemnity insurance in respect of any contractor's designed portion and prove proof of a policy on request.

## JOINT CODE OF PRACTICE: FIRE PREVENTION ON CONSTRUCTION SITES (6TH EDITION)

Clauses 6.14–6.17 of JCT(11) refer to the Joint Fire Code. The code applies to activities carried out prior to and during the procurement, construction and design process, not the completed structure, and should be read in conjunction with all current legislation.

The object of this code is the prevention of fires on construction sites. It is claimed that the majority of fires can be prevented by designing out risks, taking simple precautions and by adopting safe working practices. All parties involved must work together to ensure that adequate detection and prevention measures are incorporated during design and contract planning stages; and that the work on site is undertaken to the highest standard of fire safety thereby affording the maximum level of protection to the building and its occupants. The code is voluntary, but if applied the contractor and employer are bound to comply with it.

### *Bonds*

A bond may be thought of as a guarantee of performance. The JCT(11) includes the provision to execute three forms of bond:

- **Advanced payment bond** – advanced payments are not as common in the UK as, say for example, in France, where they are commonplace. In the event that the contractor requires an advanced payment from the employer, prior to work commencing on site, an advanced payment bond must guarantee to repay the advance in the case of default by the contractor, JCT(11) clause 4.8 and Schedule 6.1.
- **Bond for off-site materials and/or goods** – there may be occasions where, for example, it may be necessary to purchase in advance, materials and goods and to store them off site. The goods and materials in question may be expensive or delicate or both and therefore it is inappropriate to store them on site. Nevertheless the contractor applies for payment and in these circumstances a bond to cover the cost of the materials or goods should they be damaged or lost. JCT(11) Schedule 6.2 applies.
- **Retention bond** – retention, normally at 3% is deducted from all interim payments made to the contractor and held by the employer until practical completion and final account stages. If the contractor provides a retention bond then the retention deduction may be waived. JCT(11) clause 4.19 refers and Schedule 6.3.

In addition other common forms of bond are performance bonds and tender bonds.

- **Performance bond** – a performance bond is required to guarantee the performance of the contract during the works. In value terms it is usually equal to up to 10% of the value of the contract. The purpose of the performance bond is to reimburse the client in the event that the contractor does not proceed diligently.
- **Tender bond** – a bond may be required by a client to ensure that contractors who express an interest in submitting a bid for a project are bona fide. The bond fund may be used in the event that a contractor either fails to submit a bid or fails to enter into a contract after being selected.

### *Guarantees: collateral warranties*

#### Generally

A warranty is a term of a contract, the breach of which may give rise to a claim of damages but not the right to treat the contract as repudiated. It is therefore a less important term of the contract, or one which is collateral to the main purpose of the contract, the breach of which by one party does not entitle the other to treat their obligations as discharged.

Undertakings may be given that are collateral to another contract that is running side-by-side. They may be independent of the other contract because they cannot be fairly incorporated, or the rules of evidence hinder their incorporation, or because the main contract is defective in some way. A warranty is a term of a contract, the breach of which may give rise to a claim of damages. A transaction between two parties may be of particular concern or affect the performance of a third party. A collateral contract may be entered into between the third party and one of the original parties. This may be a useful device for avoiding privity of contract.

Increasingly there are a number of parties with financial stakes in the success of a construction project, e.g. funders, tenants, and purchasers who are not party to the building and other associated contracts. In the event of a third party suffering loss arising from the construction project, in the absence of a direct contract, the only remedy is a claim in the tort of negligence. However, since the 1980s the courts have severely restricted the scope of negligence claims. In order to plug this contractual gap the parties to the contract may decide to make use of collateral warranties. These are contracts in which the person or firm doing the work (the warrantor) warrants that they will properly carry out their obligations under the main contract. Therefore, in the case of defective workmanship or the like the warrantor may be sued. As with standard forms of contract there are many standard forms of collateral warranty, including those published by the JCT. It is also possible to use bespoke forms of collateral warranty for those who fear that their right to pursue a warrantor is limited by clauses in the standard forms. A classic model for the use of collateral warranties is when management contracting procurement strategy is used and the package contractors have responsibility for elements of the design. Under these circumstances, the client would procure a series of collateral warranties with the relevant package contractors. The principal disadvantage of using collateral warranties, as in the case of management contracting, is the amount of time and bureaucracy involved in procuring warranties from twenty to thirty separate organisations. It should be borne in mind that a collateral warranty should not expose the warrantor to any greater exposure than they had under the original contract and the extent of liability; usually restricted to the cost of remedial works only.

## Collateral contracts

In theory the Contracts (Rights of Third Parties) Act 1999 that came into force in May 2000 provides an alternative to collateral warranties. One of the intentions of the act was to reduce the need for collateral warranties but initial take-up of the Act was disappointing. However, when the JCT considered the provisions of the 05 suite of contracts it was decided that provisions should be incorporated for a contractual link between the main contractor, funders, etc. The majority of new 05 forms now include an obligation to provide warranties and a mechanism for invoking the Act. The new forms also allow for the calling of subcontract warranties in favour of such third parties and the employer. Both the JCT(11) (Section 7) and the JCT Design and Build(11) contain provision for collateral warranties to third parties, subcontractor collateral warranties and third party rights.

Where there is no collateral contract found a plaintiff may still sue in negligence. However, a claim may be purely economic and this may well prove fatal in establishing a duty of care. Further, claims in contract by implied terms (for example implied by the Sale of Goods Act 1979) are normally strict, that is have no defence, but a claim in negligence will require proof of fault.

## Collateral warranties between employer and subcontractor

When a subcontractor enters into a domestic contract with a main contractor there is no contractual relationship between the employer and the subcontractor. The employer could only sue the subcontractor in tort and would have to prove that a duty of care was owed. A collateral warranty between employers and subcontractors allows the employer to sue the subcontractor for any breach of the warranty's conditions, which commonly include promises on the part of the subcontractor to achieve a standard of design and workmanship as specified by the employer. The terms of the warranty may impose whatever liabilities and responsibilities the employer considers appropriate, so long as the subcontractor, being aware of such terms, is willing to tender and enter into a subcontract for the relevant work.

It is important, but nevertheless is sometimes overlooked, to ensure that both the obligation to enter into the warranty and its full wording form part of the legal obligations set down by the terms of the contract between the main contractor and the subcontractor or professional party. Warranties will also address the matter of deleterious materials, to ensure that such materials are not specified or employed in the works.

Most collateral warranties include provisions for the benefit of the warranty to be assigned by the employer to a third party, such as a purchaser or tenant. Indeed such third parties taking a legal interest in a building require such a warranty, so as to provide themselves with redress against a contractor or designer as a result of defects appearing within a period of time, commonly after 12 years from the completion of the original works.

# PART 2: CONTRACT ADMINISTRATION

## Dayworks

As previously mentioned the bills of quantities can be used by the quantity surveyor in the post-contract stage of a project as the basis for valuing variations in the contract. There are, however, some occasions when the nature of the variation is such that it is unfair or inappropriate to use the bills of quantities/pro rata pricing and in these circumstances dayworks are used. Clause 5.7 of the JCT(11) provides for pricing dayworks as a percentage addition on the Prime Cost. The use of daywork rates relieve the surveyor of having to calculate rates from basics every time daywork charges are used.

Dayworks are defined in the *Definition of Prime Cost of Daywork Carried out under a Building Contract*; a publication produced by the Royal Institution of Chartered Surveyors and the Construction Confederation. In September 2007 a major revision of the definition was published, replacing the 1975 version. There are some significant differences in the new definition.

### *Labour*

There are now two options for dealing with the prime cost of labour:

**Option A** – Percentage addition. This option is based on the traditional method of pricing labour in daywork and allows for:

- guaranteed minimum weekly earnings, e.g. standard basic rate of wages, Joint Board Supplement and Guaranteed Minimum Bonus Payment;
- all other guaranteed minimum payment, unless included with incidental costs, overheads and profit;
- differential or extra payments in respect of skill, responsibility, discomfort, inconvenience or risk, excluding those in respect of supervisory responsibility;
- payment in respect of holidays;
- any amounts that may become payable by the contractor to or in respect of operatives arising from the operation of the rules referred to;
- employer's contributions to annual holidays;
- employer's contributions to benefit schemes;
- employer's national insurance contributions;
- and contribution, levy or tax imposed by statute, payable by the contractor in their capacity as an employer.

Differential or extra payments in respect of supervisory responsibility are excluded from the annual prime cost. The time of supervisory staff, principals, foremen, gangers, leading hands and similar catagories, when working manually, is admissible under this section at the appropriate rate.

**Option B** – All inclusive rates. This option includes not only the prime cost of labour but also includes an allowance for incidental costs, overheads and profit. The all-inclusive rates are deemed to be fixed for the period of the contract. However, where a fluctuating price contract is used, or where the rates in the contract are to be index linked, the all-inclusive rates shall be adjusted by a suitable index in accordance with the contract conditions.

Option B gives the client price certainty in terms of the labour rate used in any daywork in the contract, but there is potential that the rate will be higher, as the contractor is likely to build in a contingency to cover any unknown increases in labour rates that may occur during the contract period.

For both options materials and plant are dealt with as follows.

### *Materials*

- The prime cost of materials obtained for daywork is the invoice cost after discounts over 5%.
- The prime cost of materials supplied from stock for daywork is the current market price after discounts over 5%.

### *Plant*

- Where hired for the daywork it is the invoice cost after discount over 5%.
- Where not hired for the daywork it is calculated in accordance with the RICS Schedule of Basic Plant Charges.
- Includes for transport, erection, dismantling and qualified operators.

It is up to the quantity surveyor preparing the contract documents to decide which of the above methods is most appropriate in the circumstances. Consideration should be given to the length of the contract, whether the contract is firm or fluctuating price. It should be noted that specialist trades may have their own different definitions.

## Recording dayworks

Assuming that a variation order is issued for additional works that cannot be valued using bill rates or pro rata pricing, then daywork rates are used. As this method of pricing has distinct advantages for the contractor, the recording and monitoring of dayworks must be strictly controlled. It will often be the case that a contractor will use a daywork sheet (see Figure 6.1) to record hours worked and material used carrying out additional works, in the hope that it will eventually be valued on a daywork rate basis and the final account stage and surveyors should be alert to this. If there is a clerk of works on site, then it will normally be his/her responsibility to check daywork sheets submitted by the contractor. The signature of the clerk of works will simply verify that hours, materials, etc. are correct; it does not imply that the item should be valued at daywork rates.

## Fluctuations

Fluctuations, or as it is often referred to, increased costs, is the mechanism by which the contractor is reimbursed for increases in the cost of labour, materials, plant, etc. that occur during the contract period. Particularly for large contracts with a long duration the impact of increased costs can be significant. The mechanism for the recovery of fluctuations in cost is based on the parties' perception of inflation risk during the contract period. If the perception is that the risk of inflation is low, contractors are more likely to accept the risk and incorporate an allowance in the tender to cover fluctuations in price, effectively making the contract fixed price. However, in the lifetime of the author inflation has been running at more than 25% per annum and clearly in this situation the contractor could not accept the risk and instead opt for the option to reimburse the cost of fluctuations during the contract period. It is common practice for clients to ask contractors to submit two alternative tenders; one fixed price and a second based on a fluctuations basis.

This strategy allows the employer to clearly see the amount that has been included by the tenderers to allow for the risk of increased costs and to decide whether this is allowance is reasonable.

If applicable, fluctuations in price are calculated on a monthly basis and added to the monthly valuation. Note that retention is not held on fluctuations as the contractor may not add profit onto the fluctuation claim. Most standard forms of contract contain a fluctuation clause. The JCT(11) has three fluctuation options and which are detailed in Schedule 7 of the contract:

- **Option A**: Contribution, levy and tax fluctuations.
- **Option B**: Labour and materials cost and tax fluctuations.
- **Option C**: Formula adjustment.

Yorkshire Construction plc. Sheet No. 28

Contract: New Office block, Leeds AI No. 34
Date 23/05/12 SI No. 67

Description
Alter position of window W23 and build in new window.

LABOUR

| Name | Trade | M | T | W | Th | F | Sat | Total | Rate | Add% | £ |
|---|---|---|---|---|---|---|---|---|---|---|---|
| F Blogs | Bklayer6 | | 7 | 5 | 5 | 4 | | 27 | £9.72 | 70 | 446.1 |

Total labour £446.1

MATERIALS/PLANT

| | No | Unit | Rate | Add% | £ |
|---|---|---|---|---|---|
| Materials | | | | | |
| Cement and sand | 1 | $m^3$ | 12.00 | 10 | 13.20 |
| Window | 1 | No | 250.0 | 10 | 275.00 |
| Plant | | | | | |

Total materials/plant £288.20
Tota £734.35

Signed Yorkshire Construction
G. Blogs
Date 30/05/12

Client's Representative
H. Blogs
Date 30/05/12

**Figure 6.1** Daywork sheet

## Option A

One of the common features of fluctuation provision is the establishment of a base date from which any fluctuations may be calculated. The base date is agreed between the parties to the contract and entered into the Contract Particulars. Option A allows the contractor to be reimbursed for increases in the following items:

- contributions, levies and taxes payable in respect of work people on site;
- contributions, levies and taxes payable in respect of work people off site; producing materials or goods for the site;
- duties and taxes on material, electricity, fuels, etc.

When this option is used, only increases in taxes and levies, such as VAT are reimbursed. Increases in the hourly rate of labour, etc. and increases in materials due to market conditions are not reimbursable. This approach is sometimes referred to as partial fluctuations and transfers more risk to the contractor.

**Option B**

Sometimes referred to as full fluctuation this option transfers more risk to the client, as he/she will be responsible for paying increases that occur in the following circumstances:

As Option A, plus:

- increases in rates of wages for workpeople on and off site as well as site staff in accordance with appropriate wage-fixing bodies;
- transport charges;
- materials, electricity and fuels.

To assist the calculation and agreement of fluctuations a list of basis materials is prepared by the contractor that details the prices current at the date the tender was prepared.

From the contractor's point of view the main drawback of using Options A and B and in particular B is that the amount of time and hours required to compile the fluctuations claim is considerable. In addition the quantity surveyor has to check the claim at each valuation. In order to streamline the process a third option is available:

**Option C: formula adjustment**

### *The formula*

For the purposes of the formula, construction work is divided into 42 work categories.

The formula for adjustment of the value of work allocated to work categories under Part I of these Rules for Valuation Periods up to and including the Valuation Period in which occurs the date of practical completion is as follows:

$$C = \frac{V\ (Iv - Io)}{Io}$$

where:

C = the amount of the adjustment for the work category to be paid to, or recovered from, the contractor.

V = the value of work in the work category for the valuation period.

Iv = the index number for the work category for the month during which the mid-point of the valuation period occurred.
Io = the index number for the work category for the base month.

There are separate formulae for specialist engineering works such as:

- electrical installations;
- heating, ventilating and air conditioning installations and sprinkler installations;
- lift installations;
- structural steelwork installations;
- catering equipment installations or those for which specialist formulae index numbers are published in the monthly bulletin.

The following items are excluded from the formula adjustment:

- amounts for work valued as daywork under the conditions;
- amounts for articles manufactured outside the United Kingdom which the contractor is required by the employer in the contract documents to purchase and import or have imported for direct incorporation into the works and which require no processing prior to such incorporation and which are specifically identified in the contract documents;
- amounts for work which is valued under the Conditions at a fair valuation on rates and prices current when the work is carried out or on some basis which is not referable to the level of rates, prices or amounts in the documents referred to in rules 24*a*, *b* and *c* unless made subject to formula adjustment by the operation of rule 25;
- amounts for unfixed materials and goods for which the contractor is entitled to payment;
- amounts for direct loss and/or expense ascertained under the Conditions;
- any other amounts payable by the employer which are based on the actual costs incurred by the contractor;
- amounts excluded from formula adjustment by the operation of rules 14 and 15;
- amounts for design work where separately identified in any CDP.

## Conclusion

Of all the standard forms of contract in use it would appear that the JCT (11) suite of contracts makes the most provision for the recovery of fluctuations, whereas at the other end of the spectrum the GC/Works/1 has no fluctuation provision, the assumption being that the contract will be fixed price. Of the other popular forms of contract, the NEC3, FIDIC and ICE 7th Edition, the quantity surveyor needs to specifically amend the contract conditions to include a fluctuations supplement.

## Time and the contract

The construction process is fraught with risk and uncertainty that can impact on the time taken to complete the contract; inclement weather, design changes, delay on the part of the consultants, etc. can conspire to extend the agreed contract period. For the contractor this can present a problem as the JCT(11) form of contract, for example, has provision for the employer to claim damages from the contractor if the works are not completed on time as stated in the Contract Particulars. The JCT(11) form of contract is specific in relation to time with the following dates being entered into Part 1 of the contract:

- date for possession of the site (Part 1 and 2.4);
- date for completion of the works (Part 1 and 2.3);
- amount of liquidated damages (Part1 and 2.32.2).

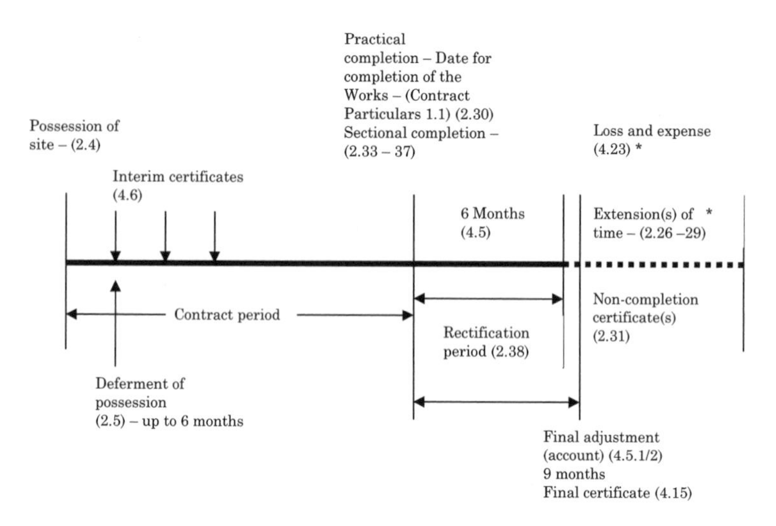


**Figure 6.2** Project time line; based on JCT(11)

In the case of possession and completion there is provision to enter several dates to allow for contracts where sections of the site and completed works are to be handed over separately in stages. Given the above, without the provision to amend or extend, for example, the completion date, then contractors would need to accept a high degree of risk and include a sum in their tender to cover the possibility of incurring damages and financial penalties. The English legal system is based on the doctrine of jurisprudence, that is to say that individual cases are decided by the courts based on the facts of each case and on decisions that have been delivered in the past. The system has the reputation in construction circles of being something of a lottery. By contrast, some European countries have a codified legal system where remedies are pre-determined and consequently the forms of contract reflect this. It is thought that the JCT(11) has evolved in its present format because of the UK legal system, based on the principal that the contract should contain every possible event that could occur during a construction project together with the remedy, instead of leaving the courts to decide. Some feel that creates a poor starting point, as it almost presumes that there will be delays during the contract period and perhaps contributes to the 'them and us' ethos that has for so long been a characteristic of UK construction.

The principal issues of time are dealt with under the following headings in the JCT(11) form of contract as illustrated in Figure 6.2:

- possession of the site and commencement of the works;
- postponement of the works;
- extension of the contract time;
- interim valuations;
- completion of the contract/practical completion;
- rectification period;
- final adjustment.

### *Possession of the site and commencement of the works*

It is obviously important for the contractor to have prompt access to the site, this is because all other time-related events such as interim payments, completion and damages for non-completion are related to this point. The date when the contractor may gain access to the site is usually stated in the invitation to tender letter and once access to the site has been granted the contractor is required to begin and work 'regularly and diligently'. The date for the possession of the site is entered into the Contract Particulars. Once the contractor has taken possession of the site and commenced work the contract is deemed to have come into existence, even if the contract itself has not been signed.

### *Postponement of the works*

There is provision in the JCT(11) for the employer to defer possession of the site to the contractor by up to six weeks. Any such delay will usually result in the contractor applying for an extension of time with an addition of costs, (see below). Any delay beyond six weeks will result in the contract becoming void.

### *Extension of time*

Extension of time provisions is required when works are delayed for the following reasons:

- to relieve the contractor from having to pay damages for completing the works later than the agreed completion date;
- without an extension of time provision, time would be considered to be 'at large' by the courts, which would make it difficult for the employer to claim liquidated damages from the contractor in the case of delayed completion.

Delays can occur for the following reasons:

- Delay on behalf of the contractor – the contractor fails perhaps due to a combination of lack of expertise, labour etc. and so fails to complete the work on time. Under these circumstances the contractor has no real excuse for failure to complete and can expect to pay the full amount of the damages stated in the contract.
- Delay caused by the employer or his/her agents and directly employed contractors – work fails to be completed on time because the architect or some other member of the design team has failed to provided information requested by the contractor on time, or contractors employed directly by the employer disrupt the main contractor's work programme. Under these circumstances the contract would be extended and in some cases would be classified as a 'relevant event' giving grounds for a claim for additional costs from the contractor.
- Delay caused by circumstances outside the control of either the contractor or the employer, for example exceptionally inclement weather, force majeure, and will normally result in an extension time being granted.

**It is a common misconception that extensions of time give an automatic right for a claim for loss and expense by the contractor; this is not the case.**

If it becomes apparent that a delay has occurred in the works then the contractor must notify the architect in writing. The notification must include the following information:

- the cause of the delay and whether in accordance with clause 2.29 the cause is a 'relevant event'. Relevant events are defined as follows:
  - variations issued by the architect;
  - instructions issued by the architect concerning:
    - discrepancies or divergence between documents such as the contract drawings and the contract bills
    - any instructions and the contract conditions
    - any drawings or other documents
    - the CDP documents;
  - postponement of the works;
  - instructions for the expenditure of a provisional sum for previously undefined works;
  - discovery of antiquities;
  - non-provision of a quotation by contractor when requested by the architect;
  - opening up works for inspection, unless the works or materials subsequently turn out to be faulty;
  - late possession of the site;
  - execution of works that have been included in the contract documents as provisional, that subsequently proves to be inaccurate in the nature or extent of work required;
  - suspension of works by the contractor for non-payment by the employer;
  - impediment by the employer, architect, quantity surveyor or any other person employed by the employer;
  - works carried out by a Statutory Undertaker;
  - exceptionally inclement weather;
  - civil commotion, threats of terrorism, strikes, lock outs;
  - UK government legislation that affects the works;
  - force majeure; which is usually interpreted as an 'act of God' that no one could have reasonably foreseen.

Once the cause of delay has been identified as a relevant event the architect has to decide whether the events qualify for an extension of time. If they do, then the architect issues an extension of time within twelve weeks of the application by the contactor. The architect is required to note the extension of time that has been allocated to each relevant event when issuing an extension of time certificate. If the works are delayed and a relevant event is not identified as the cause, then the contractor may be liable to pay liquidated damages as set out in the Contract Particulars, for details see 'Completion of the contract/practical completion' below.

**Interim valuations**

Most standard forms of contract have a provision to pay the contractor on a stage payment or quantum meruit basis, as the work proceeds. Without this provision the contractor would have to wait until the end of the contract before receiving payment and this would clearly be unworkable from a cash flow perspective. The stage payments or interim accounts are prepared usually on a monthly basis by the contractor's surveyor and the client's quantity surveyor. Until recently it had been common practice in the construction industry for the contractor not to release money to domestic subcontractors until the contractor received payment, a system referred to as 'pay when paid'. This practice caused many problems and disputes particularly between contractors and domestic subcontractors, as domestic subcontractors are not informed when the contractor receives payment and the contractor could with hold payment for weeks or longer. In 1996 the Housing Grants, Construction and Regeneration Act (referred to as the Construction Act) made pay when paid clauses unenforceable. The Construction Act contained the following provisions for all construction contracts:

- pay when paid clauses are unenforceable;
- payment by instalments for all contracts over 45 days duration;
- the contractor is to be informed when payment is due as well as the amount;
- the contractor is to be informed if the client is to withhold payment;
- the contractor has the right to withdraw from the site if not paid within a specified period.

When preparing an interim valuation the following items may be included, if appropriate:

- preliminaries as included in the bills of quantities;
- measured works as included in the bills of quantities;
- value of variations and extra works;

- work carried out by nominated subcontractors and suppliers;
- materials on site;
- materials off site;
- fluctuations;
- approved loss and expense claims.

The clauses dealing with payments on account in JCT(11) are:

### *Preliminaries*

A proportion of the preliminaries is included with the interim payment, depending on the nature of the item. For valuation purposes, preliminaries are grouped into various types:

- Offices, canteens and storage sheds. Items under this category will be assumed to have been priced on the following basis:
  - set-up costs – paid as part of Valuation No. 1;
  - hire cost per week – paid at a weekly rate during the contract period;
  - heating and cleaning – paid at a weekly rate during the contract period;
  - clearing away – paid at the end of the contract.

  For valuation purposes the contractor will be expected to provide details of the above costs. The amount included in an interim certificate for this and similar items are not related to value of the works.
- Scaffolding. This item will not be required until way into the contract period but will have costs similar to offices and canteens with initial costs, adaptation costs and removal costs.
- Water for the works. Items such as this are valued as a percentage of the contract sum and will be paid in relation to the interim amount.
- Site supervision. Items such as site supervision will be assumed to have been priced on a time-related basis. The amount to be included in valuations is calculated by dividing the total by the number of months.

### *Measured works*

The contractor's surveyor and the client's quantity surveyor will make a visual estimation of the amount of the works that have been completed on the day of the valuation. By using the bills of quantities as the reference, the value of the work can be calculated. Including and paying for work as part of the valuation does not indicate that the work has been properly executed, that responsibility remains with the architect. Usual practice is for the contractor's surveyor to prepare the valuation for the client's quantity surveyor to check. It is important that the figures are checked by the quantity surveyor as any over payment could prove to be embarrassing if the contractor goes out of business. The checking process is particularly important as the project reaches the final stages; the final account should indicate an amount due to the contractor, not an amount due from the contractor!

### *Variations*

It is common for the architect to issue variations to the contract works. As these variations are measured and agreed then the value of them should be added to the valuation. These measured and agreed variations will form the basis of the final account.

### *Nominated subcontractors and suppliers*

Nominated subcontractors and suppliers are appointed as a result of the adjustment of prime cost sums. Shortly before the valuation date the subcontractors and suppliers will submit their own interim valuations to the main contractor and these are included in the interim payment. Nominated subcontractors' and suppliers' payments are paid to the main contractor, who then pays them in turn, less any agreed discounts. The nominated subcontractors and suppliers are informed of the amount due and the main contractor must pay them within a prescribed period; if the main contractor withholds payment, then there is provision for the employer to pay them directly. The employer also has the authority to request proof that payment has been made. Prime cost sums, as previously described also have the provision for the main contractor to include provision for profit and attendance; these items should also be adjusted and included.

### *Materials on site*

As the contract progresses the main contractor and subcontractors will have materials stored on site prior to their inclusion in the works. Such materials should be included in the valuation provided that:

- they are in accordance with the specification;
- they are intended for the works and are not being stored for another contract;
- they are not delivered before they are required;
- they are properly stored;
- they are the property of the contractor.

As part of the valuation process the contractor's surveyor prepares a list of materials and these are then priced at current rates or invoices. Once paid for the materials become the property of the employer.

### *Materials off site*

Sometimes it is inappropriate for large, delicate or expensive items of plant to be stored on site. These items may be included in the valuation provided that the client's quantity surveyor visits the place where they are stored and is satisfied that the materials are labelled for delivery to the site.

### *Fluctuations*

Many standard forms of contract include the provision to calculate and include for increases in the cost of labour and materials, for example the formula approach included in the JCT(11).

### *Approved loss and expense claims*

As a project progresses there may be claims by the contractor for loss and expense. If the claim is approved by the architect and valued and agreed then any sums due should be included too.

### *Retention*

Most standard forms of contract include the provision for a percentage of interim payments to be withheld by the client as an incentive to complete the works timeously. This percentage, usually between 3% and 5%, is deducted from the interim valuation up until practical completion, when half is released back to the contractor and the percentage is reduced by half. The final percentage or moiety, being paid on the settlement of the final account. The client has no obligation to invest the retention money, but may be required to keep it in an identified bank account. Retention is also deducted from payments due to nominated subcontractors. Only fluctuations are not subject to retention.

### *Completion of the contract/practical completion*

Completion of the works is said to be when the works are practically complete. This is the point at which the architect considers that the works are complete, sufficient for the client to take possession and use the building, although it may be that there are still some minor defects of works to complete. It is the point at which the works can be used safely by the employer for the purposes for which they were designed. It is possible to arrange for sectional completion of the works, where for example a self-contained portion of the work is handed over to the employer, while work continues on the remaining sections. In cases such as these, there may be several periods of final measurement, release of retention, etc. and they require careful monitoring by the quantity surveyor, particularly where nominated subcontractors are also involved.

The issue of a certificate of practical completion is significant as:

- Practical completion is the start of the period of final measurement for the final account. The process for the preparation of the final account is described in Chapter 7. JCT(11) allows a total of nine months for the quantity surveyor to receive the necessary information from the contractor and prepare the final account.
- The first moiety of the retention fund is released to the contractor. Half of the retention held on interim certificates is handed back to the contractor. It is interest-free, the employer having no obligation to invest it and any valuations that take place between practical completion and the final adjustment (final account) will only attract a retention percentage of half that stated in the Contract Particulars.
- The employer takes over the responsibility for insuring the works. Once the works are handed over to the client, he/she becomes solely responsible for insuring the works as the contractor's liability ceases.
- Once the certificate of practical completion has been issued the contractor can no longer be charged liquidated damages.
- Practical completion is the start of the Defects Liability or Rectification Period.

### *Defects Liability Period/Rectification Period*

The Defects Liability Period, or as it is referred to in JCT(11), the Rectification Period, runs from the issue of the Certificate of Practical Completion for a period of six months, this may be longer in the case of mechanical and electrical installations. The purpose of this period is for any defects or snags that come to light once the building is handed over to be rectified by the contractor to the satisfaction of the architect and the client. The architect has 14 days after the expiry of the Rectification Period to present the contractor with a list of defects or snags after which the contractor has a reasonable time to make good the defect. Assuming that the defects are rectified the architect then issues a Certificate of Making Good after which the second moiety of the retention is released, even though the final account may not yet be agreed.

### *Final adjustment (final account)*

The final adjustment is more traditionally referred to as the final account. Although the period for preparing the final account is stated in the contract, in practice it can often take longer for the final figure to be prepared. This can be due to a number of reasons, such as change of personnel, lack of information, etc. but, perhaps the fact that there is no penalty for the late preparation of the final account is the greatest contributory factor. In addition pre-contract work often takes precedence.

### *Damages for non-completion*

If the contractor fails to complete the works on time without the issue of an extension of time certificate then a breach of contact will have taken place and the contractor will become liable to pay damages at the rate entered into the Contract Particulars. It is important that the sum entered is a reasonable pre-estimate of the financial costs incurred by the employer by the non-completion of the works and not an intimidating over estimate. Consider a new hotel complex being constructed in time for the opening of the 2012 London Olympics; if the project was not in fact completed until after the event had taken place then the employer would suffer considerable loss of income and should therefore be entitled for financial recompense. This is dealt with in clauses 2.30–2.32 in the JCT(11). Items in the sum claimed by the employer could include such items as loss of income, costs in connection with transferring guests to other accommodation, additional professional fees, etc. As considerable as an amount for damages may be in such a circumstance, it has to be justifiable and must not be regarded as punitive or seen as a punishment for the contractor, as such a sum will not be enforced by the courts. Over time various tests have been suggested to determine whether the rate of damages constitute a penalty. It will held to be a penalty if:

- the sum stipulated is extravagant and unconscionable in amount in comparison with the greatest loss that could conceivably be proved to have followed from a breach;
- a single lump sum is payable by way of compensation on the occurrence of one or more or all of several events, some of which may be considered serious and others trifling.

Interestingly, if the amount of loss entered into the Contract Particulars for Liquidated Damages is exceeded in the event of a delayed completion then the liquidated damages amount will not normally be increased, although see also *Bath and North East Somerset District Council v Mowlem plc* (2004) EWCA Civ 115. Clearly, the amount of liquidated damages is one of the many factors that a contractor should taken into account when deciding whether or not to submit a bid for the project. If a contractor wishes to pass to a subcontractor the risk under the main contract of paying liquidated damages, they may choose to do so either by a subcontract liquidated damages clause or by giving notice of the provisions of the main contract.

Clause 2.32 of the JCT(11) allows the employer to withhold or deduct liquidated damages or require the contractor to pay liquidated damages at the rate stated in the Contract Particulars. From the quantity surveyor's view point, if there is a chance of liquidated damages being applied, caution should be taken when preparing the interim valuations and final account as it is obviously easier to withhold damages from monies due to a contactor rather than go cap in hand to recover them from monies already paid!

## Insolvency

Insolvency is a generic term, covering both individuals as well as companies. If you cannot pay your business debts when they become due, or if the assets of your business are less than your debts, your business is insolvent. The debt involved may not necessarily be a large sum; any outstanding debt in excess of £750.00 can be used as the basis of a winding-up partition. Unless you pay those debts quickly, then the insolvency will lead to bankruptcy or winding up. Bankruptcy applies to individuals such as sole traders and those that have given personal guarantees for loans. Winding up and liquidation apply to companies. Becoming bankrupt may involve restrictions, but the situation is now less onerous for individuals whose businesses have failed through no fault of their own. Most are discharged from this process within 12 months although there can be longer-term effects on their credit rating. Insolvency rules differ slightly in different parts of the United Kingdom and there are separate departments that deal with insolvency in Scotland and Northern Ireland. On average, during the period 2001–2011, in England and Wales there were approximately 13,000 company liquidations, (with a peak in 2009) and 25,000 individual insolvencies per annum and these are to some extent a barometer of the country's economic health. In Scotland and Northern Ireland the number of insolvencies and liquidations, year on year, vary more widely. There are a number of pieces of legislation governing insolvency including The Insolvency Act 1986 and 2000.

### *Liquidation of the contractor*

The unique structure of the UK construction industry has been discussed previously and it is unfortunate that this structure, with a proliferation of sole traders, means that construction often heads the insolvency league. The factors that make construction companies vulnerable to insolvency are:

- the large number of sole traders, often set up during times of plentiful workload with little regard to how the business will cope in leaner times. The majority of work is now carried out by subcontractors, who will be squeezed by the main contractors when times get tough and this may force them into liquidation;
- the practice of winning work via competitive tendering with unsustainable profit margins, in the hope that profit can be generated during the contract through extra works or claims for loss and expense;
- the volatile nature of demand for construction works, making it hard to secure continuity of work;
- taking on too much work and being unable to carry it out because of insufficient cash flow to purchase materials or hire labour.

Section 8 of the JCT(11) is devoted to insolvency and termination of the contract and contains a major change to the JCT(98) version of the contract. For, if the contractor becomes insolvent, the employer can terminate the contract immediately by notice. The onus is therefore on the employer to act swiftly and decisively in the instance of the insolvency of the contractor. This is a major change to JCT(98) where determination occurred immediately without any action from the employer in the majority of cases, however, doubts were cast on the validity of the JCT(98) provision, as it seemed to infringe a fundamental principle of bankruptcy law. The contractor is also required to inform the employer if he/she is approaching insolvency. No further payments need be made to the contractor until the total financial implications are known. All domestic subcontractors are automatically determined on the termination of the main contract.

Of course it is not just the contractor who may become insolvent, the employer or the professional advisors may also become insolvent and this will be discussed later.

Consider the following scenario: having borrowed against a business plan that has not worked, a building contractor finds that it is suffering cash flow problems. In an effort to survive, the company reports its problems to its bank and the bank asks for more information on the problems the company faces. Struggling with the problem the directors find it difficult to produce the information required by the bank. Often the accountancy and reporting systems of small contractors are not robust and a lot of time is needed to work out where the company is going, what the depth of the problems is and the necessary reporting to the bank is delayed.

The company's problems may be manifested in the following ways:

- the company's agreed overdraft is constantly at the limit;
- cheques that are written by the company are returned by the bank due to insufficient funds because of cash flow problems as money is not being received by the company.

After several weeks or months of problems the contractor's bank may become concerned and ask the contractor to reduce their exposure to risk by either:

- Asking that the directors increase their guarantees and securities. The majority of small building contractors and subcontractors have their overdrafts secured against second charges on their director's homes or other assets. Additional guarantees may be required in order for continued bank support. It should be noted that the attitude of banks towards levels of exposure and debt vary considerably according to the state of the construction market and the economy in general. It is not uncommon for funds to be agreed when times are good only to be withdrawn at short notice, as happened during the so-called 'credit crunch' of 2008 and the market downturn in 1990. This withdrawal of funding by banks can happen very quickly as they seek to protect their shareholders!
- Assuming that it is not possible for the contractor to furnish additional securities then the bank will ask for investigating accountants to look at the business. Normally, this is a large firm of accountants who have their own insolvency practitioners. During the period the business continues to trade. The insolvency practitioners will ascertain whether:
  - the business is viable;
  - the business is stable and has a long-term future;
  - the bank's exposure is sufficiently covered in the event of failure.

  If the insolvency practitioner considers that the company is in serious risk of failure and that the bank will lose money then he/she will usually recommend to the bank that they appoint an administrator or a receiver.

**An administrator** is appointed and the company goes into administration, when it is thought that the business may be rescued and may survive as a going concern. In these circumstances the company will be run on a day-to-day basis by the administrator. Note: once a company has gone into liquidation an administration order cannot be made but, once an administration order has been issued, no liquidation or winding-up procedures can be commenced; in effect giving all concerned a breathing space.

**A receiver** is a civil servant in the Insolvency Service and as an officer of the court he/she will be notified by the court of the bankruptcy or winding-up order. The receiver will then be responsible through his/her staff for administering the initial stage, at least, of the insolvency case. This stage includes collecting and protecting any assets and investigating the causes of the bankruptcy or winding up.

Liquidation is the process of selling the company's assets in order to raise money to clear the contractor's debts. Once sold, the proceeds will be distributed in strict order as follows:

- secured creditors, with fixed charges such as banks and finance houses;
- liquidator's fees;
- preferential creditors, that rank equally for payment:
  - HM Revenue and Customs (PAYE and VAT)
  - Social security payments
  - Pension scheme payments
  - Employees' salaries for up to four months;
- secured creditors without a fixed charge;
- unsecured creditors;
- dividends, if applicable;
- repayment of shareholders' capital, if applicable;
- surplus assets apportioned among shareholders, if applicable.

As can be seen, this is a very long list and given that the contractor has probably gone into liquidation because of debt or lack of funds, the creditors towards the end of the list have little chance of recovering all that is owed to them. This factor should be borne in mind by the quantity surveyor if he/she suspects that a contractor is about to go into liquidation and will be discussed in greater detail later.

### *How should the quantity surveyor deal with insolvency of the contractor?*

As mentioned previously, although a contractor can stop trading, literally overnight, there are usually tell tale signs that all is not well and if these are noticed then the quantity surveyor must be prepared to act quickly if the worst happens. The advance warning of financial problems are:

- subcontractors complain of delayed payment or non-payment;
- the progress of the work may slow as the contractor finds it difficult to purchase material, builders merchants may agree to supply materials on a cash-only basis instead of a monthly credit account;

- attempts to maximise the interim valuations and include items and materials ahead of their incorporation of the works. It is vital that the quantity surveyor does not over certify at interim valuations, as recovery of the money may prove to be problematic.

Assuming that the quantity surveyor arrives at the office one morning to discover that a contractor on one of his projects has ceased trading, he/she must take the following immediate steps:

- Secure the site, if necessary by employing a security firm. This action is necessary as subcontractors and other creditors may try to remove material and plant from site; some of these materials may have been paid for and therefore be the property of the employer.
- Once the site is secure an accurate assessment must be made as to the state of the works and how much is required for completion. This is usually done by quantity surveyors appointed by the receiver.
- Stop all payments to the contractor.
- A check should be made to ensure that the works continue to be insured.
- If the contract required the contractor to take out a performance bond, the underwriters should be informed.

The way forward will depend on the stage that the works have reached and the need for a quick completion. If the works are substantially complete then perhaps the best way forward is to offer the remaining work to another contractor on a cost plus or fixed fee basis. If on the other hand work had only recently started then the original bill of quantities, assuming that there was one, can be used as a basis to obtain new bids. The most awkward situation is where the project is significantly advanced, with several months work to complete. In this situation the most accurate approach is to prepare an addendum bill of quantities, this can then be used to negotiate with another contractor or as the basis of obtaining bids.

The final account for a project where the original contractor has ceased trading is more complicated than normal. Firstly, it is necessary to prepare a hypothetical account, as if liquidation had not occurred, then calculate the actual cost comprising the costs of the original contractor together with the costs associated with completion of the project. Any surplus or shortfall will result in a payment or recovery by the client.

### *Employer's liquidation*

Of course it isn't just the contractor that could cease trading, the employer can also face financial problems and in these circumstances the contractor needs protection. The legal process will be as described earlier depending on whether the client is an individual or a company. The signs of potential difficulties are:

- the employer fails to pay interim certificates on time;
- the employer refuses to pay for portions of the works on the grounds that there are defects that are unfounded.

Contractors would be advised to carry out credit checks on employers, particularly those that they have not worked for previously to try to avoid financial problems.

Clause 8.10 of JCT(11) states that in the event of employer insolvency the contractor may give notice to the employer that the contract is terminated. The notice is best given to the liquidator and is sent by recorded delivery. Clause 8.12 sets out the entitlement of the contractor once termination has occurred as follows:

- the contractor is entitled to remove from site any temporary buildings, plant, tools, and equipment belonging to the contractor;
- the contractor is allowed to remove from site all goods and site materials.

Within a time limit prescribed in clause 8.12.3 an account must be prepared that details the following items:

- the total value of work properly executed;
- any sums in respect of direct loss and expense;
- the reasonable costs associated with the removal of temporary buildings, materials, etc.;
- the cost of materials that were ordered prior to liquidation and not yet delivered to site, but nevertheless have to be paid for.

What happens to retention?

The JCT(11) seeks to protect the retention fund by establishing that the employer's interest in the monies is fiduciary as trustee on behalf of the contractor. The contractor may request that retention monies are held in a separate account in the joint names of the contractor and the employer to protect the contractor's interests in the event of the insolvency of the contractor.

### *Termination*

In addition to liquidation of one of the parties, Section 8 also sets out the other circumstances when the contract can be terminated. For the employer grounds for termination include:

- suspending work;
- failure to proceed regularly and diligently;
- refusal to comply with written instructions;
- sub-letting or assigning work without consent;
- non-compliance with CDM regulations.

For the contractor grounds for termination includes:

- failure to pay certificates when due;
- obstructing the issue of instructions or certificates;
- obstructing the sub-letting or assignment of work.

Termination of the works should only be taken as a last resort as the consequences have such an impact on the eventual completion and cost of the works could be far reaching. Better the devil you know!

**Loss and expense claims**

Often one of the most contentious parts of the final account is a claim submitted by the contractor for loss and expense incurred while carrying out the works. A claim is a method of recompensing the contractor for proven loss and expense that is not covered in any other way. There is evidence to suggest that some contractors win work on very slender profit margins, secure in the knowledge that there will be the opportunity to submit a claim for loss and expense due to matters materially affecting the regular progress of the work. To many, compiling a claim for loss and expense is something of a black art and it is true to say that traditionally many claims have been submitted by a contractor on the poorest of backup information and records, only to be agreed by the client! Section 4 of the JCT(11) form of contract makes provision for the contractor to submit a loss and expense and stipulates that the contractor must inform the architect as soon as it is apparent that loss and expense has been incurred and that all necessary support data should be provided. The following items are considered to be 'relevant matter' and therefore form the basis of a claim:

- variations;
- postponement of the works;
- adjustment of provisional sums for undefined work;
- opening up work for inspection that subsequently prove to be in accordance with the contract;
- discrepancy and divergence between the contract documentation (quantity surveyors beware), bills of quantities and drawings are specifically mentioned;

- suspension of works by the contractor for non-payment by the employer;
- execution of works that have been included in the contract documents as provisional, that subsequently proves to be inaccurate in the nature or extent of work required;
- any impediment by the employer, architect, quantity surveyor or any other person employed by the employer.

The claim is based on the period certified by the architect as being a justified extension of time in accordance with the terms of a contract. When preparing a claim a contractor will consider the following items during the extension period:

- site costs, the costs associated with keeping the site operational;
- head office overheads for the period of claim;
- finance charges: any additional expenses, associated with funding the project for the extended period;
- fluctuation in accordance with the contract for the extension period;
- costs associated with uneconomic working; the contractor may have to carry out the works in an out-of-sequence way due to a relevant event, as listed previously;
- loss of profit caused by the impact of the relevant event. It has been hotly debated as to whether a claim for loss of profit is admissible and must be carefully prepared and documented if it is to be successful.

Perhaps two of the most important watchwords when preparing a claim are transparency and records.

## Dispute resolution

England and Wales are Common Law countries with an adversarial court system and construction law exists as a recognised legal specialism.

Unfortunately, all too often in the construction industry, the parties to the contract reach a point where a dispute cannot be resolved and in these circumstances the traditional approaches to finding a solution have been litigation and arbitration. The difficulty with both of these approaches is that they are very expensive and time consuming and cannot be commenced until after the contract has finished or work has stopped. It has long been the view that litigation in particular is a bit like playing Russian roulette with little certainty over the outcome or the quantum of the award. Very serious consideration should be given by all parties before going down this route. In England and Wales, the Technology and Construction Court ('TCC') formally known as the Official Referees' Court is a specialist court, which deals principally with technology and construction disputes. The full range of work undertaken by the TCC is set out in CPR Part 60 and the accompanying practice direction. TCC judgements which may be of interest to practitioners are accessible on the British & Irish Legal Information Institute (BAILII) website. Mr Justice Jackson is the judge in charge of the TCC and he is available to manage and try TCC cases either in London or at any of the court centres mentioned above.

In an attempt to simplify matters, an approach known as alternative dispute resolution (ADR) has been introduced and the past few years have seen ADR pass from the realms of quirky to mainstream. Following the recommendations of the Latham Report and the UK Housing Grants and Regeneration Act 1996, fundamental changes have taken place in the processes and procedures of dispute resolution. JCT(11) still provides for litigation to be the default mechanism for dispute resolution, however it allows the parties to give serious consideration to mediating any dispute, particularly if the project is in its early stages and there are sound reasons for avoiding adversarial processes. Two wholly new clauses have been added to Section 9 – Settlement of Disputes:

- **9.1 Mediation**: this clause is a statement that the parties will give consideration to mediation to settle disputes. It is not an obligation that they must do so, or that they must even consider doing so. Mediation is a voluntary non-binding process in which a neutral mediator assists the parties to attempt to achieve a negotiated settlement. Mediation thus retains the characteristic of negotiation that gives the parties full control over how their dispute is to be resolved, but the mediator is there to help them through any impasses that may occur, and help them explore various options for settlement. The approach of the mediator is generally to try to get the parties to look at the dispute differently; to look forward rather than backward; to consider where their future interests and needs lie, rather than their strict legal entitlements and wants. The mediator also has the potential, from what he learns from the parties during the mediation, to expand the settlement options available to the parties. In addition to meeting the parties together, the mediator also holds separate private meetings with each party, known as caucuses. All of the proceedings in mediation are private and confidential. It is usually stipulated in the mediation agreement that the mediator may not be called as a witness in any subsequent proceedings, but a special confidentiality applies to matters disclosed in caucus. The mediator undertakes not to divulge them to the other party unless expressly permitted by the party who has made the disclosure. In that way, so long as the mediator has gained the trust and confidence of the parties, he can get a better understanding of each party's situation, motivation and aspirations, and with this knowledge, may discover novel ways in which each party may satisfy the interests of the other. The mediator is however merely the catalyst. It is ultimately for the parties to decide whether any of the avenues suggested by the mediator will satisfy their interests sufficiently to resolve the dispute.

One of the strengths of mediation is that the solution to the dispute may be very far removed from the respective contractual rights and obligations of the parties. So long as the proposed resolution is seen to be of equivalent value by the recipient party, it will often be accepted in settlement of the dispute. Thus a monetary debt may be discharged by the provision of a useful service instead, which is an outcome not possible where the decision-making process is delegated to a neutral third party. Such novel solutions are also unlikely in negotiation due to the inherent unwillingness of opponents to share their personal and business information with each other, but they may be more candid with a mediator in whom they have confidence.

A typical mediation will commence with the mediator explaining the process to the parties and getting them to explain to him and each other their views of the issues that need to be resolved. Instead of seeking to enforce the parties' respective rights and obligations as he perceives them, the mediator will attempt to build a picture of the parties' interests, needs and available resources, to see if the available resources of one might satisfy the needs of the other, and vice versa, to the extent necessary to resolve the dispute. The mediator in caucus will also conduct reality checks with each party separately to test the viability of their initial positions, and gently try to persuade them away from their positional bargaining stances towards a more principled negotiation approach.

- **9.2 Adjudication**: the SBC/Q 2011 adopts the Scheme for Construction Contracts (England and Wales) Regulations 1998.

  An adjudicator considers the submissions of the parties, and the evidence adduced, and issues a binding interim decision within a fixed period of time. The adjudicator's decision is binding, and must be implemented, until such time as the dispute is finally decided by litigation, by arbitration (if the contract so provides), or by agreement. Thus the loser in adjudication must pay what is ordered to the winner, even if it is the loser's intention to refer the dispute afresh to litigation or arbitration.

Under the statutory adjudication system introduced by the Housing Grants, Construction and Regeneration Act 1996, a construction adjudicator in the UK has 28 days from the date the dispute is referred to give his/her decision as to how that dispute should be resolved on an interim basis. This prompt decision-making process has become very popular in the UK, and is now the principal method of resolving disputes within the UK construction industry. Although interim in nature, the adjudicator's decision is frequently taken as finally resolving the dispute.

The adjudicator is usually immune from suit by either of the parties, but is not immune from being sued by third parties who may be adversely affected by his/her decision. Thus adjudicators often seek an indemnity from the parties in respect of third-party actions.

- **9.3 Arbitration** is a private system of dispute resolution with statutory backing that is frequently used as an alternative to litigation in connection with construction disputes, where it is considered desirable that the skills of the decision-maker should be more heavily biased towards relevant technical issues, rather than purely legal issues. However, to discharge their duties satisfactorily, good arbitrators need a high degree of competence in both technical and legal matters. Arbitration is similar to litigation in the sense that the award of the arbitrator is usually final and binding on the parties. Depending on the jurisdiction within which the arbitration is held, the grounds for subsequent appeal to the courts are generally very limited. Enforcement of the arbitrator's award is usually through the courts of whatever jurisdiction the winner chooses to enforce it, which may not necessarily be the jurisdiction where the arbitration was held. Thus arbitration has considerable attraction in international disputes, where the loser in the arbitration may not hold sufficient assets in the national jurisdiction where the arbitration was held. The parties to arbitration may have their differences settled in private, and the existence and result of the arbitration will usually remain confidential so long as there is no need to enforce the award in open court.

  Arbitration is consensual. It cannot be imposed on the other party, as litigation can, without the prior agreement of the other party and has the potential to be quicker and cheaper than litigation, but the realisation of this potential rather depends on the procedure adopted by the arbitrator, the parties and the parties' legal representatives, if any. There is no obligation on the parties to employ legally qualified representatives, but they frequently do. In such circumstances, there may be pressure from the party representatives for a procedure that mimics litigation. Unless this is controlled, the flexibility and potential economy of arbitration over litigation will be lost. There is generally no set procedure in arbitration, unless it is a scheme arbitration tailored to a particular type of dispute. More usually it is up to the parties and the arbitrator between them to customise the procedure to suit the needs of the particular case.

In arbitration there is not, nor should there be, any such thing as judgement in default. Notwithstanding a party's refusal to participate in the arbitration when it arises, the claimant is, or should be, always required to satisfy the arbitrator on the balance of probability that its claim is valid. To the extent that it fails to do that, the claim should fail, whether or not the respondent participates. The arbitrator is expected to apply the rules of natural justice which involves firstly having no personal interest in the outcome of the case, and secondly allowing each party a reasonable opportunity to state its case and to know and answer the case against it. This second requirement places a duty on the arbitrator to give equal attention to each party's submissions, and not to rely on any special knowledge of his own, of which the parties may be unaware, without first putting it to the parties and allowing them to comment.

As previously indicated, the arbitrator's award is enforceable through the courts. When asked to enforce, the court is not permitted to interfere with the contents of the award, and will confine its enquiries to whether the award was made pursuant to a valid arbitration agreement by a validly appointed arbitrator. Thus the court will generally not attempt to check whether the award accords to its own view of the issues, and will simply enforce, or not enforce, according to its view of the validity of the arbitration agreement and the validity of the arbitrator's appointment.

Thus an award may be challenged on the basis that the arbitrator did not have the necessary jurisdiction to make the award he has made. In very exceptional circumstances, a party may also, but only with leave of the court, appeal against the award on grounds of an error of law within the award.

## PART 3: SITE ORGANISATION

### Welfare facilities

Construction site workers need adequate toilet and washing facilities, a place to warm up and eat their food and somewhere to store clothing. The minimum welfare facilities are set out in the Construction (Health, Safety and Welfare) Regulations 1996.

## Planning the site layout

When planning the layout of the site the following factors should be taken into account:

- Security.
- Space available.
- Site accommodation and welfare facilities.
- Temporary services: water, drainage, power – 110, 240 and 415 volt, if required.
- Access to the site, impact on traffic.
- Storage of and security materials; millions of pounds a year is wasted due to the inadequate storage of materials. Also materials storage should be planned to avoid the necessity of double or even triple handling. Subcontractors' requirements must also be considered and provided for.
- Plant: depending on the nature of the contract and the degree of mechanical plant being used provision should be considered to allow the free movement of dumper trucks, excavators etc. as well as bases for tower cranes.
- The sequence of site operations.

The layout of the site will depend on the location of the site. On a cramped city centre site, surrounded by busy roads and other buildings, it is often necessary to stack the site accommodation to reduce its footprint. If possible the site accommodation should be placed on areas that will eventually be landscaped and as close to the site entrance as possible. In the case of refurbishment contracts then it may be possible to use part of the existing building for accommodation.

Legislation governing the provision of site accommodation and welfare facilities are:

- Construction (Health, Safety and Welfare) Regulations 1996;
- Offices Shops and Railway Premises Act 1963;
- Health and Safety at Work Act 1974;
- Fire precautions (Factories, Offices, Shops and Railway Premises) Order 1989.

## Planning and programming

There are a number of techniques open to a contractor to prepare a programme for the various activities involved in the construction of a new building. The principles behind programming are to plan and illustrate:

- when activities start and finish;
- what activities must be completed before another can start;
- what activities can be carried out simultaneously;
- what activities cannot start until another is finished.

These techniques vary in their complexity and sophistication as shown in Figure 6.3.

The bar chart is arranged with the contract period in weeks running vertically with the breakdown of activities arranged horizontally down the left-hand side. The weeks where the activities are planned to take place are then shaded in as shown. The advantages with this technique are:

- easy to understand and widely used;
- illustrates the relationship between activities and where activities overlap;
- easy to compile.

The disadvantages are:

- simplistic presentation that does not allow the interrelationship of complex processes to be shown.

## Critical path analysis and PERT charts

Critical path analysis (CPA) and PERT were developed in the 1950s to control large defence projects. They act as the basis both for preparation of a

| Year | 2012 | | | | | | | |
|---|---|---|---|---|---|---|---|---|
| Month | October | | | | November | | | |
| Activity Week | 6 | 13 | 21 | 28 | 3 | 10 | 17 | 24 |
| Clear site | ■ | ■ | | | | | | |
| Excavation | | ■ | ■ | ■ | ■ | | | |
| Brickwork | | | | ■ | ■ | ■ | ■ | ■ |

**Figure 6.3** Bar chart or Gantt chart

schedule and of resource planning. They enable the commercial manager to monitor achievement of project goals and help see where remedial action needs to be taken to get back on course. The benefit of using CPA over Gantt charts is that CPA formally identifies tasks which must be completed on time for the whole project to be completed on time (these are the tasks on the critical path) and also identifies tasks which can be delayed for a while, if resources need to be redeployed to catch up elsewhere. The disadvantage of CPA is that the relation of tasks to time is not as immediately obvious as the Gantt charts and this can make them more difficult to understand. A CPD is prepared as follows:

- List all activities in the plan.
- For each activity show the:
  - earliest start date;
  - estimated length of time;
  - whether parallel or sequential;
  - if sequential, show which stage they depend on.
- Plot the activities as a circle and arrow diagram:

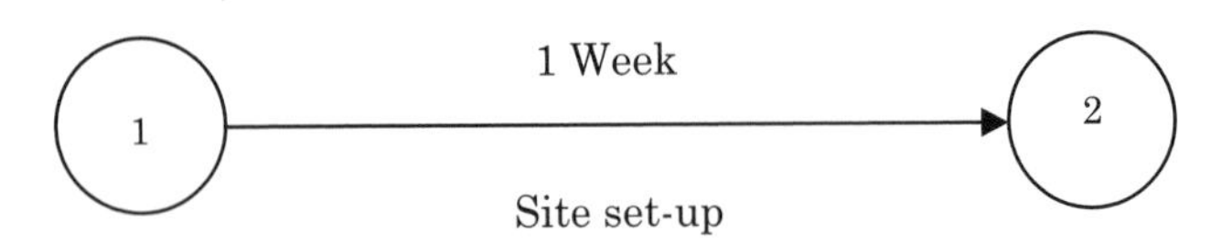


  The above shows the start event (circle 1) and the completion of the site set-up task (circle 2). The arrow between them shows that the activity of carrying out the site set-up should take one week.
- Where one activity cannot start until another activity has been completed start the arrow for the dependent activity at the completion event circle of the previous activity as follows:

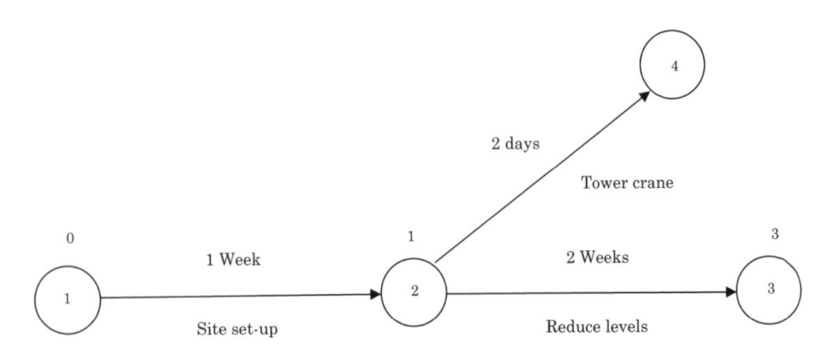

Here the activities of 'Tower crane' and 'Reduce levels' cannot be started until 'Site set-up' has been completed.

Also note:

- within CPA activities are referred to by numbers in circles at each end;
- activities are not drawn to scale;
- the numbers above the circles are the earliest possible time that this stage in the project will be reached.

PERT is a variation to CPA that takes a slightly more sceptical view of time estimates made for each project. To use it, estimate the shortest possible time each activity will take, the most likely length of time and the longest time that might be taken, if the activity takes longer than expected. Then use the following formula to calculate the time to use for each project stage:

$$\frac{\text{shortest time} + 4 \times \text{likely time} + \text{longest time}}{6}$$

This helps to bias time estimates away from the unrealistically short time scales often assumed.

# 7

# Final account

## FINAL ACCOUNT

The final account stage of a contract is the process during which the quantity surveyor determines the final cost of a project, based on the following documents:

- the form of contract
- original priced bill of quantities
- variations
- drawings
- agreed contractor claims.

For public sector projects in particular the final account acts as the final part of the audit trail, allowing all financial transactions relating to the contract to be clearly traced with the contract bills of quantities being the starting point; it can be a lengthy and time consuming process. For a private sector client it may not always be necessary to produce a fully detailed final account as long as the client is convinced that the final project has delivered value for money. For some public sector clients the final accounts documents are subjected to audit. The processes described in the following pages are those required by most public sector clients, but may be adapted for use in the preparation of final accounts for private clients too.

The final account should be prepared during the contract period, as some of the detail required will have been used during the preparation of variations and interim certificates and not left until the contract is complete. One reason for doing this is that it is better to measure and price variations during the currency of the contract while the facts are still fresh in people's minds, before work becomes covered up and before changes of personnel make the accurate preparation of a final account more difficult and time consuming. Contracts will typically specify a fixed period following practical completion of the work within which the account must be prepared.

The final account is composed of the following:

- Statement of Final Account;
- Final account summary;
- Adjustment of Prime Cost Sums;
- Adjustment of Provisional Sums;
- Adjustment of Provisional Items;
- Adjustment of variation account – see Section 7;
- Adjustment for fluctuations – see Section 6;
- Adjustment for contractors' claims (if applicable) – see Section 6.

and is prepared in accordance with the conditions of contract. For example in the JCT (11) Standard Form the clause dealing with the preparation of the final account or final adjustment is clause 4.5. After the architect has issued the Certificate of Practical Completion the contractor must supply the quantity surveyor with all the necessary documentation in order that the final account (adjustment) can be prepared. The quantity surveyor then has an additional three months to prepare the final account (adjustment), there is no stipulated penalty stated in the contract if these deadlines are not met.

## THE VARIATION ACCOUNT

Section 5 of the JCT(11) deals with variations to the contract.

After a contract has been signed it cannot be changed or varied by the parties. However, given the nature of the construction process, with all of its inherent risk and uncertainty, most standard forms of construction contract include the provision for variations or alterations to the works.

The JCT(11) form of contract clause 5.1 defines the term variation as follows:

> [T]he alteration or modification of the design, quality or quantity of the Works including:
>
> - the addition, omission or substitution of any work;
> - the alteration of the kind or standard of any of the materials or goods to be used in the works;
> - the removal from the site of any work executed or materials or goods that are not in accordance with the contract.

In addition the contract allows for variations or alterations to obligations and restrictions imposed on the contractor such as access to the site, working hours, etc.

The variation account will probably be the largest, in terms of documentation, in a final account. The number and nature of change orders or variations issued during the contract period will vary considerably according to the type of contract. For example, contracts with little or no pre-contract planning or documentation or refurbishment contracts can expect to generate a substantial number of variations whereas comparatively uncomplicated projects based on fully detailed documentation may generate very few.

The procedure for issuing a variation order using JCT(11) is as follows:

- All variations must be in writing and issued by the architect/contract administrator. If the contractor objects to carrying out the variation then he/she has seven days to object. Any objection must be in writing. If the contractor fails to comply with the instruction within seven days then the architect/contract administrator has the right to get another contractor to complete the variation and contra charge the contractor.
- In addition to architect's instructions, variations may also be given to the contractor in the form of an oral instruction or a site instruction, given by the clerk of works. Both of these forms of variation have to be confirmed in writing by the architect/contract administrator in order to become an official variation.
- Some variations will be issued by way of a drawing together with a covering architect's instruction and there may be occasions where several revisions of the same drawings may be issued. In circumstances such as these it is important that all drawings are carefully logged in a drawing register and only the latest revision is used to measure and assess the variations/alterations. Note that to be valid drawings must be signed by the architect or accompanied with a covering letter – see *Myers v Sarl* (1860).
- The architect may not omit work, which has been measured in the bills of quantities and priced by the contractor, in order to give it to others to carry out.
- Any errors or omissions that come to light during the contract period, are in the case of JCT(11) clause 2.14/15, dealt with as a variation order.

The valuing of variations is generally divided into two operations as follows.

**Measurement and pricing of variations**

The JCT(11) – clause 5.4 gives the contractor the right to be present when variations are measured. In practice measurement of variations can take place on site by the employer's quantity surveyor who then passes it to the contractor for checking and agreement, or by the contractor as part of the interim valuation process, which has to be checked and agreed by the employer's quantity surveyor. Once agreed variations are included in the final account and included in interim certificates.

JCT(11) – clause 5.6 sets out the rules for valuing variations, which are a set of procedures in descending order of preference, namely:

- where the additional works are of similar character to items in the original contract bills then the bill rates are used;
- where the additional works are different in so much as they are carried out under different conditions or there is a significant change in the quantities, then the bill rates are used in order to prepare a fair price, usually by the build up of pro rata prices – see below;
- if neither of the above two approaches is appropriate then work shall be valued at fair rates and prices;
- finally, under certain circumstances, defined in clause 5.7 dayworks may be used as the basis for valuation.

No allowances for any effect on the regular progress of the work caused by the issue of a variation order can be built into the pricing. Any such claims have to be the subject of a claim for loss and expense – see Chapter 6.

From the quantity surveyor's view point it is best to measure and value variations a quickly as practical so that increases/decreases can be built into financial statements and final accounts. From a logistical point of view it is good practice to group similar items together, for example work to the substructure or drainage being measured in one omnibus item.

## Pro rata pricing

With respect to the valuation of variation the JCT(11) refers to taking into account similarity, character, conditions and quantity of work involved and this can be problematic at times.

One of the provisions in the JCT(11) form is to price variations, where no direct price is available, at rates that are based on the bill rates. This is known as pro rata pricing and as with approximate quantities there are a number of approaches.

As explained in Chapter 2, bill rates are an amalgam of materials, labour, plant, profit and overheads, therefore when attempting a pro rata calculation a known built-up rate is broken down into its constituent parts in order that it can be adjusted. If too many of the elements that constitute a rate have to be adjusted then it is better to build up a new rate from basic principle as described in Chapter 5.

One of the advantages of having a bill of quantities is that the degree of detail contained in the document can be used as a basis for the valuing of variations during the post-contract stage of a project. As discussed in the previous pages bill rates are composed from the following:

- labour
- plant
- profit and overheads.

The technique of pro rata pricing involves disassembling a bill rate and substituting new data in order to calculate a new rate that can be used for pricing variations.

### *Example*

The client's quantity surveyor is meeting with the contractor to price a variation that has been measured previously. It is not possible to directly use the existing bill rates to price an item; therefore a pro rata price is built up. As the contractor is present it may be possible to refer back to the original price build-up and this will make the process easier. There are occasions, however, when the quantity surveyor works alone and therefore appropriate assumptions have to be made. The approach is to analyse the original bill rate as follows:

- deduct profit and overheads;
- analyse materials, labour and plant costs;
- adapt costs to suit new item; and
- add back profit and overheads.

BILL OF QUANTITIES ITEM

50mm thick cement and sand (1:3) level screed over 600mm wide in one coat to concrete – £13.58/m$^2$

VARIATION ACCOUNT ITEM

38mm thick cement and sand (1:3) level screed over 600mm wide in one coat on concrete – cost/m$^2$?

ANALYSIS OF BILL RATE

| | | | |
|---|---|---|---|
| 50mm thick cement and sand screed | £13.58 | | |
| Less profit and overheads 15% | £ 2.04 | | |
| Net cost | £11.54 | | 11.54 |
| Deduct materials | | | |
| $1m^3$ cement = 1400kg cement @ £85.00 per tonne | | 119.00 | |
| Unloading 1 hour/tonne @ £13.20 | | 18.48 | |
| $4m^3$ sand = 6400 kg @ £10.00 per tonne | | 64.00 | |
| | | 201.48 | |
| Add shrinkage 25% | | 50.37 | |
| Cost per $5m^3$ | | 251.85 | |
| Cost per $m^3 \div 5$ | | 50.37 | |
| Mixing | | | |
| Assume 200 litre mechanical fed mixer @ £20.00 per hour; output $4m^3$ per hour – cost per $m^3$ | | 5.00 | |
| Cost per $m^3$ | | 55.37 | |
| Cost per $m^2$ – 50mm thick | | 2.52 | 2.52 |
| Cost of labour for 50mm thick screed | | | £9.29 |
| Add materials | | | |
| Cost per $m^3$ as before | | 55.37 | |
| Cost per $m^2$ – 38mm thick | | | £2.10 |
| | | | £11.39 |
| Labour costs | | | |
| Labour costs for a 38mm thick screed will be approximately 25% cheaper as it will be possible to place a 38mm screed more quickly than a 50mm thick: £9.29 × 0.25 | | | £2.33 |
| | | | £9.06 |
| Add Profit and overheads 15% | | | £1.36 |
| Cost per $m^2$ for 38mm thick cement and sand (1:3) screed | | | **£10.42** |

## FINAL ACCOUNT STANDARD FORMAT

There follows an example of standard approach that may be used for the preparation and presentation of a final account. Important points to note are:

- All omissions are entered in grey whereas all additions are in black.
- The account is presented as a series of several self-contained accounts:

  - Variations
  - Prime Cost Sums
  - Provisional Items
  - Provisional Sums

  the totals of which are carried to the final account summary.

- Each item in the various accounts is numbered and contains the architect's instruction(s) that authorised the item to be carried out/adjusted.
- When, during the preparation of the final account, items are omitted from the original bill of quantities, the original bill item should be lined through in red as shown below and referenced to the final account item(s):

£

K Additional builder's work in connection with Mechanical Installation

Provide the Provisional Sum of £3,000.00 for additional work in connection with mechanical installation.

3000 00

*Provisional Sums Account*
*Item No. 1*

- The above entry indicates to the auditors that this item has been omitted from the bill(s) of quantities and adjusted/added back as Item 1 of the Provisional Sums Account.
- All items such as provisional sums, provisional items and prime cost sums must be adjusted even in the event that there is no item to add back as the sums have not been expended. In this case the item will be omitted and nil added back resulting in an overall saving.
- Provisional items are those items of measured works included in the bills of quantities and marked as PROVISIONAL.

STATEMENT OF FINAL ACCOUNT

for

REFURBISHMENT OF TENANTS' MEETING HALL
TONBRIDGE

Architect

Gardiner & Partners
6 Derby Walk
Tonbridge TN4 8HN

Borough Technical Services

P.S. Brookes FRICS
Technical Services Group
67 Uxbridge Road
Tonbridge TN5 6JK

Contractor

J. Harris & Co. Ltd
37 Newton Terrace
Tonbridge TN3 8GH

26 June 2012

I/we the undersigned hereby certify that the gross total value of the final account for this contract has been agreed in the sum of £2,645,363.78
Two million six hundred and forty five thousand three hundred and sixty three pounds and seventy eight pence

and that payment of this gross amount shall be in full and final settlement of this account, subject to any adjustments required following the Local Authority's audit and liquidated and ascertained damages which the employer may deduct and that I/we have no further claims on this contract.

Signed...................................................................................

For and on behalf of ..............................................................

..............................................................

Date..................

FINAL ACCOUNT SUMMARY

for

REFURBISHMENT OF TENANTS' MEETING HALL – TONBRIDGE

| | Omissions £ | Additions £ | £ |
|---|---|---|---|
| **Contract sum** | | | 2,670,000.00 |
| Less Contingencies | | | 15,000.00 |
| | | | 2,655,000.00 |
| From Prime Cost Sums summary | 18,325.00 | 16,899.00 | |
| From Provisional Sums summary | 13,300.00 | 2,689.00 | |
| From Provisional Items summary | 3,191.44 | 61.70 | |
| From variation account summary | 75,839.04 | 73,672.78 | |
| Fluctuations | – – | 896.78 | |
| Agreed claim | – – | 6,800.00 | |
| | 110,655.48 | 101,019.26 | |
| | 101,019.26 | | |
| | £9,636.22 | | 9,636.22 |
| | | | £2,645,363.78 |
| Less amount paid in interim certificates nos 1–12 | | | £2,642,876.00 |
| **Balance due** | | | **£2,487.78** |

Therefore in this example the sum of £2,487.78 is due for payment to the contractor in full and final settlement and the statement of final account can be signed.

REFURBISHMENT OF TENANTS' MEETING HALL – TONBRIDGE

Prime Cost Sums account summary

| No | Description | Omission £ | Addition £ |
|---|---|---|---|
| 1 | Electricity main | 8,525.00 | 8,557.00 |
| 2 | Steel cladding | 7,000.00 | 4,556.00 |
| 3 | Fire alarm installation | 2,000.00 | 2,986.00 |
| 4 | Electricity installation | 800.00 | 770.00 |
| | Prime Cost Sums account carried to final account summary | £18,325 | £16,869 |

---

REFURBISHMENT OF TENANTS' MEETING HALL – TONBRIDGE

Provisional Sums account summary

| No | Description | Omission £ | Addition £ |
|---|---|---|---|
| 1 | Contingencies | 10,000.00 | – |
| 2 | Roofing repairs | 500.00 | 650.00 |
| 3 | Entrance porch | 2,000.00 | 1,489.00 |
| 4 | Exit signs | 800.00 | 550.00 |
| | Provisional Sums account carried to final account summary | £13,300.00 | £2,689.00 |

REFURBISHMENT OF TENANTS' MEETING HALL – TONBRIDGE

Provisional Items account summary

| No | Description | Omission £ | Addition £ |
|---|---|---|---|
| 1 | Softwood noggins | 41.44 | 61.70 |
| 2 | Excavation | 2,000.00 | – |
| 3 | Drainage | 500.00 | – |
| 4 | External works | 650.00 | – |
| | Provisional Items account carried to final account summary | £3,191.44 | £61.70 |

REFURBISHMENT OF TENANTS' MEETING HALL – TONBRIDGE

Variation account summary

| No | Description | Omission £ | Addition £ |
|---|---|---|---|
| 1 | Remeasurement of substructure | 52,231.04 | 48,789.78 |
| 2 | Remeasurement of external works | 4,700.00 | 6,897.00 |
| 3 | Remeasurement of drainage | 6,908.00 | 8,986.00 |
| 4 | Etc. | 12,000.00 | 9,000.00 |
| | Variation account carried to final account summary | £75,839.04 | £73,672.78 |

REFURBISHMENT OF TENANTS' MEETING HALL – TONBRIDGE

Daywork account

An allowance for carrying out work and valuing on a daywork basis is usually included in the bills of quantities as a provisional sum(s). Therefore in the final account any allowance for dayworks will be omitted in the provisional sums account and if authorised, any work will be added back in the appropriate account as follows in this example:

| | Omissions | |
|---|---|---|
| OMISSIONS<br>VARIATION ACCOUNT | £ | p |
| ITEM No. 24 | | |
| A.I. No 23 | | |
| ADDITIONAL PLASTERWORK | | |
| OMIT | | |
| NIL | – | – |
| ADDITIONAL PLASTERWORK<br>Carried to variation account summary | £– | – |

| | Additions | |
|---|---|---|
| ADDITIONS | £ | p |
| VARIATION ACCOUNT | | |
| ITEM No. 24 | | |
| A.I. No 23 | | |
| ADDITIONAL PLASTERWORK | | |
| ADD | | |
| Additonal plasterwork as DWS No. 34/9 | 259.98 | |
| ADDITIONAL PLASTERWORK<br>Carried to variation account summary | **£259.98** | |

| | Omissions | |
|---|---|---|
| OMISSIONS | £ | p |
| VARIATION ACCOUNT | | |
| ITEM No. 1 | | |
| A.I. Nos 9, 10, 11, 34, 79 | | |
| SUBSTRUCTURE | | |
| OMIT | | |
| BQ Page 3/1/A–N | 33,000 | 89 |
| " " 3/2//A–L | 2,456 | 38 |
| " " 3/3/A–L | 10,786 | 45 |
| " " 3/4/A–K | 5,987 | 32 |
| SUBSTRUCTURES<br>Carried to variation account summary | £52,231.04 | |

| | Additions | |
|---|---|---|
| ADDITIONS | £ | p |
| VARIATION ACCOUNT | | |
| ITEM No. 1 | | |
| A.I. Nos 9, 10, 11, 34, 79 | | |
| SUBSTRUCTURE | | |
| ADD | | |
| Include here the remeasured and priced dimensions for the substructure as agreed with the main contractor. | | |
| Assumed as: | 48,789.78 | |
| SUBSTRUCTURES<br>Carried to variation account summary | £48,789.78 | |

| | Omissions | |
|---|---|---|
| OMISSIONS | £ | p |
| PRIME COST SUMS | | |
| ITEM No. 1 | | |
| A.I. No 3 | | |
| ELECTRICITY MAINS | | |
| OMIT | | |
| The PC sum of £8,500.00 for work to electricity mains | | |
| As 5/35/F | 8,500 | 00 |
| Allow for profit | | |
| As 5/35/G | – | – |
| Allow for attendance | | |
| As 5/35/H | – | – |
| Allow for special attendance | | |
| As 5/35/J | 25 | 00 |
| ELECTRICITY MAINS<br>Carried to prime cost sums summary | £8,525 | 00 |

| | Additions | |
|---|---|---|
| ADDITIONS | £ | p |
| PRIME COST SUMS | | |
| ITEM No. 1 | | |
| A.I. No 3 | | |
| ELECTRICITY MAINS | | |
| ADD | | |
| Include the sum of £8,532.00 as Southern Electricity's account No. C445 dated 24 April 2011 | 8,532 | 00 |
| Allow for profit | – | – |
| Allow for attendance | – | – |
| Allow for special attendance | | |
| As 5/35/J | 25 | 00 |
| ELECTRICITY MAINS<br>Carried to prime cost sums summary | £8,557 | 00 |

* Note the account ref C445 should be included with this item in the final account and referenced to this item.

Note that when pricing the bills of quantities the contractor chose not to price the items of Profit and General attendance here and therefore they remain unchanged in the final account. Special attendance was priced but will only be adjusted if the amount or the nature of the works changes considerably.

| | Omissions £ | p |
|---|---|---|
| OMISSIONS | | |
| PROVISIONAL SUMS | | |
| ITEM No. 1 | | |
| A.I. No 19 | | |
| CONTINGENCIES | | |
| OMIT | | |
| The provisional sum of £10,000.00 for contingencies | | |
| As1/48/H | 10,000 | 00 |
| CONTINGENCIES Carried to provisional sum summary | £10,000 | 00 |

| | Additions £ | p |
|---|---|---|
| ADDITIONS | | |
| PROVISIONAL SUMS | | |
| ITEM No. 1 | | |
| A.I. No 19 | | |
| CONTINGENCIES | | |
| ADD | | |
| Nil | – | – |
| CONTINGENCIES Carried to provisional sum summary | £– | – |

* Note there is no addition to the Contingencies item as this sum has been expended over a number of items in the final account where additions exceed omissions.

| OMISSIONS | Omissions £ | p |
|---|---|---|
| PROVISIONAL ITEMS | | |
| ITEM No. 1 | | |
| Softwood noggins | | |
| OMIT | | |
| BQ ref 3/34/A | 41 | 44 |
| Softwood noggins<br>Carried to provisional items summary | £41 | 44 |

| ADDITIONS | Additions £ | p |
|---|---|---|
| PROVISIONAL ITEMS | | |
| ITEM No. 1 | | |
| Softwood noggins | | |
| ADD | | |
| 50 × 75mm Sawn softwood noggins | | |
| As 3/44/A | £61 | 78 |
| Softwood noggins<br>Carried to provisional items summary | £61 | 78 |

* Note that an Architect's Instruction is not necessarily needed to adjust a provisional item.

# Useful links and contacts

**Association of Consultant Architects**
60 Godwin Road
Bromley
Kent BR2 9LQ
www.acarchitects.co.uk/
Tel/Fax: +44(0)20 8466 9079

**British Property Federation**
5th Floor
St Albans House
57–59 Haymarket
London SW1Y 4QX
www.bpf.org.uk/
Tel: +44(0)20 7828 0111
Fax: +44(0)20 7834 3442

**BSI British Standards Group**
389 Chiswick High Road
London W4 4AL
www.bsi-global.com/
Tel: +44(0)20 8996 9001
Fax: +44(0)20 8996 7001

**Building Cost Information Service (BCIS)**
12 Great George Street
London SW1P 3AD
www.bcis.co.uk/
Tel: +44(0)20 7695 1500
Fax: +44(0)20 7695 1501

**Building Research Establishment (BRE)**
Bucknalls Lane
Watford WD25 9XX
www.bre.co.uk/
Tel: +44(0)19 2366 4000

**Building Research Establishment (Scotland)**
Scottish Enterprise Technology Park
East Kilbride
Glasgow G75 0RD
Tel: +44(0)13 5557 6200

**Chartered Institute of Building**
Englemere
Kings Ride
Ascot
Berks SL5 7TB
www.ciob.org.uk/
Tel: +44(0)13 4463 0700
Fax: +44(0)13 4463 0777

**Chartered Institution of Building Services Engineers**
222 Balham High Road
London SW12 9BS
www.cibse.org/
Tel: +44(0)8675 5211
Fax: +44(0) 8675 5449

**Construction Confederation**
55 Tufton Street
London SW1P 3QL
www.constructionconfederation.co.uk/
Tel: +44(0)87 0898 9090
Fax: +44(0)87 0898 9095

**Construction Skills**
Eastleigh House
1st Floor
Upper Market Street
Easteigh SO50 9FD
www.cskills.org/
Tel: +44(0)84 4844 0046
Fax: +44 (0)84 4844 0045

**Dept. for Business Enterprise & Regulatory Reform (BERR)**
Ministerial Correspondence Unit
1 Victoria Street
London SW1H 0ET
www.berr.gov.uk/
Tel: +44(0)20 7215 5000
Fax: +44(0)20 7215 0105

**Europa – The European Union on-line**
http://europa.eu/
Tel: 00800 6 7 8 9 10 11

**European Committee for Standardization (CEN)**
Avenue Marnix 17
Brusselswww.cen.eu/
Tel: +32 2550 0811
Fax: +32 2550 B-1000

**Federation of Master Builders (FMB)**
Gordon Fisher House
14–15 Great James Street
London WC1N 3DP
www.fmb.org.uk/
Tel: +44(0)20 7242 7583
Fax: +44(0)20 7404 0296

**Health and Safety Executive (HSE)**
Rose Court
2 Southwark Bridge
London SE1 9HS
www.hse.gov.uk/
Tel: 0845 345 0055

**HM Revenue & Customs**
www.hmrc.gov.uk/

**Institute of Clerks of Works**
Equinox
28 Commerce Road
Lynch Wood
Peterborough PE2 6LR
www.icwgb.org/
Tel: +44(0)17 3340 5160
Fax: +44(0) 17 3340 5161

**International Federation of Consulting Engineers (FIDIC)**
World Trade Center II
Geneva Airport
Box 311
29 route de Prés-Bois
CH-1215 Geneva 15
www.fidic.org/
Tel: +41(0)2 2799 4900
Fax: +41(0)2 2799 4901

**Institution of Chemical Engineers (IChemE)**
Davis Building
176–189 Railway Terrace
Rugby CV21 3HQ
www.icheme.org/
Tel: +44(0)17 8857 8214
Fax: +44(0)17 8856 0833

**Institution of Civil Engineers**
One Great George Street
London SW1P 3AA
www.ice.org.uk/
Tel: +44(0)20 7222 7722

**Institution of Structural Engineers**
11 Upper Belgrave Street
London SW1X 8BH
www.istructe.org/
Tel: +44(0)20 7235 4535
Fax: +44(0)20 7235 4294

**Insolvency Service**
21 Bloomsbury Street
DX 120875 Bloomsbury 6DX
London WC1B 3QW
www.bis.gov.uk/insolvency
Tel: 0845 602 9848

**Joint Contracts Tribunal Ltd**
4th Floor,
28 Ely Place
London EC1N 6TD
www.jctltd.co.uk
Fax: +44(0)20 7404 7387

**National Building Specification (NBS)**
The Old Post Office
St Nicholas Street
Newcastle upon Tyne NE1 1RH
www.thenbs.com/
Tel: +44(0)19 1244 5500
Fax: +44(0)19 1232 5714

**National Home Energy Rating (HNER)**
National Energy Centre
Davy Avenue
Knowlhill
Milton Keynes MK5 8NA
www.nher.co.uk/
Tel: +44(0)19 0867 2787

**Office of Government Commerce (OGC)**
Rosebery Court
St. Andrew's Business Park
Norwich NR7 0HS
www.ogc.gov.uk/
Tel: +44(0)84 5000 4999

**ProCure21**
Dept. of Health
Room 3N11
Quarry House
Quarry Hill
Leeds LS2 7UE
www.procure21plus.nhs.uk/
Tel: +44(0)11 3254 5851
Fax: +44(0)11 3254 6691

**Royal Institute of British Architects**
66 Portland Place
London W1B 1AD
www.architecture.com/
Tel: +44(0)20 7580 5533
Fax: +44(0)20 7255 1541

**Royal Institution of Chartered Surveyors**
12 Great George Street
London SW1P 3AD
www.rics.org.uk/
Tel: +44(0)870 333 1600
Fax: +44(0)207 334 3811

**Society of American Value Engineers (SAVE)**
136 South Keowee Street
Dayton
OHIO 45402
USA
www.value-eng.org/
Tel: (937) 224 7283
Fax: (937) 222 5794

**United Nations Commission on International Trade**
Law (UNCITRAL)
UNCITRAL Secretariat
Vienna International Centre
P.O. Box 500
A-1400 Vienna
Austria
www.uncitral.org/
Tel: +43-(1) 2 6060 4060/4061
Fax: +43-(1) 26060 5813

**Whole Life Cost Forum**
www.wlcf.co.uk/

# Further reading

## 1 The quantity surveyor and the construction industry

Badke, E. (2008) 'Are you measuring up?' *RICS Construction Journal*, June–July, pp.15–17.
Cartlidge, D. (2011) *New Aspects of Quantity Surveying Practice,* Third Edition, Spon Press, Oxford.
Dalziel, B. and Ostime, N. (2008) *Architect's Job Book.* RIBA Publications.
Harvey, J. (2004) *Urban Land Economics,* Sixth Edition, Palgrave Macmillan.
Morton, R. (2007) *Construction UK: Introduction to the Industry*, Blackwell Publishing.
Office of National Statistics (2010) *Annual Construction Statistics.* HMSO.
RICS (2006) *The Future of Surveying Education,* The Royal Institution of Chartered Surveyors.
RICS (2012) *RICS New Rules of Measurement: Order of Cost Estimating and Elemental Cost Planning,* Second Edition, RICS Books.
RICS (2012) *RICS New Rules of Measurement: Bill of Quantities for Works Procurement,* First Edition, RICS Books.
Thompson, F.M.L. (1968) *Chartered Surveyors: The Growth of a Profession*, Routledge & Kegan Paul.

## 2 Forecasting costs and values

Cartlidge, D. and Mehrtens, I. (1982) *Practical Cost Planning*, Hutchinson.
Issac, D. (2003) *Property Finance,* Second Edition, Palgrave Macmillan.
Millington, A.F. (2002) *Property Development*, Estates Gazette.
Myres D. (2004) *Construction Economics: A New Approach,* Spon Press.
Office of Government Commerce (2003) *Procurement Guide No. 7: Whole Life Costing and Cost Management*, HMSO.
Stone, P.A. (1968) *Building Design Evaluation: Costs-in-Use,* E & F.N. Spon.

## 3 Measurement and quantification

Davidson, J. and Hambleton, P. (2006) *SMM7 Questions and Answers*, RICS Books.
Keily, P. and McNamara, P. (2003) *SMM7 Explained and Illustrated*, RICS Books.

Lee, S., Trench, W. and Willis, A. (2011) *Willis's Elements of Quantity Surveying*, Eleventh Edition, Blackwell Publishing.
RICS (2012) *RICS New Rules of Measurement: Bill of Quantities for Works Procurement*, First Edition, RICS Books.

## 4 Procurement

Cartlidge, D. (2004) *Procurement of Built Assets*, Butterworth Heinemann.
Cartlidge, D. (2006) *Public Private Partnerships in Construction*, Taylor & Francis.
Flyvbjerg, B. (2003) *Mega Projects and Risk*, Cambridge.
Hughes, W., Hillebrandt, P. and Greenwood, D. (2006) *Procurement in the Construction Industry*, Taylor & Francis.

## 5 Pricing and tendering

Brook, M. (2008) *Estimating and Tendering for Construction Work*, Butterworth Heinemann.
Construction Faculty (2005) *E-Tendering*, RICS Books.
CIOB (2009) *Code of Estimating Practice*, Seventh Edition, Wiley Blackwell.
Finch, R. (2011) *NBS Guide to Tendering for Construction Projects*, RIBA Publishing.

## 6 Contract procedure

Chappell, D. (2006) *Construction Contracts Questions and Answers*, Taylor & Francis.
Davison, J. (2006) *JCT 2005 What's New?* RICS Books.
Hacket, M., Robinson, I. and Statham, G. (2006) *Aqua Group Guide to Procurement Tendering and Contract Administration*, Blackwell Publishing.
RICS Quantity and Construction Professional Group Board (2012) *Black Book*, RICS Publishing.
Wallace, D.I. (1994) *Hudson's Building and Engineering Contracts Main Work and Supplement*, Sweet and Maxwell.

## 7 Final account

Ashworth, A. and Hogg, K. (2007) *Willis's Practice and Procedure for the Quantity Surveyor*, Twelfth Edition, Blackwell Publishing.

# Useful measurement rules and conventions

**Abbreviations**

| | |
|---|---|
| aggregate | agg |
| as before | ab |
| as described | ad |
| asphalt | asph |
| brickwork | bkwk |
| centres | c /c |
| coat | ct |
| common brickwork | cb |
| cement | ct |
| concrete | conc |
| damp-proof course | dpc |
| damp-proof membrane | dpm |
| deduct | ddt |
| ditto | do |
| excavate | exc |
| fair face | ff |
| foundations | fdns |
| galvanised | galv |
| ground level | GL |
| half-brick wall | hb wall |
| hardwood | hw |
| height | ht |
| horizontal | hoz |
| joint | jt |

| | |
|---|---|
| kilogram | kg |
| mild steel | ms |
| not exceeding | ne |
| pipe | pi |
| pointing | ptg |
| rainwater pipe | rwp |
| reinforced concrete | rc |
| reduced level | RL |
| reinforcement | reinf |
| softwood | sw |
| thick | thk |
| vertical | vert |

**Conversion tables**

SI (International System of Units) has been used in the construction industry since the 1970s. Previous, to this a hybrid decimal system based on 12 was used in both measurement and bill of quantities preparation. Despite the used of SI units it is still common for prices and rental values to be quoted in imperial units, for example; £32 per square foot.

| | **Metric** | **Imperial** | **Imperial** | **Metric** |
|---|---|---|---|---|
| Length | 1mm | 0.03937 in | 1 in | 25.4mm |
| | 1m | 3.281 ft | 1 ft | 0.3048m |
| | 1m | 1.094 yd | 1 yd | 0.9144m |
| Area | $1mm^2$ | $0.00153\ in^2$ | $1\ in^2$ | $645.2mm^2$ |
| | $1m^2$ | $10.764\ ft^2$ | $1\ ft^2$ | $0.0929m^2$ |
| | $1m^2$ | $1.196\ yd^2$ | $1\ yd^2$ | $0.8361m^2$ |
| | 1 hectare | $11,960\ yd^2$ | 1 hectare | $10,000m^2$ |
| Mass | 1 kg | 2.205 lb | 1 lb | 0.4536kg |
| | 1 tonne | 0.9842 tons | 1 ton | 1.016 tonnes |

## Areas

When calculating irregular areas or volumes, the approach should be to divide the total area or volume into more easily measured regular areas such as triangles. For example:

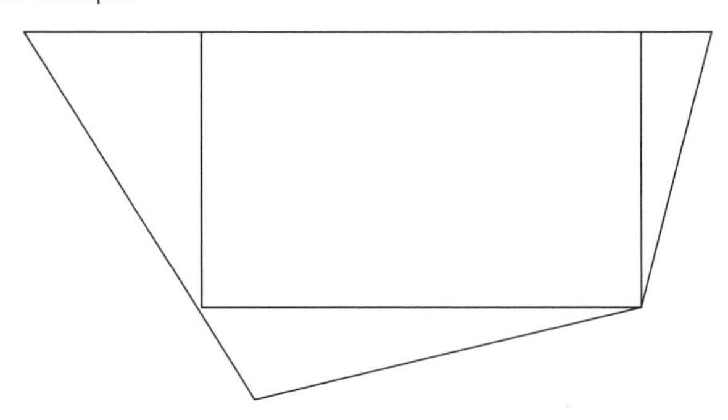

## Roof members and coverings

Calculation of coverings

The length of the rafter C can be determined using Pythagoras:
$C = \sqrt{B^2 + A^2} = 2.83$

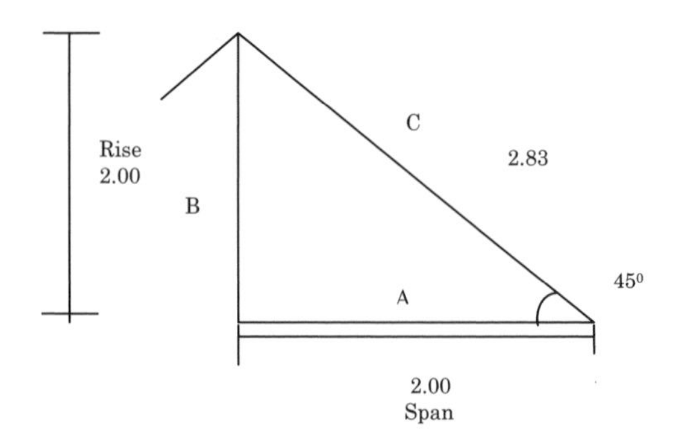


The span is taken to the extreme projection of the roof and includes eaves projections, etc.

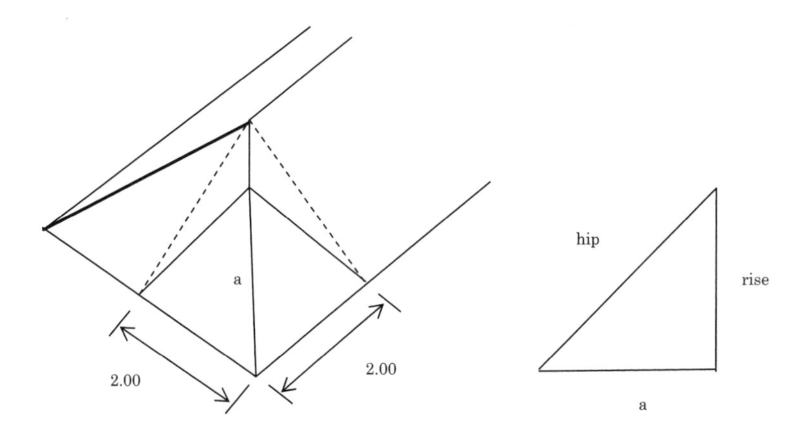


The length of the hip also has to be calculated as follows:

To calculate a: $a = \sqrt{2.00^2 + 2.00^2} = 2.83$

Therefore length of hip $= \sqrt{2.83^2 + 2.00^2} = 3.47$

# Index